BIRDS OF THE ROCKY MOUNTAINS

Happy Birthday 1992
Fred

BIRDS *of the*
ROCKY MOUNTAINS

with particular reference to
National Parks in the
Northern Rocky Mountain Region

by
Paul A. Johnsgard

University of Nebraska Press
Lincoln and London

First Bison Book printing: 1992
Most recent printing indicated by the last digit below:
10   9   8   7   6   5   4   3   2   1

Library of Congress Cataloging-in-Publication Data
Johnsgard, Paul A.
Birds of the Rocky Mountains: with particular reference to national parks in
the Northern Rocky Mountain region / by Paul A. Johnsgard.
p.   cm.
Originally published: Boulder, Colo.: Colorado Associated University
Press, 1986.
Includes bibliographical references and index.
ISBN 0-8032-7574-9
1. Birds—Rocky Mountains.   2. National parks and reserves—Rocky
Mountains.   I. Title.
QL683.R63J64   1992   91–42200
598.2978—dc20   CIP

Reprinted by arrangement with Paul A. Johnsgard

∞

Contents

List of Figures

List of Plates

Preface and Acknowledgments

The Rocky Mountain region has fascinated me ever since I first travelled to Glacier and Yellowstone national parks as a teenager, and saw for the first time such wonderful birds as ospreys, dippers, and Lewis's woodpeckers. At various times in my adult life I have also felt compelled to return to the mountains again and again. Because of my Nebraska home, these more recent trips have most often been to Rocky Mountain National Park and Grand Teton National Park, both of which are only about a day's drive away. Grand Teton National Park has been my special favorite, and was the subject for an earlier book on the region's natural history. Shortly after writing that book its publisher, Colorado Associated University Press, asked me if I would be interested in writing a book on the birds of Rocky Mountain National Park. I rejected that idea immediately, inasmuch as I had no interest in doing a book on such a restricted area, but countered with the proposal of a book covering the entire region of the northern Rockies, and encompassing all the U.S. parks north to the Canadian border. This idea was accepted and I began work on the book in 1982. It soon became apparent that I could add most of the Alberta and British Columbia parks without much additional effort, and thus I moved the boundaries of map coverage north to the 52d parallel, or the boundary of Banff and Jasper national parks. Since the avifauna of Jasper is almost identical to that of Banff I decided to deal with them collectively, pointing out any significant differences in the text. The western boundary of Idaho became my western limits, the eastern boundaries of Montana and Wyoming my eastern limits, and the 40th parallel in Colorado my southern limits. The resulting total area of coverage in this book is some 353,000 square miles, or about 70% of the area covered in my earlier (1979) book on the breeding birds of the Great Plains. A total of 354 species are included in this book, compared with 325 in the earlier one (which was restricted to breeding species only).

These two books are somewhat complementary both in their geographic coverage and utility, for the earlier book has sections on breeding biology and characteristics of nests, eggs, incubation periods, etc., which for reasons of space have not been included here. At the suggestion of the publisher, this book includes short "identification" sections that should help facilitate field identification for persons already somewhat familiar with bird groups, but is not nor cannot be considered a substitute for a good color-illustrated field guide. This book should be especially useful to residents of or visitors to Montana and Wyoming, since both of these states lack "state bird books," and to a lesser degree persons in Idaho and Colorado, for which states bird books have been written but are now out of print. I hope that my book will be of supplementary value to *The Birds of Alberta* by Salt and Salt (1976), which provides similar distribution maps covering all of that province as well as the other two prairie provinces. Because of the high levels of visitation to the national parks of the Rocky Mountain region, particular attention has been paid to the status of each species in these parks, based on published or unpublished park records and additional information accumulated by me in the course of my research. Rather surprisingly, considerable information "gaps" still exist for nearly all these parks, in spite of the great attention they have received, and I would appreciate receiving information that updates or modifies the materials presented here.

An important component of the book was the inclusion of latilong data from montane portions of Montana, Wyoming, and northern Colorado. For permission to use Montana data I am indebted to the late P. D. Skaar, and for similar data from Wyoming and Colorado I appreciate permission from the editors of the *Wyoming Avian Atlas* and the *Colorado Bird Distribution Latilong Study*. However, I made some modifications of these sources, in part based on personal information and in part demanded by standardization of coding symbols, and thus the data are not invariably identical to these sources. It might be further noted that the reader may occasionally find apparent discrepancies in the text among the status symbols indicated for a particular park, that park's latilong data, and the general range map. In part these result from the fact that the boundaries of parks and latilongs never coincide. Furthermore, the range maps are usually more "generous" in estimating a species' range than is indicated by available data from a latilong or particular park. Such a situation reflects my belief that, based on its general distributional and habitat characteristics, the species may eventually be found there. More rarely, the mapped breeding range of a species may not include a park for which one or more historic breeding records exist; this is sometimes the case for species whose ranges have retracted or otherwise changed in recent years. Finally, the diffi-

culties inherent in drawing range maps covering some of the topographi-
cally most complex areas in North America need hardly be mentioned.

Many biologists assisted me in various ways, including the Chief Natu-
ralists of several parks, specifically Glenn Kaye, Clyde Lockwood,
George Robinson, and Patrick Smith. Other biologists provided unpublished data, including Greg Beaumont, Charles Chase III, David Cos-
tello, Kenneth Diem, Richard Follet, Richard Hutto, B. Riley
McClelland, Suzanne Murray, Ronald Ryder, David Shea, and Paul
Wright. Dr. Diem also provided a critical reading of the entire manuscript. Photographs were offered or provided me by Hans Aschen-
brenner, Kenneth Fink, Alan Nelson, Mardy Murie, and Ed Schulen-
berg.

I also wish to thank the U.S. National Park Service and the University of Wyoming for providing me with research space at the Jackson
Hole Biological Station in 1983, and for similar accommodations in 1975
and 1976 at the station's earlier location.

Lastly, I would be remiss if I were not to thank my various field
companions, especially Tom Mangelsen and my son Scott, for sharing
many wonderful days in the mountains with me, surrounded by some
of the most glorious landscapes on earth.

*This book is dedicated to my new grandson Scott;
may he love these mountains as one would a favorite book,
and may the life therein offer him its manifold lessons.*

Introduction

The Rocky Mountains represent the longest and in general the highest of the North American mountain ranges, extending for nearly two thousand miles from their origins in Alaska and northwestern Canada southward to their terminus in New Mexico, and forming the continental divide for this entire length. As such, these mountains have provided a convenient corridor for northward and southward movement of both plants and animal life, but on the other hand have produced important barriers to eastern and western plant and animal movements. These effects result not only from their height and physical nature, but also from their manifold effects on such things as precipitation, humidity, temperature, and other important climatic factors affecting plant and animal life.

The bird life of the Rocky Mountains is surprisingly uniform, in spite of their great latitudinal spread and the equally wide altitudinal variations that occur in the region. Thus, a bird-watcher in Banff or Jasper national parks will encounter the vast majority of the same breeding species in the coniferous zones of those areas as one who is observing birds nearly a thousand miles to the south in Rocky Mountain National Park, although particular bird species would occur at considerably different altitudes. This is a result of the horizontal zonation patterns of organisms, which individually distribute themselves along the slopes of mountains within vertical bands that conform to their limits of physiological stress and their biological requirements for food, cover, reproduction sites, and the like. Because of these biotic interactions, complex plant and animal communities have evolved through time. The observant naturalist can soon learn to recognize these communities on the basis of one or two "dominant" plant species that represent biological reflections of the overall physical and biotic environment of that particular altitude and latitude. In earlier biological literature these aggregations of interacting organisms were often called "life zones," based on temperature and moisture characteristics, and were given a variety of names that tended to have geographic connotations. More recently, the life zone concept has been replaced with one based

1

primarily on the native plant life, inasmuch as it is believed that such plant "communities" provide an immediate biotic index to the climate, as well as to the soil characteristics and to the recent history of the environment, such as disturbance by burning, grazing, and the like. Specifically, a succession of natural plant communities are believed to develop and replace earlier developmental ("seral") stages in a relatively orderly fashion, until a "climax" or essentially permanent and self-regenerating stage is reached. By mapping large regional areas on the basis of actual potential climax plant communities, an insight into their climates, soils, and biotic productivities can be attained. Many such areas, because of periodic or continuous disturbance, probably may never reach this ultimate climax stage, but instead remain in various earlier transitional stages. In the Rocky Mountain region repeated forest fires have caused much of the area to become vegetated by fire-adapted communities, such as lodgepole pine or aspen, for example, and in other areas timbering may have caused regeneration of a second-growth forest that likewise represents a non-climax community type.

Because of both the physical characteristics of the communities (height, density, etc., of the vegetation), and their biological characteristics (food availabilities, presence of competing or otherwise interacting organisms such as predators and prey species), birds and other organisms occur in greatly differing levels of abundance in differing communities. Those plants which exert the greatest ecological effects in a community (the "dominants"), are obviously likely to play important roles in the distribution and abundance of many bird species. However, particular birds may be dependent upon certain non-dominant plants or even other features of the environment. Thus, for example, three-toed woodpeckers favor recently burned forest areas, where an abundance of wood-boring beetles are attracted to dead trees. Indeed, many other species of birds require dead snags, rotted trees, or other relatively uncommon or unique features of a community in order to find all the requirements for performing all of the aspects of their particular ecological "profession," or niche. Thus, ecological communities that have a great diversity of plant life also tend to have a diversity of bird life, for in such communities there are far more opportunities for diverse niches to be supported.

One of the major roles of national parks is to insure the existence of protected examples of all the major habitat types and ecological communities typical of a particular region; in a sense they are natural repositories of ecological diversity. Even though most established national parks are areas of outstanding natural beauty, as indeed was the case with all the national parks included in this book, it was also recognized by their founders that their natural integrity should remain inviolate forever, and that they should be allowed to continue to support the broadest possible range of plant and animal life. For an ecologist, a national park might be more exciting because it supports a rare species of fern, orchid, or bird, than because it has awesome peaks, high waterfalls, or stunning geysers. Or, it may well be the only place left in that part of North America that has not felt the effects of mining or oil drilling machinery, or heard the sounds of chain saws or timber axe in recorded time. It is in part for this reason, that the parks are the last and best vestiges of a pristine North America, that I have decided to concentrate on them in writing this book. Furthermore, they are visited each year by millions of people wanting information on the distribution, abundance, and identification of birds and other organisms, and to varying degrees they have been studied by biologists sufficiently that relatively complete bird lists are available now for most parks.

I have selected an area of the central and northern Rocky Mountains that includes a total of eight national parks in a relatively confined area, namely between 40° and 52° north latitude, or specifically from northern Colorado to the arctic watershed divide in southern Alberta and adjacent British Columbia. Four Canadian national parks (Banff, Yoho, Kootenay, and Watertown Lakes) and four U.S. parks (Glacier, Yellowstone, Grand Teton, and Rocky Mountain) fall within this area, in addition to a national monument, several national wildlife refuges, and numerous national forests and provincial or state parks.

I have adopted a tri-level approach to the information presented in this book. First, a seasonal abundance and breeding chart for each species is provided in the back matter for each national park mentioned. Secondly, the distributional status for each species is regionally presented in "latilong" form. Latilongs are geographic areas that are defined by quadrants of latitude and longitude, each block measuring approximately 50 × 70 miles (in the area concerned), and provide convenient methods of illustrating bird distributions in areas where more "fine-grained" analyses are impossible. Groups of latilongs were chosen that include all four U.S. national parks, and also include a maximum of typical Rocky Mountain terrain outside the parks. Thirdly, regional distribution maps show each species' apparent breeding or residential distribution (in the case of breeding species) or non-breeding status, within the northern Rocky Mountain regions, here chosen to include all of Montana, Idaho, and Wyoming, and adjacent parts of Colorado and Alberta.

The list of species to be included in this book was developed using information available on all of the included national parks as well as Dinosaur National Monument. All species reliably reported from any of these areas are included in the book. A few additional species have also been included that have not yet been definitely reported from any of the parks, but which perhaps do occasionally occur, for a total of 354 species.

It might be of interest to compare the species coverage of this book with that of the various states and provinces that are variably encompassed. Idaho has a total avifaunal list of 305 species (Burleigh, 1972), including 197 total known breeders (Idaho Fish and Game Information Leaflet No. 12, undated). This book includes all known Idaho breeders except for the mountain quail (which is local from Latan to Owyhee counties), and the locally introduced (Lemhi County) Gambel's quail. Wyoming has a total list of 377 bird species, of which 227 are considered to be breeders (Oakleaf et al., 1982). This book includes all known Wyoming breeding birds, except for the barn owl (local in southern Wyoming) and the Scott's oriole (local in southwestern Wyoming). Montana has a total list of 378 species, of which 237 are known breeders (Skaar, 1980). This book includes all the state's known native breeders except for the piping plover (local in northeastern Montana). Alberta has a total list of 333 species, and 247 breeding species (Salt and Salt, 1976). This book includes all of the montane breeding avifauna, and 97% of the province's total breeding species. Colorado has a total list of 416 species, and approximately 260 breeding species. This book includes all of the typical montane breeding avifauna, and about 90% of the state's total breeding species.

Latilong and Abundance Coding

The methods of using latilongs for plotting bird distribution was devised by Skaar for use in Montana, and initially was applied to the area around Bozeman, and later (1975, 1980) to the entire state. Skaar argued that, although the area encompassed by latilong varies with latitude owing to the gradual narrowing of the distances between lines of longitude toward the poles, the encompassed areas of latilongs along the Montana–Canada border are, for example, only 5.4% smaller than ones on the Montana–Wyoming border. Thus, for all practical purposes, those used in this book can be considered approximately equal in size.

Skaar's pioneering work in Montana was followed by similar analyses for Wyoming (Oakleaf et al., 1979, 1982), and for Colorado (Kingery and Graul, 1978; Chase et al., 1982). As a result, most of the Rocky Mountain area south of Canada and under consideration in this book has been subjected to latilong analysis, with the exception only those portions of the Rocky Mountains occurring in eastern Idaho.

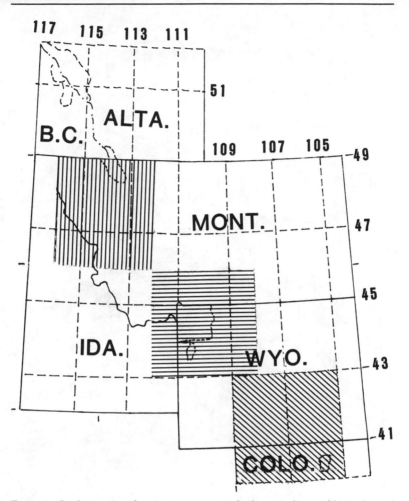

Figure 1. Outline map of region encompassed, showing lines of latitude and longitude, and (shaded) latilong groupings selected for detailed information presentation.

6

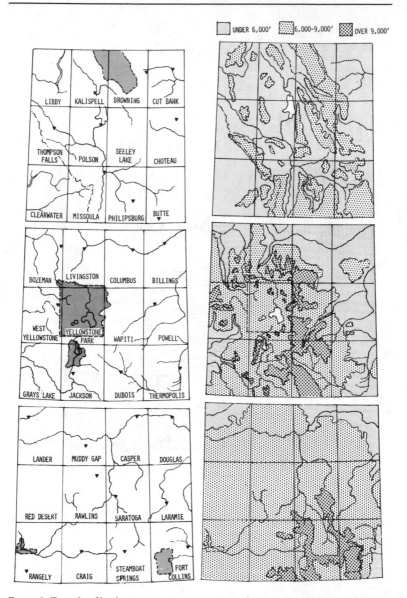

UNDER 6,000'   6,000-9,000'   OVER 9,000'

LIBBY   KALISPELL   BROWNING   CUT BANK

THOMPSON FALLS   POLSON   SEELEY LAKE   CHOTEAU

CLEARWATER   MISSOULA   PHILIPSBURG   BUTTE

BOZEMAN   LIVINGSTON   COLUMBUS   BILLINGS

WEST YELLOWSTONE   YELLOWSTONE PARK   WAPITI   POWELL

GRAYS LAKE   JACKSON   DUBOIS   THERMOPOLIS

LANDER   MUDDY GAP   CASPER   DOUGLAS

RED DESERT   RAWLINS   SARATOGA   LARAMIE

RANGELY   CRAIG   STEAMBOAT SPRINGS   FORT COLLINS

Figure 2. Details of latilong groupings summarized in species accounts, showing (left) latilong names and locations of features for which each latilong is named, and (right) selected contour intervals for areas within latilong groupings.

Because of this available useful data base, I requested and obtained permission from the appropriate authors to extract latilong data from a sample group of latilongs that center on the U.S. national parks and extend diagonally southward from northwestern Montana to northern Colorado along the continental divide. Three quadrants, each measuring four latilongs wide by three latilongs high, or approximately 180–210 miles east–west by 207 miles north–south, were thus selected (Fig. 1). Within these three quadrants, two latilongs fall entirely within the state of Idaho, but otherwise all the latilongs fall partially or entirely within the states of Montana, Wyoming, and Colorado. Names applied to these latter latilongs (Fig. 2) are the same as those used in the previously mentioned studies, while new names were devised for the two strictly Idaho-restricted latilongs. Data for these two latilongs, which I have named Clearwater and Bear Lake, were assembled from various sources, but primarily from information obtained on the birds of Clearwater National Forest and Bear Lake National Wildlife Refuge. In each case, the limits of the national forest and the national wildlife refuge do not entirely correspond with latilong limits, and thus some possibilities for errors of latilong attribution exist.

Unfortunately, no consistent method of coding latilong information has yet been formalized, and the methods used for Wyoming, Montana, and Colorado all differ to some extent from one another. I settled for using a system that is not entirely like any one of the three, but which concentrates on reproductive status (breeding or non-breeding) and seasonal occurrence. The symbols used are as follows:

R = breeding permanent resident
r = resident, breeding unproven
S = breeding summer resident
s = summer resident, breeding unproven
M = migrant, excepting wintering visitors
W = migrants that sometimes or regularly overwinter
V = vagrant, out of normal range
X = extirpated from area
? = inadequate or conflicting information

Latilong codes in this book thus differ to varying degrees from those found in the individual state references, and in some cases decisions were necessarily made (especially as to "vagrant" vs "migrant" status) of a subjective nature. There are also a few cases in which the status for a given latilong differs from that reported in individual state summaries; this is the result of additional or different information available to me that seemed to warrant making such deviations. A different set of symbols, provided in a chart following the species accounts, has been used to indicate relative seasonal or overall abundance for each of the national parks. These are as follows:

A = abundant
C = common
U = uncommon
O = occasional
R = rare
V = vagrant (accidental)
? = inadequate information
* = breeds (or has bred) in area

Habitats and Ecological Distributions

All bird species, by virtue of variably specialized niche adaptations, are most abundant in or may even be completely restricted to particular habitats. It is thus particularly important that birders pay attention to the habitat in which they are observing birds, for this very often provides important clues as to the species of birds that are most likely to be encountered. Except for such specialized habitats as rocky outcrops, mud flats, and other special substrates, the majority of habitats are most easily described in terms of their dominant plants. Furthermore, as noted earlier, there is a rather remarkable consistency in the vertical stratification or zonation patterns throughout the central and northern Rocky Mountains. Thus, there is a sequence of community types that typically occurs sequentially from the plains and foothills upwards toward the mountain tops. The exact altitudes at which a particular habitat type occurs varies greatly, depending both on latitude and also on such local effects as directional exposure, protection from winds, soil characteristics, and the like. However, as shown in Table 1, most vegetation zones in the Rocky Mountains occur in broad belts ranging in altitudinal width from a few hundred feet to about two thousand feet.

Although many bird watchers may not be interested in making such fine botanical distinctions as separating Douglas fir forest from, for example, Engelmann spruce, subalpine fir forest, in many cases this is not actually necessary for useful habitat identification. For example, many birds seemingly respond to forest habitats in terms of rather broad life-form characteristics (e.g., mature hardwood forest, open woodlands, coniferous forest, timberline thickets, etc.) and thus detailed botanical recognition may not be necessary. Nevertheless, habitats are identified in the text as accurately as possible, and thus some familiarity with the usual vegetational zonation patterns in the Rocky Mountains can be very useful in finding particular birds.

Table 1. *Vertical Distribution of Major Plant Communities in the Central and Northern Rocky Mountains*

Physiography	Vegetation Zones	Traditional "Life-Zones"	Approximate Average Altitude in Feet		
			Colorado	Wyoming, S. Idaho	Montana, S. Alberta
Alpine	Tundra	Arctic–Alpine	11,000–13,000+	10,300–12,000+	6600–8000+
Subalpine (timberline)	Engelmann spruce, Subalpine fir	Hudsonian	10,000–11,000	9500–10,300	6000–6600
Montane	Climax Phase Douglas fir Western redcedar, Western hemlock Ponderosa pine Seral Phase Lodgepole pine Quaking aspen	Canadian	8000–10,000	7500–9500	4500–6000
Foothills and Mesas	Pinyon, Juniper Oak, Mountain mahogany Sagebrush scrub Saltbush, Greasewood	Transition	6000–8000	5500–7500	4000–4500
Plains and Valleys	Shortgrass Plains Riparian Deciduous Forest	Upper Sonoran	under 6000	under 5000	under 4000

For example, in the tundra zone from Colorado north to Jasper National Park, the most typical breeding bird species are the white-tailed ptarmigan and the rosy finch. In the southern Rockies of Colorado, the rosy finch is of the brown-capped race, while in the central Rockies of Wyoming it is of the black-bodied race, and farther north it is the gray-headed form. In the subalpine zone below timberline such species as the Brewer's sparrow, pine siskin, red and white-winged crossbill, and white-crowned sparrow are typical breeding forms, and slightly lower in the montane coniferous forest a great number of breeding species are found. Among those that are particularly associated with the mesic or typical montane forest are the spruce grouse, goshawk, Cooper's hawk, sharp-shinned hawk, great gray owl, boreal owl, pileated woodpecker, three-toed and black-backed woodpeckers, Williamson's sapsucker, gray and Steller's jays, Clark's nutcracker, Wilson's, yellow-rumped, and MacGillivray's warblers, winter wren, golden-crowned and ruby-crowned kinglets, mountain chickadee, red-breasted nuthatch, Townsend's solitaire, varied, hermit, and olive-backed thrushes, mountain bluebird, and dark-eyed junco. In the moister western red cedar, western hemlock forests of northwestern Montana, the chestnut-backed chickadee is especially characteristic, while in the drier and more park-like ponderosa pine forests such species as the blue grouse, band-tailed pigeon, calliope hummingbird, American robin, black-billed magpie, Lewis' woodpecker, and pygmy nuthatch are more likely to be found. The birds of lodgepole pine forests are in general very much like those of the Douglas fir forests and other coniferous montane forest communities, while the aspen forests typically have a somewhat more diversified breeding avifauna, including such species as the ruffed grouse, flammulated owl, pygmy owl, yellow-bellied sapsucker, tree swallow, eastern bluebird, and warbling vireo.

The woodlands of pinyon pine, juniper, oak, and mountain mahogany, which are best developed to the south and west of the region under consideration here, carry into the region a distinctive group of birds such as the common poor-will, saw-whet owl, pinyon jay, gray flycatcher, plain titmouse, blue-gray gnatcatcher, western bluebird, Bewick's wren, black-throated gray warbler, and Grace's warbler. The drier and lower sagebrush scrub and less widespread alkaline-associated saltbush and greasewood communities likewise have a few highly distinctive breeding species, such as sage grouse, mountain plover, sage thrasher, Brewer's sparrow, and sage sparrow. The birds of the shortgrass plains beyond the foothills typically have wide breeding distributions on the Great Plains, while those of riparian deciduous forest usually have their affinities with the deciduous forest communities that occur widely over eastern North America. For example, the yellow-billed cuckoo, eastern screech-owl, red-headed and red-bellied woodpeckers, least flycatcher, eastern phoebe, blue jay, eastern bluebird, brown thrasher, warbling vireo, red-eyed vireo, chestnut-sided warbler, bay-breasted warbler, indigo bunting, fox sparrow, rufous-sided towhee, orchard oriole, and Baltimore oriole are all likely to be encountered in mature riparian woodlands as well as in deciduous forests much farther to the east.

Even in national parks not all the habitats are pristine; historical and recent forest fires have placed much of the area of Yellowstone Park in various stages of vegetational succession dominated by lodgepole pine, for example, and ranching activities in Grand Teton National Park have influenced grassland and shrub succession in non-forested areas. Browsing of elk has greatly influenced aspen distribution and survival, and damming of streams by beavers has resulted in the formation of unique beaver-pond communities, with an interesting and diverse associated plant, bird, and mammal life. Indeed, such species as the trumpeter swan and sandhill crane are largely dependent upon beaver activity in the Grand Teton–Yellowstone area for the production and maintenance of suitable breeding habitat.

Man's effects on the environment are apparent everywhere throughout the Rocky Mountains, as lumbering, agriculture, mining, energy development, road-building, and other familiar symbols of modern civilization have left their marks on the landscape. In general, the influence of man is to reduce environmental diversity, by eliminating either unwanted ("weeds," "pests," predators, etc.) or "worthless" species, in favor of more economically desirable uses for the land. As a result, some bird species have become extremely rare throughout the region, even in national parks. These include the peregrine falcon (now being reintroduced). This and other species have in some areas been locally extirpated, such as the flammulated owl and the Cassin's kingbird from Rocky Mountain National Park. Others, such as the trumpeter swan and sandhill crane, are gradually responding to protection and management, and are moving back into areas from which they have been absent for many decades. Yet others have benefited greatly from man's activities, and have become extremely abundant in and around human activity centers. These include such introduced species as the house sparrow and European starling, and various native species including the American robin, common grackle, and brown-headed cowbird.

Climate, Landforms, and Vegetation

The entire area under consideration is characterized by continental climate, with great seasonal and daily changes in temperature, and fairly short and cool summers. Most of the precipitation is orographic in nature; that is, it is related to topography, with the heaviest precipitation levels typically occurring on the western slopes, and the eastern slopes and valleys often showing reduced precipitation or "rain-shadow" effects (Fig. 3). In general, the higher precipitation levels are in the northwestern portion of the area, in northern Idaho and northwestern Montana, where moist winter air from the Pacific Northwest spills inland to produce the lush western red cedar, western hemlock forests of the west-facing slopes. The driest parts of the area are in the Snake River basin of southern Idaho, the Bighorn Basin of Wyoming, and the Red Desert region of southeastern Wyoming, where annual precipitation is sometimes under ten inches.

The landforms of the region are, of course, dominated by mountainous topography (Fig. 4). These consist of three relatively discrete regions: the northern Rockies, extending from Canada south into western Montana and Idaho as far as the Snake River basin; the central Rockies, centering on the Yellowstone Plateau of northwestern Wyoming and adjacent portions of southern Montana and extreme eastern Idaho; and the southern Rockies of Colorado and extreme southern Wyoming. These are generally relatively high mountains, with the highest elevations being 12,294 feet in Alberta, 12,665 feet in Idaho, 12,850 feet in Montana, 13,785 feet in Wyoming, and 14,431 feet in Colorado. The Rocky Mountains form the continental watershed throughout, separating the Great Plains to the east from the Great Basin to the west, and providing the headwaters for such major river systems as the Snake, Colorado, and Missouri. Only in central Wyoming does the continental divide drop below 7,000 feet. There, in the Red Desert area, it separates and encloses the arid and alkaline Great Divide Basin before rising and leaving the state in the Sierra Madre range. Besides the primary Rocky Mountain chain, there are a number of smaller subsidiary ranges, including the Bighorn Mountains of Wyoming, and several smaller groups of mountains in eastern Wyoming and Montana (Fig. 5).

The geologic forces that shaped the area of the northern and central Rocky Mountains are complex, but the mountains are largely the result of folding and thrust-faulting of sedimentary layers starting in late Cretaceous times some seventy million years ago. Lateral pressures on these layers caused folding, buckling, and faulting to occur, with large areas being lifted upwards and subsequently eroded away. After the Cre-

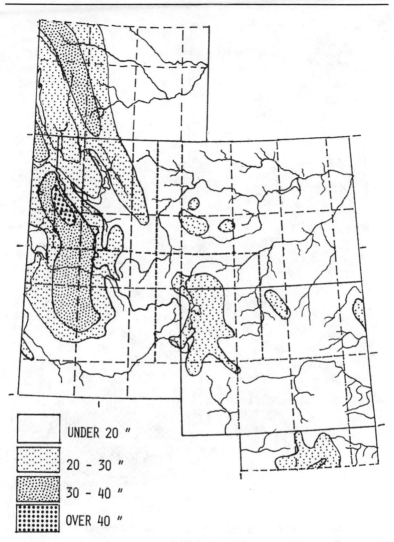

Figure 3. Outline map of region, showing annual precipitation patterns and river drainages.

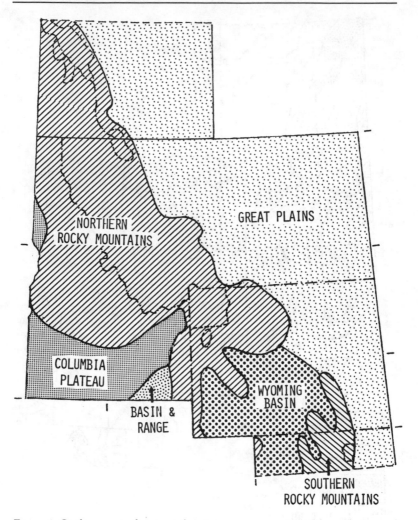

Figure 4. Outline map of region, showing state or provincial boundaries and boundaries of major physiographic features.

taceous layers had been eroded away, progressively earlier layers of Mesozoic and Paleozoic deposits were exposed, until finally the Precambrian core levels were locally exposed. By early Cenozoic times the mountains were perhaps as rugged as the present-day Rockies, but were generally much lower and far more subtropical in climate. Folding and thrust-faulting actions terminated by about the end of the Eocene some sixty million years ago, but during various later parts of the Cenozoic several periods of vulcanism resulted in the deposition of great lava plateaus in the Columbia Plateau, as well as tremendous depositions of wind-blown ash, raising mountain valleys and filling basins. In mid-Cenozoic times the entire Rocky Mountain area began to rise gradually, until the range reached its present-day elevation. This lifting was countered by continued erosion, especially during Pleistocene times, when actions of ice, water, and winds removed several thousands of feet from the exposed strata, and left evidence of such glacial handiwork as U-shaped valleys, cirques, and morainal lakes.

The cold climate of Pleistocene times brought with it a host of northern plant and animal species, which moved variably southward along the Rocky Mountain chain. Subsequently, many of the less mobile species became isolated on high peaks, as the climate ameliorated and a period of drying and warming began. This, of course, was much more true of plant life than of the more mobile animals, but to some degree can also be observed in such locally variable species as the rosy finches, with their distinctive isolated populations from Canada southward. In any case, tundra areas were isolated in post-Pleistocene times and became restricted to the highest mountains, while montane forests and progressively more arid-adapted vegetational communities established varied patterns of geographic distribution reflecting altitude, precipitation, soils, and other environmental factors (Figs. 6–10).

The bird watcher should be aware of the major vegetational zones in the Rocky Mountains, both as a means of predicting the occurrence of particular birds and also as a way of more fully appreciating the complex ecological interactions visible in the region. The major plant communities, and some of their associated botanic characteristics, are as follows:

ALPINE TUNDRA: Areas above timberline where trees are absent or confined to exceptionally protected locations, and dominated by perennial herbs and shrubs. Summers are very short, and very few breeding birds occur here.

SUBALPINE ZONE (timberline zone): This area of generally low and often twisted trees ("krummholz") is typically dominated by subalpine fir (*Abies lasiocarpa*) and Engelmann spruce (*Picea engelmannii*), although

16

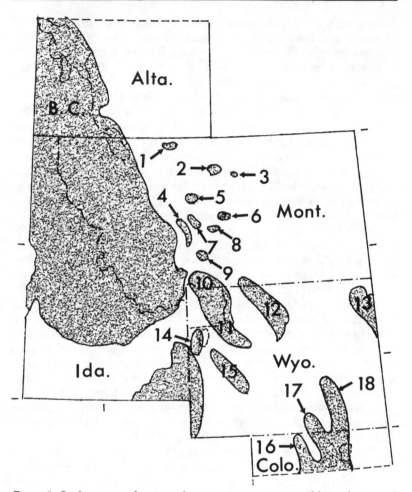

Figure 5. Outline map of region, showing state or provincial boundaries and locations of major and minor mountain ranges. Numbered ranges include 1: Sweetgrass Mts. 2: Bearpaw Mts. 3: Little Rocky Mts. 4: Big Belt Mts. 5: Highwood Mts. 6: Judith Mts. 7: Little Belt Mts. 8: Big Snow Mts. 9: Crazy Mts. 10: Snowy Mts. 11: Absaroka Mts. 12: Big Horn Mts. 13: Black Hills. 14: Teton Mts. 15: Wind River Mts. 16: Park Mts. 17: Medicine Bow Mts. 18: Laramie Mts.

in some areas of Wyoming and Colorado the white-barked pine (*Pinus albicaulis*) is a major timberline species, and somewhat farther south limber pine (*Pinus flexilis*) is a characteristic timberline species as well as occurring on drier foothill areas.

DOUGLAS FIR CLIMAX: Dominated by Douglas fir (*Pseudotsuga menziesii*), sometimes in dense, single-species stands, but also often sharing dominance in the Central Rockies with blue spruce (*Picea pungens*) or Engelmann spruce.

WESTERN RED CEDAR, WESTERN HEMLOCK CLIMAX: On the moist western slopes of Glacier Park, and elsewhere in the northern Rockies, this distinctive community, dominated by these massive and beautiful forest giants, is locally found. The western red cedar (*Thuja plicata*) has fluted trunk, fern-like foliage, and grayish bark. The western hemlock (*Tsuga heterophylla*) is a similarly beautiful and important timber tree, and few of these forests remain except in protected areas such as parks.

PONDEROSA PINE: This forest is extremely widespread throughout the Rockies, often forming the lower edge of the montane coniferous forest, and frequently extending out into the high plains in scattered groves on mesas or other favored sites. The dominant, and sometimes only, tree is ponderosa pine (*Pinus ponderosa*), which typically grows in open rather than dense groves, with considerable grassy or shrubby cover between the trees. In eastern Wyoming the limber pine is an important component of this forest type; in western Wyoming and Colorado various junipers replace it to some degree.

LODGEPOLE PINE: Vast areas of the lower and middle portions of the montane forest are covered by lodgepole pine (*Pinus contorta*) in the central and northern Rockies; for example, most of Yellowstone Park is dominated by such forests, which typically occur following fire. The stands are usually very dense, with little undergrowth, and do not support a diverse breeding bird population.

QUAKING ASPEN: Aspen (*Populus tremuloides*) groves occur widely in the central and northern Rockies, either as a successional community following fire or logging, or as an apparent climax community in low hillsides too dry to support coniferous forests. It is an easily recognized community type, and often is rich in bird life.

PINYON–JUNIPER: On foothills and other areas below the coniferous forest a low forest composed of various species of junipers (*Juniperus monosperma*, *J. scopulorum*, *J. occidentalis*, etc.) and arid-adapted pines (*Pinus monophylla*, *P. edulis*, etc.) locally occurs. It is poorly represented in our area, but extends north to the Snake River of Idaho.

OAK–MOUNTAIN MAHOGANY: Like the last community type, this is also an arid-adapted community better developed in the southern

18

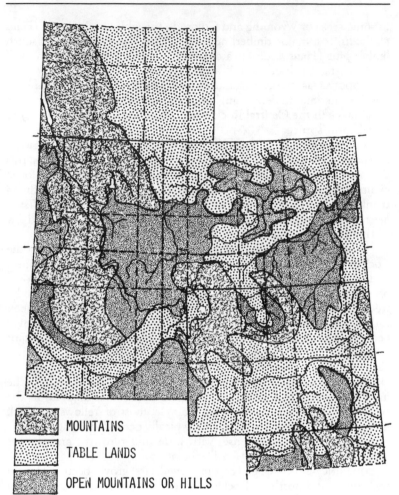

Figure 6. Outline map of region, showing distribution of table lands, open mountains or hills, and mountains.

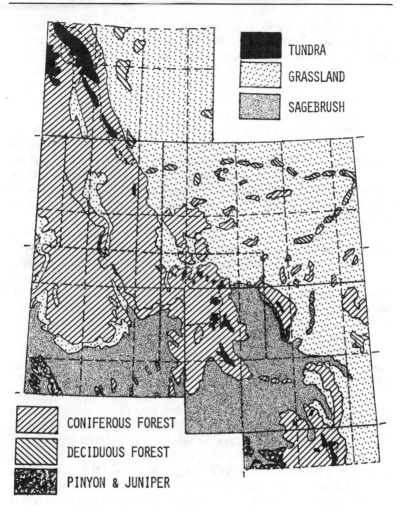

Figure 7. Outline map of region, showing distribution of natural vegetation community types.

Rockies than in our area. It is largely limited to Colorado and extreme southern Idaho. It consists of several species of low oaks (*Quercus gambelii* primarily) and mountain mahogany (*Cercocarpus parviflorus, C. ledifolius,* etc.), as well as other low shrubs such as serviceberry (*Amelanchier utahensis*). These shrubs typically grow in clumps, separated by grassy areas, forming chaparral-like community types. Like the pinyon–juniper community, it supports a distinctive bird life.

SAGEBRUSH SCRUB: Over vast areas of the intermountain west the land is dominated by sagebrush, especially big sagebrush (*Artemisia tridentata*). In some areas the sage shares dominance with various grasses, but its distinctive silvery gray color allows for ready identification. In recent years much of the sagebrush lands have been converted to agriculture through irrigation, and such sage-adapted species as the sage grouse have suffered accordingly.

SALTBUSH, GREASEWOOD: In the Great Divide Basin area of Wyoming, and locally elsewhere, the alkaline soils allow only for the growth of this distinctive community type. The vegetation is scattered, shrubby, and bunch-like, with the dark green color of the greasewood (*Sarcobatus vermiculatus*) strongly contrasting with the more grayish shadscale (*Atriplex canescens*) and saltbush (*A. confertifolia*). The bird life of these communities is similar to that of sagebrush scrub, but is generally low in species diversity and abundance.

SHORTGRASS PLAINS: The vast grassy plains lying to the east of the Rocky Mountains support such attractive species as long-billed curlew, upland sandpiper, and several species of grassland sparrows such as grasshopper sparrow, lark bunting, vesper sparrow, McCown's longspur and chestnut-collared longspur. They are dominated by numerous species of low, perennial grasses such as grama (*Bouteloua* spp.) and buffalo grass (*Buchloë dactyloides*), as well as other taller grass species in protected or ungrazed areas.

RIPARIAN DECIDUOUS FOREST: The upper reaches of the Yellowstone, Missouri, Platte, and other major rivers of the Great Plains bring west into the region an important biota that is especially rich in eastern bird life, as noted earlier. The major trees are cottonwoods, alders, and willows, which sometimes attain considerable height, depending on amounts and seasonality of water availability.

Typical Rocky Mountain Avifauna

On the basis of their widespread occurrence in the national parks under consideration here (reported from at least five of the eight), a collective list of fifty-three "typical" Rocky Mountain birds can be com-

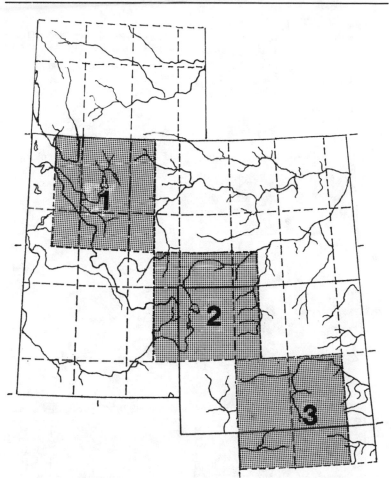

Figure 8. Outline map of region, showing river drainages and (shaded) latilong groupings.

22

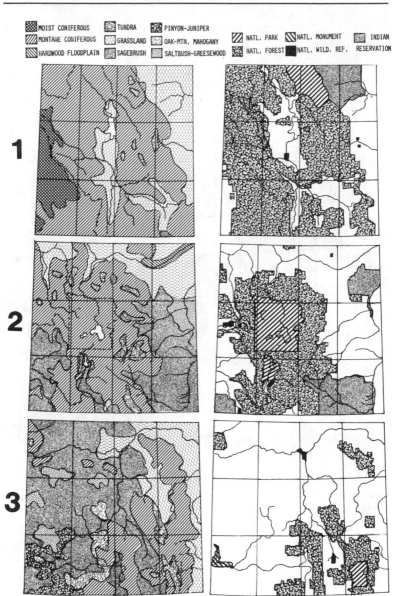

MOIST CONIFEROUS TUNDRA PINYON-JUNIPER

MONTANE CONIFEROUS GRASSLAND OAK-MTN. MAHOGANY

HARDWOOD FLOODPLAIN SAGEBRUSH SALTBUSH-GREESEWOOD

NATL. PARK NATL. MONUMENT INDIAN

NATL. FOREST NATL. WILD. REF. RESERVATION

1

2

3

Figure 9. Details of latilong groupings indicated in Figs. 1 and 8, showing natural vegetation community types (left) and patterns of federal land usage (right).

piled, which provides a nuclear group of birds that might be observed while visiting almost any of the Rocky Mountain parks.

Water-related Species	Forest-related Species	Widespread and Other Species
Mallard	Blue Grouse	Killdeer
Blue-winged Teal	Ruffed Grouse	Common Nighthawk
Common Merganser	Great Horned Owl	Willow Flycatcher
Osprey	Northern Pygmy Owl	Tree Swallow
Spotted Sandpiper	Yellow-bellied Sapsucker	Violet-green Swallow
Common Snipe	Three-toed Woodpecker	Rough-winged Swallow
Belted Kingfisher	Olive-sided Flycatcher	Cliff Swallow
American Dipper	Gray Jay	Barn Swallow
Water Pipit	Steller's Jay	American Robin
Common Yellowthroat	Clark's Nutcracker	Yellow Warbler
Lincoln's Sparrow	Common Raven	McGillivray's Warbler
	Black-capped Chickadee	Chipping Sparrow
	Mountain Chickadee	Vesper Sparrow
	Red-breasted Nuthatch	Savannah Sparrow
	Brown Creeper	Brewer's Blackbird
	Ruby-crowned Kinglet	Rosy Finch
	Mountain Bluebird	
	Townsend's Solitaire	
	Hermit Thrush	
	Yellow-rumped Warbler	
	Wilson's Warbler	
	Western Tanager	
	Fox Sparrow	
	White-crowned Sparrow	
	Dark-eyed Junco	
	Pine Grosbeak	

Widespread Western Avifauna

Besides the "typical" Rocky Mountain avifauna just listed, there are also a considerable number of species that are rather generally distributed through the Rocky Mountain region, even though they may not occur in the majority of the parks covered here. The following is a list of fifty of these, exclusive of those species included in the immediate previous listing.

Water-related Species	Forest-related Species	Widespread and Other Species
Great Blue Heron	Sharp-shinned Hawk	Northern Harrier

Water-related Species	Forest-related Species	Widespread and Other Species
Canada Goose	Cooper's Hawk	Swainson's Hawk
Northern Pintail	Northern Goshawk	Red-tailed Hawk
Green-winged Teal	Downy Woodpecker	Golden Eagle
Sora	Hairy Woodpecker	American Kestrel
Wilson's Phalarope	Common Flicker	Prairie Falcon
Yellow-headed	Western Wood-pewee	Rock Dove
Blackbird	House Wren	Mourning Dove
Red-winged Blackbird	Golden-crowned	Western Kingbird
	Kinglet	Eastern Kingbird
	Veery	Horned Lark
	Swainson's Thrush	Black-billed Magpie
	Cedar Waxwing	Loggerhead shrike
	Solitary Vireo	European Starling
	Warbling Vireo	Lazuli Bunting
	Orange-crowned	Western Meadowlark
	Warbler	Common Grackle
	American Redstart	Brown-headed Cowbird
	Black-headed Grosbeak	American Goldfinch
	Rufous-sided Towhee	House Sparrow
	Northern Oriole	
	Cassin's Finch	
	Red Crossbill	
	Pine Siskin	

Individual Park Specialties

Nearly every one of the Rocky Mountain parks, because of its location or special ecological attributes, offers the bird watcher a few species not found, or at least not so easily found, in any of the other parks. The following is a list of the relatively unique breeding species that might be looked for in particular when visiting one of these localities during summer, although some of them are relatively rare.

U.S. SITES

Dinosaur Natl. Mon.	Rocky Mountain Natl. Park	Grand Teton Natl. Park
Turkey Vulture	Band-tailed Pigeon	Trumpeter Swan
Plain Titmouse	Common Poorwill	Sandhill Crane
Bushtit	Pygmy Nuthatch	Sage Grouse
Brown Towhee	Western Bluebird	Black Rosy Finch
Black-throated Gray	Virginia's Warbler	
Warbler	Brown-capped Rosy	
	Finch	

Dinosaur Natl. Mon.	*Rocky Mountain Natl. Park*	*Grand Teton Natl. Park*

Canyon Wren
House Finch
Lesser Goldfinch

Yellowstone Natl. Park	*Glacier Natl. Park*
White Pelican	Hooded Merganser
Double-crested	Vaux's Swift
Cormorant	Chestnut-backed
Caspian Tern	Chickadee
Pinyon Jay (rare)	LeConte's Sparrow
Sage Thrasher (rare)	Chestnut-collared
	Longspur

CANADIAN SITES

Waterton Lakes Natl. Park	*Yoho Natl. Park*
Sharp-tailed Grouse	Barred Owl
Black-billed Cuckoo	Rusty Blackbird
Ruby-throated Humming- bird	
Brown Thrasher	
Ovenbird	
Indigo Bunting	
LeConte's Sparrow	

Banff Natl. Park	*Jasper Natl. Park*
Alder Flycatcher	Greater Yellowlegs
White-throated Sparrow	Willow Ptarmigan
Golden-crowned Sparrow	Palm Warbler
Purple Finch	Rusty Blackbird
	(Plus all Banff specialties)

Substrate-dependent Avifauna

A considerable number of species that breed in the Rocky Mountain region do so under special ecological conditions, and their nest-site or foraging requirements tend to dictate local distributions more so than do vegetational characteristics. These include the following species:

Cliff-nesting Species
Turkey Vulture
Ferruginous Hawk
Peregrine Falcon
Prairie Falcon
Rock Dove
White-throated Swift
Black Swift
Say's Phoebe

Fish-dependent Species
Common Loon
Grebes (especially Western)
American White Pelican
Double-crested Cormorant
Great Blue Heron
Hooded Merganser
Common Merganser
Osprey
Bald Eagle
Caspian Tern
Common Tern
Belted Kingfisher

Nest in Pre-existing Holes
Wood Duck
Barrow's Goldeneye
Bufflehead
Hooded Merganser
Common Merganser
American Kestrel
Flammulated Owl
Western Screech Owl
Northern Pygmy Owl
Barred Owl
Boreal Owl

Northern Saw-whet Owl
Vaux's Swift
Violet-green Swallow
Tree Swallow
Chickadees (all species)
Brown Creeper
House Wren
Western Bluebird
Mountain Bluebird
European Starling
House Sparrow

Self-excavating Hole-nesters
Woodpeckers (all spp.)
Nuthatches (all spp.)

Nest in Banks or Burrows
Belted Kingfisher
Burrowing Owl
Bank Swallow
Rough-winged Swallow

Nest on Low Islands
American White Pelican
Ring-billed Gull
California Gull
Common Tern
Caspian Tern

Nest on Human-made Structures
Rock Dove
Barn Swallow
Cliff Swallow
Eastern Phoebe
Say's Phoebe

Nest along Mountain Streams
Harlequin Duck
American Dipper

Synopsis of Major Bird-watching Areas

Colorado

ROCKY MOUNTAIN NATIONAL PARK. This national park (Fig. 10) encompasses about 417 square miles of Colorado's magnificent Front Range, and altitudes range from Longs Peak, 14,255 feet above sea level, to 7,800 feet at Estes Park. Tourists can drive over the continental divide, at 12,183 feet, on Trail Ridge Road, the highest road in any national park. About one-third of the park is above 11,000 feet in elevation, and there are over 50 square miles of tundra vegetation present. As a result, this park includes many areas of alpine tundra that are easily accessible; there are also large tracts of montane coniferous forest, mainly of ponderosa pine, plus Engelmann spruce and subalpine fir at subalpine levels, and associated birds. Besides a locally available park checklist of birds, there is also an excellent (but out of print) booklet on the birds of this park ("Birds of Rocky Mountain National Park," A. Collister, 1970, Museum Pictorial of the Denver Museum of Natural History). For more information, contact the Superintendent, Estes Park, Colorado 80517.

DINOSAUR NATIONAL MONUMENT. This national monument is located on the Colorado–Utah border, and comprises 320 square miles. The vegetation and topography are not typical Rocky Mountain, but instead are an extension of the basin and range topography of Utah and Nevada. As a result, the bird life is distinctly arid-adapted, and includes many species otherwise occurring only to the west and south. There is no published checklist as of this writing, but the list for Brown's Park National Wildlife Refuge (see below) is probably applicable, and a preliminary mimeographed bird list for the Monument is also available. Species listed as breeding in the Monument in the text of this book include those reported for Brown's Park. For further information, contact the Superintendent, Box 128, Jensen, Utah 84035.

BROWN'S PARK NATIONAL WILDLIFE REFUGE. This is a refuge of more than 13,000 acres in extreme northwestern Colorado along the Green River and adjoining Dinosaur National Monument. It consists mostly of mountain meadows and rocky slopes, bluffs and marshy habitats as well as the Green River itself. A bird checklist is available. For more information, contact the Refuge Manager, Greystone Route, Maybell, Colorado 81640.

ARAPAHO NATIONAL WILDLIFE REFUGE. This 12,814-acre refuge in north-central Colorado is located near the Continental Divide, at about

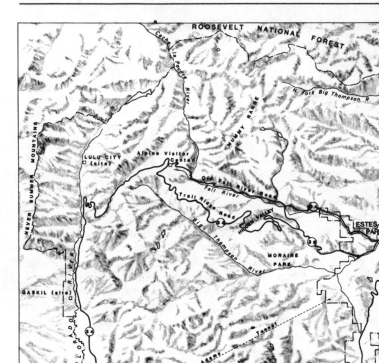

Figure 10. Outline map of Rocky Mountain National Park

8,300 feet elevation, in an arid rain-shadow area only about 60 miles away from Rocky Mountain National Park. It is located in the North Park region of Colorado, with "park" indicating a mountain meadow or relatively treeless area in an otherwise generally forested region. A bird checklist is available. For further information, contact the Refuge Manager, Box 457, Walden, Colorado 80480.

Wyoming

GRAND TETON NATIONAL PARK. This national park (Fig. 11) includes Jackson Hole and the adjacent Teton Range, and consists of more than 480 square miles of high mountains (maximum height 13,766 feet), coniferous forests, and sage-covered plains between 6,000 and 7,000 feet. All the roads are limited to the lower elevations, but many hiking trails extend into the mountain forests and alpine zones, where many birding opportunities exist. A checklist of Grand Teton birds has been published by the Grand Teton Natural History Association and the National Park Service, and another covering all of Jackson Hole has been produced by the Wyoming Game and Fish Department (260 Buena Vista, Lands, Wyoming 82520). A more extended coverage, illustrated with color photos of many of the commoner species, is "Birds of Yellowstone and Grand Teton National Parks," by D. Follett, Yellowstone Library and Museum Association in cooperation with the National Park Service. For more information, contact the Superintendent, Moose, Wyoming 83012. For information, including a checklist of 173 bird species that have been observed on the adjoining National Elk Refuge, contact the Refuge Manager, P.O. Box C, Jackson, Wyoming 83001.

YELLOWSTONE NATIONAL PARK. This, the oldest national park (Fig. 12), is also the largest south of Canada, covering more than 3,400 square miles. Its highest point (Eagle Peak) is 11,358 feet. Most of the roads are in excess of 7,000 feet, with passes as high as 8,850 feet. The majority of the land area of the park is covered with dense lodgepole pine forests, with Engelmann spruce and subalpine fir at the higher elevations, and some fairly extensive areas of sage-dominated grasslands between 5,000 and 7,500 feet. Yellowstone supports the only nesting colony of white pelicans in any national park, and is also notable for its nesting populations of bald eagles, ospreys, and trumpeter swans. Besides the booklet mentioned in the account of Grand Teton National Park, there is also a checklist published by the Yellowstone Library and Museum Association. For more information, contact the Superintendent, Mammoth Hot Springs, Wyoming 82190.

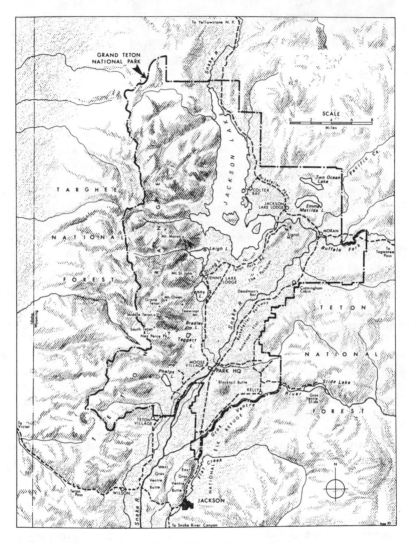

Figure 11. Outline map of Grand Teton National Park.

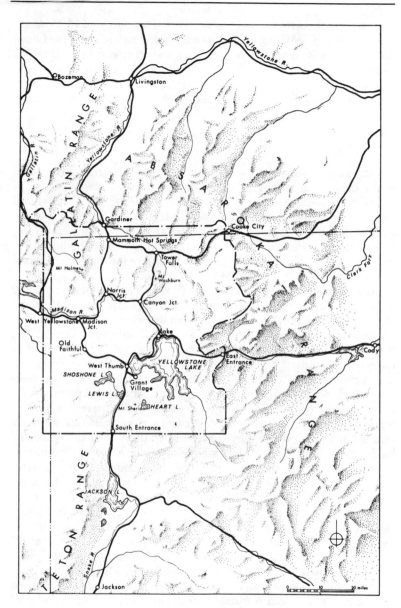

Figure 12. Outline map of Yellowstone National Park.

SEEDSKADEE NATIONAL WILDLIFE REFUGE. This refuge of 14,455 acres in southwestern Wyoming lies at about 6,000 feet elevation, in the Green River Valley. It is dominated by sagebrush and other arid-adapted plants, and has an associated bird fauna, together with bluff-associated and riparian woodland birds, although no refuge checklist is yet available. Further information can be obtained from the Refuge Manager, P.O. Box 67, Green River, Wyoming 82935.

Montana

RED ROCK LAKES NATIONAL WILDLIFE REFUGE. This refuge is located in the Centennial Valley of southwestern Montana, directly west of Yellowstone National Park. It encompasses some 63 square miles and its elevations range from 6,000 to nearly 10,000 feet. Its centerpiece is a 12,000-acre marsh that is a primary breeding area for trumpeter swans, as well as more than 20 other species of waterfowl. A checklist of 209 refuge birds (144 breeders) is available. For more information, contact the Manager, Box 15, Lima, Montana 59739.

GLACIER NATIONAL PARK. This park (Fig. 13) consists of nearly 1,600 square miles in northwestern Montana, and includes some of the most spectacular glacial topography to be seen anywhere south of Canada. Logan Pass at 6,664 feet lies at the crest of the Continental Divide. The park ranges in altitude from 10,243 feet to slightly over 3,000 feet at its eastern boundary. Slightly more than a third of the park lies in the alpine zone. Below 7,000–8,000 feet the park is largely covered by montane coniferous forest, especially of lodgepole pine, Engelmann spruce, and subalpine fir, but on the lower western slopes there is a moister phase dominated by western red cedar and western hemlock. A major bird attraction is provided by the concentrations of several hundred bald eagles (444 in 1977) at McDonald Creek each October. In addition to an excellent early analysis of the bird and mammal life in the park, "Wild Animals of Glacier National Park," 1918, by V. and F. M. Bailey, U.S. Government Printing Office, Washington, D.C.), there is a more recent "Birds of Glacier National Park," 1964, by L. P. Parratt, Special Bulletin No. 9, Glacier Natural History Association, and U.S. National Park Service. A checklist of Glacier National Park birds is also available from the Glacier Natural History Association. For more information, contact the Superintendent, West Glacier, Montana 59936.

NATIONAL BISON RANGE. This area of steep hills and narrow canyons lies southwest of Glacier National Park in the Flathead Valley, and consists of nearly 30 square miles of grassland and forest habitats. A list of 187 bird species includes breeding golden eagles, wood ducks, and

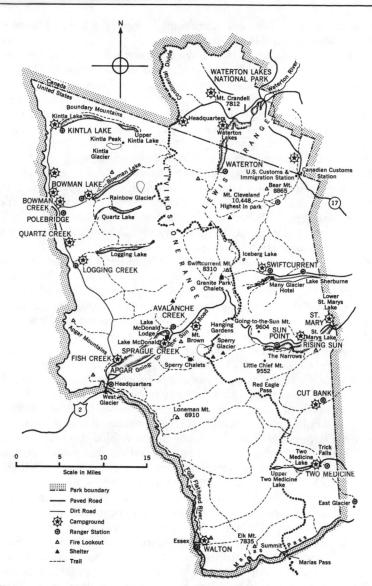

Figure 13. Outline map of Glacier and Waterton Lakes National Parks. From *A Guide to the National Parks*, Volume I by William H. Matthews III, copyright 1968 by William H. Matthews III. Reprinted by permission of Doubleday & Company, Inc.

hooded mergansers. For more information, contact the Refuge Manager, Moiese, Montana 59824.

NINEPIPE AND PABLO NATIONAL WILDLIFE REFUGES. These refuges are northeast of the National Bison Range, and collectively consist of about seven square miles of water, marsh, and upland grasslands. A collective bird checklist of 188 species includes five species of nesting grebes and 16 species of nesting waterfowl. For more information, contact the Refuge Manager, Moiese, Montana 59824.

CHARLES M. RUSSELL NATIONAL WILDLIFE REFUGE. This refuge consists of some 1,700 square miles of grassland habitats, in addition to Fort Peck Reservoir. A bird checklist of 236 species is available, and includes such breeding species as ferruginous hawk, golden eagle, prairie falcon, piping plover, and mountain plover. For more information, contact the Refuge Manager, P.O. Box 110, Lewistown, Montana 59457.

MEDICINE LAKE NATIONAL WILDLIFE REFUGE. This refuge in northeastern Montana consists of marshes and grasslands, with many typical prairie nesting species. The bird checklist includes 221 species, and has such prairie breeders as Sprague's pipit, Baird's sparrow, LeConte's sparrow, and chestnut-collared longspur, plus some colonial marsh nesters such as double-crested cormorant, American white pelican, and California and ring-billed gulls. For more information, contact the Refuge Manager, Medicine Lake, Montana 59247.

BENTON LAKE NATIONAL WILDLIFE REFUGE. This refuge of more than 12,000 acres consists mostly of grasslands and associated wetlands, and its bird life consists of a variety of shorebirds, waterfowl, and marsh birds, as well as a considerable diversity of raptors. It is located about 10 miles north of Great Falls, Montana. Breeding colonies of eared grebes and Franklin's gulls are among the conspicuous marsh birds, and there is a large population of breeding ducks. A bird checklist is available that lists 175 species, including 59 breeders. For more information, contact the Refuge Manager, Benton Lake National Wildlife Refuge, P.O. Box 450, Black Eagle, Montana 59414.

BOWDOIN NATIONAL WILDLIFE REFUGE. This refuge of some 15,000 acres is similar to the preceding one in that it consists mostly of grasslands and associated wetland habitats. It is located about 7 miles east of Malta, Montana. Colonial nesting marsh birds include American white pelicans, double-crested cormorants, eared grebes, and California, Franklin's, and ring-billed gulls. A bird checklist is available that lists 207 species, including 96 breeders. For more information, contact the Refuge Manager, P.O. Box J, Malta, Montana 59538.

METCALF NATIONAL WILDLIFE REFUGE. This rather small (2,700-acre) refuge is located in the Bitterroot Mountains of Montana, about 25 miles south of Missoula. It is centered along the Bitterroot River, and includes coniferous woodland as well as open field and aquatic habitats. For more information, contact the Refuge Manager, P.O. Box 257, Stevensville, Montana 59870.

Idaho

GRAY'S LAKE NATIONAL WILDLIFE REFUGE. This refuge of some 15,000 acres in southeastern Idaho is located about 35 miles north of Soda Springs, and supports the densest known breeding population of greater sandhill cranes. It also is the location of the egg-transplant experiment involving whooping cranes, and thus it is the best location in the region for seeing this rare species. A bird list is available that totals over 160 species, including 69 breeding species. For more information, contact Refuge Manager, Box 837, Soda Springs, Idaho 83276.

CAMAS NATIONAL WILDLIFE REFUGE. This is a little-visited refuge of more than 10,000 acres, located about 4 miles northwest of Hamer, Idaho. It consists of a diverse array of habitats ranging from wetlands to prairie, irrigated meadows, and sagebrush. For more information, contact the Refuge Manager, Hamer, Idaho 83425.

KOOTENAI NATIONAL WILDLIFE REFUGE. This small refuge of less than 3,000 acres in extreme northern Idaho is located about five miles west of Bonners Ferry. It consists of montane forest and marshland habitats, and the refuge checklist lists 218 species of birds, including 85 breeders. For more information, contact the Refuge Manager, Star Route 1, Box 160, Bonners Ferry, Idaho 83805.

Alberta

WATERTON LAKES NATIONAL PARK. This national park (Fig. 13) is contiguous with Glacier National Park (U.S.), and shares many of the same habitat types and topographic characteristics. It covers 203 square miles, and has a maximum elevation of 8,833 feet. A bilingual bird checklist is available. For more information, contact the Superintendent, Watertown Park, Alberta T0K 2M0.

BANFF NATIONAL PARK. This national park (Fig. 14) encompasses some 2,564 square miles along the western border of Alberta, and is largely a high montane park, with a maximum elevation of about 11,500 feet. A considerable part of the park is above timberline, and several large glaciers are found within its boundaries. A bilingual bird checklist covering both Banff and Jasper national parks is available. For more information, contact the Superintendent, Box 900, Banff, Alberta T0L 0C0.

JASPER NATIONAL PARK. This national park (Fig. 14) consists of some 4,200 square miles of high mountain scenery along the western border of Alberta, directly north of Banff National Park. Its highest point is 12,294 feet, the highest point in Alberta, but some of the river valleys are only slightly above 3,000 feet, providing an enormous vertical habitat range. A bilingual checklist of birds is available, and a birder's guide to the park has also been recently published (*Alberta Naturalist* 11:134–140, 1981). For more information, contact the Superintendent, Box 10, Jasper, Alberta T0E 1E0.

British Columbia

YOHO NATIONAL PARK. This national park (Fig. 14) encompasses 507 square miles of montane habitats, and has a maximum elevation of 10,346 feet. It lies along the western border of Banff National Park, and shares many of the same species with that park. A bilingual bird checklist is available. For more information, contact the Superintendent, Field, British Columbia V0A 1G0.

KOOTENAY NATIONAL PARK. This national park (Fig. 14) encompasses 543 square miles, and is contiguous with Yoho National Park on the north and with Banff National Park on the east. Its highest elevation is 10,511 feet, on the western edge of the park, but much of the area along the highway is at elevations of under 5,000 feet. A checklist of the park's birds is available. For more information, contact the Superintendent, Radium Hot Springs, British Columbia V0A 1M0.

GLACIER NATIONAL PARK AND MT. REVELSTOKE NATIONAL PARK. These two national parks (Fig. 14) occur west of the main chain of mountains forming the Continental Divide, and are not dealt with in detail here. Their topography and bird life are very similar to those of Yoho and Kootenay parks. A preliminary bird checklist for Glacier National Park is available. For more information, contact the Superintendent, Box 350, Revelstone, British Columbia V0E 2S0.

Figure 14. Outline map of Canadian national and provincial parks and wilderness areas in the region covered by this book. Map by the author.

38

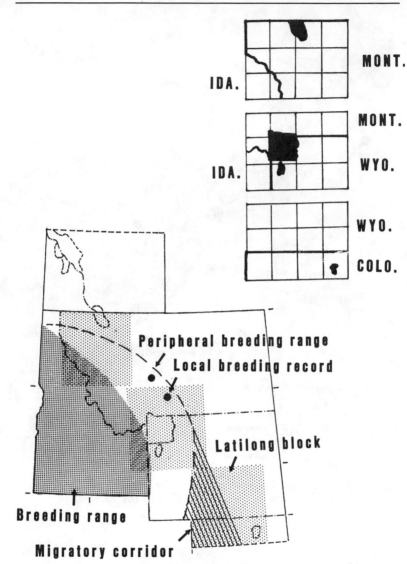

Figure 15. Hypothetical range map, indicating meanings of graphic symbols, and showing (light shading) latilong blocks summarized in text accounts.

Checklist of Birds of the Rocky Mountain Parks

	UNITED STATES PARKS				CANADIAN PARKS			
	Rocky Mountain	Grand Teton	Yellow-stone	Glacier	Waterton Lakes	Kootenay	Yoho	Banff/ Jasper
Red-throated Loon				V	V			V
Arctic Loon		V		V			V	
Common Loon*	M	S	S	S	s	s	s	S
Pied-billed Grebe*	s	S	S	s	S			S
Horned Grebe*		M	S	s	M	M	M	s
Red-necked Grebe*		M	V	s	S		M	S
Eared Grebe*		M	S	s	M	M	M	M
Western Grebe*	M	S	s	S	M	M	M	M
American White Pelican*	s	s	S	s				V
Double-crested Cormorant*		s	S	S	V			
American Bittern*	s	S	S	V	M		M	S
Least Bittern*	V							
Great Blue Heron*	M	S	S	S	s	s	M	V
Great (Common) Egret*		V	V					
Snowy Egret*	M	M	M					
Cattle Egret*	V	V						
Green-backed (Green) Heron*								V

Symbols: R = breeding resident; r = resident, breeding unproven; S = breeding summer resident; s = summer resident, breeding unproven; M = migrant, including wintering visitors; V = vagrant, out of normal range; X = extirpated from area; * = species also included in the Birds of the Great Plains (Johnsgard, 1979).

Checklist of Birds of the Rocky Mountain Parks

	UNITED STATES PARKS				CANADIAN PARKS			
	Rocky Mountain	Grand Teton	Yellowstone	Glacier	Waterton Lakes	Kootenay	Yoho	Banff/Jasper
White-faced Ibis*	M	M	V					
Tundra (Whistling) Swan	M	M	M	M	M	V	M	R
Trumpeter Swan*		R	R	M				V
Greater White-fronted Goose		V						
Snow Goose	V	M	M	M	M	V	M	M
Ross' Goose	V		V	V				
Canada Goose*	M	R	R	R	S	M	S	S
Wood Duck*	V	M	S	S	M		M	S
Green-winged Teal*	s	R	R	s	s	S	S	S
American Black Duck*			V					
Mallard*	r	R	R	R	S	S	S	R
Northern Pintail*	M	R	R	r	M	s	s	S
Blue-winged Teal*	S	S	S	s	S	S	s	S
Cinnamon Teal*	V	S	S	s	S		M	s
Northern Shoveler*	S	S	S	s	M	M	M	M
Gadwall*	s	R	R	S	S		M	M
Eurasian Wigeon		V	V	V				
American Wigeon (Baldpate)*	M	S	S	s	s	M	M	S
Canvasback*	M	S	M	M	M		M	M
Redhead*	M	S	S	M	M		M	M
Ring-necked Duck*	M	S	S	S	M	s	S	S
Greater Scaup	M	V	M	V	M		S	S

Species								
Lesser Scaup*	M	s	S	M	M	M	S	S	S
Harlequin Duck	S	S		S	S	S	S	S	S
Oldsquaw	M		M	M	M	M	M	M	M
Black Scoter	V				V	V			
Surf Scoter	V	V	V						
White-winged Scoter*	V	V	M	M	M	M	M	M	M
Common Goldeneye	R	R	R	M	M	M	M	M	M
Barrow's Goldeneye	R	R	R	R	R	R	s	S	S
Bufflehead*	R	R	R	M	M	S	s	S	S
Hooded Merganser*	M	M	M	S	S	M	M	M	M
Common Merganser*	R	R	R	R	S	R	S	S	S
Red-breasted Merganser	M	M	M	S	M	S	M	M	M
Ruddy Duck*	M	S	S	s	M	M	M	M	M
Turkey Vulture*	s	M	M	M	M				V
Osprey*	S	S	S	S	S	s	M	S	S
Bald Eagle*	M	R	R	R	M	M	s	M	R
Northern Harrier (Marsh Hawk)*	r	R	R	s	S	M	M	M	S
Sharp-shinned Hawk*	r	S	S	R	S	s	s	s	R
Cooper's Hawk*	R	S	S	R	S	s	s	S	S
Northern Goshawk*	R	R	R	R	R	r	s	s	R
Red-shouldered Hawk*			V	V	V				
Broad-winged Hawk*		V	V						V
Swainson's Hawk*	s	S	S	S	s	V	V	V	V
Red-tailed Hawk*	r	R	R	R	S	S	S	S	S
Ferruginous Hawk*	s	S	S	M	M	V	V	V	S

Symbols: **R** = breeding resident; **r** = resident, breeding unproven; **S** = breeding summer resident; **s** = summer resident, breeding unproven; **M** = migrant, including wintering visitors; **V** = vagrant, out of normal range; **X** = extirpated from area; * = species also included in the *Birds of the Great Plains* (Johnsgard, 1979).

Checklist of Birds of the Rocky Mountain Parks

	UNITED STATES PARKS				CANADIAN PARKS			
	Rocky Mountain	Grand Teton	Yellow-stone	Glacier	Waterton Lakes	Kootenay	Yoho	Banff/Jasper
Rough-legged Hawk	M	M	M	M	M		M	M
Golden Eagle*	R	R	R	R	R	s	S	R
American Kestrel*	R	S	S	S	S	S	S	S
Merlin (Pigeon Hawk)*	S	M	M	s	s	s	s	s
Peregrine Falcon*	R	V	S	M	V			M
Gyrfalcon		V		V				V
Prairie Falcon*	R	R	S	S	s	s		M
Gray Partridge*	V	R	V	r	V			
Chukar		r		r				V
Ring-necked Pheasant*				r	V			
Spruce Grouse*	R	R	?	R	R	R	R	R
Blue Grouse		R	R	R	R	R	S	R
Willow Ptarmigan			V	V				R
White-tailed Ptarmigan	R	R	R	R	R	R	R	R
Ruffed Grouse*		R	R	R	R	R	R	R
Sage Grouse*	R	R	r					
Sharp-tailed Grouse*		r	V	r				
Wild Turkey*					R			
Virginia Rail*	R	M	V					V
Sora*	S	S	S	S	s	s	s	S
Common Moorhen*	V							
American Coot*	R	S	S	s	M	S	M	s

Species	1	2	3	4	5	6	7
Sandhill Crane*	M			M			V
Whooping Crane*			V		S	V	
Black-bellied Plover		M	M	M	M	V	V
Lesser Golden Plover		V	V	V	V	V	V
Semipalmated Plover	M	M	S	S	S	M	M
Killdeer*	R	S	M	R	R	S	S
Mountain Plover*		M	M	M	M	V	
Black-necked Stilt*		M	M	M	M	V	
American Avocet*	s	M	M	M	M	V	V
Greater Yellowlegs	M	M	M	M	M	M	S
Lesser Yellowlegs	M	M	M	M	M	M	S
Solitary Sandpiper	M	M	M	M	M	S	s
Willet*	s	s	S	S	V	s	S
Wandering Tattler					V		
Spotted Sandpiper*	S	S	S	S	S	S	S
Upland Sandpiper*		S	S	M	S	V	V
Long-billed Curlew*	s	S	S	M	M	V	V
Hudsonian Godwit		s				V	
Marbled Godwit*	M	M	M	M	M		V
Ruddy Turnstone			V	V	V		
Black Turnstone			M	V			
Sanderling			M	M			V
Semipalmated Sandpiper	M	M	M	M		V	V
Western Sandpiper	M	M	M	M	M	M	M

Symbols: **R**=breeding resident; **r**=resident, breeding unproven; **S**=breeding summer resident; **s**=summer resident, breeding unproven; **M**=migrant, including wintering visitors; **V**=vagrant, out of normal range; **X**=extirpated from area; *=species also included in the *Birds of the Great Plains* (Johnsgard, 1979).

Checklist of Birds of the Rocky Mountain Parks

	UNITED STATES PARKS				CANADIAN PARKS			
	Rocky Mountain	Grand Teton	Yellow-stone	Glacier	Waterton Lakes	Kootenay	Yoho	Banff/Jasper
Least Sandpiper		M	M	M	M	M	M	M
Baird's Sandpiper	M	M	M	M	M	M	M	M
Pectoral Sandpiper		M	M	M	M		M	V
Stilt Sandpiper		M					M	
Short-billed Dowitcher	M	M		M	M	M	M	M
Long-billed Dowitcher	R	R		S				M
Common Snipe*	s	S	S	M	s	s	S	S
Wilson's Phalarope*		S	S	M	S	M	M	s
Red-necked Phalarope	M	M	M	M	M	M	M	s
Red Phalarope								V
Pomarine Jaeger	V							
Parasitic Jaeger	V	V			M			V
Long-tailed Jaeger						V	V	V
Franklin's Gull*	M	M	M	s	M			V
Bonaparte's Gull	M	M	M	M	M	V	M	M
Mew Gull						V	M	V
Ring-billed Gull*	r	M	s	s	s	M	M	s
California Gull*	M	M	S	s	s	M	M	s
Herring Gull	M		M	M			M	M
Sabine's Gull			V					V
Caspian Tern*		M	S	V				M
Common Tern*		M	M	M			V	M

Species							
Forster's Tern*	s	M	M	M	M	M	V
Black Tern*	M	M	S	S	S		M
Rock Dove*	S	r	r		M		R
Band-tailed Pigeon	R	M	M	M	V	M	S
Mourning Dove*		S	S	S	s	s	S
Black-billed Cuckoo*	V	V	V	s	s		
Yellow-billed Cuckoo*		V	V				
Flammulated Owl	X	V				V	
Western Screech Owl	r	r	R	R	M		V
Great Horned Owl*	R	R	R	R	R	R	R
Snowy Owl		V	V	M			V
Northern Hawk Owl			V	V			R
Northern Pygmy Owl	R	r	R	R	r		R
Burrowing Owl*		M	M				V
Barred Owl*		V		R	R	S	R
Great Gray Owl*		R	R	R	R	r	s
Long-eared Owl*	R	R	R	s	r	S	
Short-eared Owl*		R	R	s	M	M	M
Boreal Owl	V	r	V	R		r	R
Northern Saw-whet Owl*	S	R	R	R	M	r	R
Common Nighthawk*	s	S	S	S	s	s	S
Common Poor-will*	S	V	S	S	s		S

Symbols: **R**=breeding resident; **r**=resident, breeding unproven; **S**=breeding summer resident; **s**=summer resident, breeding unproven; **M**=migrant, including wintering visitors; **V**=vagrant, out of normal range; **X**=extirpated from area; *=species also included in the *Birds of the Great Plains* (Johnsgard, 1979).

Checklist of Birds of the Rocky Mountain Parks

	UNITED STATES PARKS					CANADIAN PARKS		
	Rocky Mountain	Grand Teton	Yellow-stone	Glacier	Waterton Lakes	Kootenay	Yoho	Banff/Jasper
Black Swift	S		V	S		s	s	S
Chimney Swift*	V							
Vaux's Swift		s	S	S		V	s	
White-throated Swift*	S	s	S	M				S
Magnificent Hummingbird	V							
Ruby-throated Hummingbird		V						
Black-chinned Hummingbird	M	M		V	s			
Calliope Hummingbird	S	S	S	S	S	S	s	s
Broad-tailed Hummingbird*	S	S	S	s				
Rufous Hummingbird	S	S	S	S	s	s	S	S
Belted Kingfisher*	r	R	R	R	S	S	S	R
Lewis' Woodpecker*	M	S	S	s	s			S
Red-headed Woodpecker*	s	M	M	V	s			
Acorn Woodpecker		V						
Yellow-bellied Sapsucker*	S	S	S	S	S	S	S	S
Williamson's Sapsucker	S	S	S	S	V			
Downy Woodpecker*	R	R	R	R	R	r	M	R
Hairy Woodpecker*	r	R	R	R	R	R	r	R
White-headed Woodpecker		V						
Three-toed Woodpecker	R	R	R	R	S	R	R	R
Black-backed Woodpecker*		R	R	R	S	R	M	r

Northern Flicker*	S	R	R	R	S	S	S	S	S
Pileated Woodpecker*	S	R	V	R	R	R	s	R	R
Olive-sided Flycatcher*	S	S	S	S	S	S	s	S	S
Western Wood Pewee*	S	S	s	s	S	s	s	S	S
Yellow-bellied Flycatcher								M	V
Alder Flycatcher	S	M	S	S	S				
Willow Flycatcher*	S	s	S	S	S	s	s	M	S
Least Flycatcher*	S	S	S	S	S	S	s	s	S
Hammond's Flycatcher	S	s	S	S	S			s	S
Dusky Flycatcher*	S	S	s	S	S	S	S	s	S
Western Flycatcher*	S	s	s	s	S	s		s	s
Eastern Phoebe*				S?				V	s
Say's Phoebe*	M	M	M	S?		M	s	M
Vermilion Flycatcher*	V		V		V	V		
Ash-throated Flycatcher*	X	V		V				
Cassin's Kingbird*	M	V						
Western Kingbird*	M	s	s	s	S	s	V	
Eastern Kingbird*	S	s	s	s	S	S	s	s	S
Scissor-tailed Flycatcher*				V	V			
Horned Lark*	r	r	r	R	S	M	S	M	S
Purple Martin*	V	S						
Tree Swallow*	S	S	S	S	S	S	S	S	S
Violet-green Swallow*	S	S	S	S	S	s	S	S	S

Symbols: **R**=resident, breeding unproven; **r**=resident, breeding unproven; **S**=breeding summer resident; **s**=summer resident, breeding unproven; **M**=migrant, including wintering visitors; **V**=vagrant, out of normal range; **X**=extirpated from area; *****=species also included in the *Birds of the Great Plains* (Johnsgard, 1979).

Checklist of Birds of the Rocky Mountain Parks

	UNITED STATES PARKS				CANADIAN PARKS			
	Rocky Mountain	Grand Teton	Yellow-stone	Glacier	Waterton Lakes	Kootenay	Yoho	Banff/Jasper
Northern Rough-winged Swallow*	s							
Bank Swallow*	S	S	S	S	s	S	S	S
Cliff Swallow*	S	S	S	S	s	S	s	S
Barn Swallow*	S	S	S	S	s	S	S	S
Gray (Canada) Jay*	R	R	R	R	R	R	R	R
Steller's Jay*	R	R	R	R	R	r	R	r
Blue Jay*	V			M	M			M
Scrub Jay	M							
Pinyon Jay*	M	M	s					
Clark's Nutcracker	R	R	R	R	R	R	R	R
Black-billed Magpie*	R	R	R	R	R	M	M	R
American Crow*	r	R	R	S	S	s	S	R
Common Raven*	r	R	R	R	R	R	R	R
Black-capped Chickadee*	R	R	R	R	r	R	R	R
Mountain Chickadee	R	R	R	R	r	r	R	R
Boreal Chickadee*				R	r	r	R	R
Chestnut-backed Chickadee				R	V			
Red-breasted Nuthatch*	R	R	R	R	R	R	R	R
White-breasted Nuthatch*	R	R	R	R	M	M	V	M
Pygmy Nuthatch*	R	r	r					V

Species								
Brown Creeper*	R	R	R	R	s	R	R	R
Rock Wren*	S	S	S	S	s			s
Canyon Wren*	S	S			s			
Bewick's Wren*	V	V						
House Wren*	S	S	S	S	S	R		s
Winter Wren*	M	M	S	S	S		S	S
Marsh Wren*		S	S	V	s			s
American Dipper*	R	R	R	R	R	R	R	R
Golden-crowned Kinglet*	S	r	R	R	s	s	S	S
Ruby-crowned Kinglet*	S	S	S	S	s	s	s	S
Blue-gray Gnatcatcher*	V	V						
Eastern Bluebird*	V		V					V
Western Bluebird*	S	V	S	s				V
Mountain Bluebird*	S	S	R	S	S	S		S
Townsend's Solitaire*	R	R	S	S	S	S	R	R
Veery*	M	s		S	s		V	V
Gray-cheeked Thrush								V
Swainson's Thrush*	S	S	S	S	s	S	S	S
Hermit Thrush*	S	S	S	S	s	S	S	S
American Robin*	R	S	S	S	S	S	S	S
Varied Thrush	V	V	V	S	S			S
Gray Catbird*	s	S	S	S	S	S	M	V
Northern Mockingbird*	V	V						V

Symbols: **R**=breeding resident; **r**=resident, breeding unproven; **S**=breeding summer resident; **s**=summer resident, breeding unproven; **M**=migrant, including wintering visitors; **V**=vagrant, out of normal range; **X**=extirpated from area; *****=species also included in the *Birds of the Great Plains* (Johnsgard, 1979).

Checklist of Birds of the Rocky Mountain Parks

	UNITED STATES PARKS				CANADIAN PARKS			
	Rocky Mountain	Grand Teton	Yellow-stone	Glacier	Waterton Lakes	Kootenay	Yoho	Banff/Jasper
Sage Thrasher*	S	S	S	S				S
Brown Thrasher*	M				s			
Water Pipit	S	s	S	S	s	S	S	S
Sprague's Pipit*	V	V						S
Bohemian Waxwing	M	M	M	R	M	S	M	r
Cedar Waxwing*	r	r	M	S	S	s	S	s
Northern Shrike	M	M	M	M	M	M	M	M
Loggerhead Shrike*	s	s	M	s	M			
European Starling*	R	R	R	R	S	S	R	R
Solitary Vireo*	S	M	M	S	s	s	s	S
Warbling Vireo*	S	S	S	S	s	s	s	S
Philadelphia Vireo*							V	V
Red-eyed Vireo*	s	s	M	S	s	s	s	S
Golden-winged Warbler*	V							
Tennessee Warbler*	M	M		M	s	s	M	S
Orange-crowned Warbler	S	S	S	S	s	s	S	S
Nashville Warbler*	M	V	V	M				M
Virginia's Warbler*	S							
Northern Parula*	V							

Species							
Yellow Warbler*	S	S	S	S	S	s	S
Chestnut-sided Warbler*	V	V	V	V	S	s	M
Magnolia Warbler*	V			V	V	V	M
Cape May Warbler*	V	V	V				
Yellow-rumped Warbler*	S	S	S	S	S	s	S
Black-throated Gray Warbler	V			V			V
Townsend's Warbler	M	M	M	S	S	s	S
Black-throated Green Warbler*	V	V	V		M	s	V
Blackburnian Warbler*	V	V	V	V			
Grace's Warbler	V						
Palm Warbler*	V	V	V		M	s	V
Bay-breasted Warbler*	V	V	V			s	V
Blackpoll Warbler	M			V			s
Black-and-white Warbler*	M	M	M	S	S	s	S
American Redstart*		V	S	V	S	s	S
Prothonotary Warbler*	V	S	V				
Worm-eating Warbler*	V	V					
Ovenbird	V	V	V				
Northern Waterthrush*	M	M	M	S	s	s	M
MacGillivray's Warbler*	s	S	S	S	S	s	S
Common Yellowthroat*	V	S	S	S	S	S	S
Hooded Warbler*	S		S				
Wilson's Warbler		S	S	S	S	s	S
Canada Warbler*							V
Yellow-breasted Chat*	V	V	V	V			

Symbols: **R** = breeding resident; **r** = resident, breeding unproven; **S** = breeding summer resident; **s** = summer resident, breeding unproven; **M** = migrant, including wintering visitors; **V** = vagrant, out of normal range; **X** = extirpated from area; * = species also included in the *Birds of the Great Plains* (Johnsgard, 1979).

Checklist of Birds of the Rocky Mountain Parks

	UNITED STATES PARKS				CANADIAN PARKS			
	Rocky Mountain	Grand Teton	Yellow-stone	Glacier	Waterton Lakes	Kootenay	Yoho	Banff/ Jasper
Hepatic Tanager	V							
Scarlet Tanager*	V		V					
Western Tanager*	S	S	S	S	s	S	s	S
Rose-breasted Grosbeak*	M	M						
Black-headed Grosbeak*	S	S	S	S	s	V		V
Lazuli Bunting*	s	S	M		s	s		
Indigo Bunting	V	V			s		M	
Green-tailed Towhee*	S	S	S	V				
Rufous-sided Towhee*	M	M	s	s	M	V		V
Brown Towhee*	s							
American Tree Sparrow	M	M	M	M	M		M	M
Chipping Sparrow*	S	S	S	S	s	S	S	S
Clay-colored Sparrow*	M	S			x	M	M	S
Brewer's Sparrow*	S	S	S	s	s	s	M	S
Field Sparrow*				V				
Vesper Sparrow*	S	S	S	S	S	s	M	S
Lark Sparrow*	S	M	S	M			V	
Black-throated Sparrow*		V						
Sage Sparrow	V	M	M	s	s			
Lark Bunting*	M	M	M	S	s			V
Savannah Sparrow*	S	S	S	S		s	S	S
Baird's Sparrow*								V

Species	1	2	3	4	5	6	7	8
Grasshopper Sparrow*		S	S	S	s	s	s	
LeConte's Sparrow*	V					s	s	
Fox Sparrow	S	S	S	S	S	S		S
Song Sparrow*	R	R	R	R	R	R	s	S
Lincoln's Sparrow*	S	S	S	S	s	s		S
Swamp Sparrow*		V	M	M	M	M		V
White-throated Sparrow*	M	M	M	M	M	M		S
Golden-crowned Sparrow					s	s		S
White-crowned Sparrow	S	S	S	S	S	S	S	S
Harris' Sparrow	M	M	M	V	V		M	M
Dark-eyed Junco*	R	R	S	R	R	S	R	S
McCown's Longspur*	V	V	M	M				V
Lapland Longspur		M	M	M	M	M		M
Chestnut-collared Longspur*		V	s		s	s		
Snow Bunting	M	M	M	M	M	M	M	M
Bobolink*	V	M	S	S	s	s	M	V
Red-winged Blackbird*	R	R	R	R	R	R	S	S
Western Meadowlark*	S	s	S	S	S	S	M	M
Yellow-headed Blackbird*	M	S	M	M	M	S	M	M
Rusty Blackbird	M	V	M	M	M	M		s
Brewer's Blackbird*	S	S	S	S	S	S	S	S
Common Grackle*	S	M	S	V	V	s		V
Brown-headed Cowbird*	S	S	S	S	S	S	s	S
Orchard Oriole*	V	V	V	V			V	
Northern Oriole*	s	S	S	S	S	s	M	M

Symbols: R=breeding resident; r=resident, breeding unproven; S=breeding summer resident; s=summer resident, breeding unproven; V=vagrant, out of normal range; M=migrant, including wintering visitors; X=extirpated from area; *=species also included in the Birds of the Great Plains (Johnsgard, 1979).

Checklist of Birds of the Rocky Mountain Parks

	UNITED STATES PARKS				CANADIAN PARKS			
	Rocky Mountain	Grand Teton	Yellow-stone	Glacier	Waterton Lakes	Kootenay	Yoho	Banff/ Jasper
Rosy Finch	R	R	R	R	r	r	R	R
Pine Grosbeak	R	R	R	R	r	R	R	R
Purple Finch*				S	s	s	M	S
Cassin's Finch	R	R	R	S	S		s	M
House Finch*	R	s	r		s			
Red Crossbill*	R	r	R	R	r	r	r	R
White-winged Crossbill*	M	V	M	r	r		r	R
Common Redpoll*	M	M	M	M	M	M	M	M
Hoary Redpoll		V			M			M
Pine Siskin*	R	R	R	R	s	s	S	S
Lesser Goldfinch*	M							
American Goldfinch*	s	S	S	s	s		M	s
Evening Grosbeak*	S	R	R	R	r	s	s	R
House Sparrow*	R	R	r	V	S	V	R	R

Symbols: R = breeding resident; r = resident, breeding unproven; S = breeding summer resident; s = summer resident, breeding unproven; M = migrant, including wintering visitors; V = vagrant, out of normal range; X = extirpated from area; * = species also included in the Birds of the Great Plains (Johnsgard, 1979).

Species Accounts

Red-throated Loon (*Gavia stellata*)

Identification: In breeding plumage adults of both sexes can be readily identified by their pale gray head and neck color, except for a reddish brown foreneck and vertical striping down the nape and hindneck. The back lacks the strong white spotting typical of the other two loon species that occur in the area, and the birds are also slightly smaller in size. At close range, the bill appears slightly upturned, although the upper bill profile is almost perfectly straight. Wailing calls are typical on the breeding grounds, and sometimes also are uttered during spring migration.

Status: A vagrant throughout the area, varying from rare to accidental.

Habitats and Ecology: Like the other loons, found on larger rivers and lakes that support good fish populations.

Seasonality: The nearest known area of regular breeding is in northwestern Saskatchewan, although in the early 1970s breeding was suspected in the Caribou Mountains of Alberta. There was a reported case of nesting in Yellowstone National Park in 1929, when a pair of adults with young were reported on July 15 (Komsies, 1930). Oakleaf et al. (1982) consider this as a "historical record"; if valid it represents the only known nesting of the species south of the Canadian border.

LATILONG STATUS

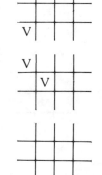

Comments: Red-throated loons as well as arctic loons apparently make direct, perhaps normally non-stop, flights from their coastal wintering areas to their tundra breeding grounds, and thus are unlikely to be seen in this area. Like other loons, they evidently also often migrate at night, and so are rarely seen in the air.

RARE MIGRANT

Suggested Reading: Palmer, 1962; Peterson, 1976.

56

Arctic Loon (*Gavia arctica*)

Identification: In breeding plumage adults of both sexes can be readily identified by their silvery gray head and hindneck, with a black foreneck, and with large patches of white spots on the back surrounded by darker black on the flanks. Immatures or birds in fall plumage are very difficult to separate from other loons, and standard field guides should be consulted. Sometimes the slightly smaller size of the arctic loon helps separate it from the common loon, as does its more sharply defined separation of dark and white areas of the head and foreneck. Separation from the red-throated loon is aided by the more slender and seemingly slightly upturned bill of the latter; in the common and arctic loons the bill tapers gradually to a sharp point.

LATILONG STATUS

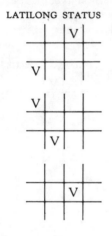

RARE MIGRANT

Status: This is a vagrant species throughout the area, and must be considered as rare to accidental.

Habitats and Ecology: To be expected on large rivers and lakes supporting good fish populations.

Seasonality: The nearest known breeding areas are in extreme northeastern Alberta, near Leland Lake. Otherwise, the species might be observed during spring and fall migration, probably during the same general period as indicated for the other loons.

Comments: Like the other loons, arctic loons are non-gregarious and highly territorial during the breeding season. Their loud, wailing calls, which are associated with this high degree of territoriality, are thus most often heard on the breeding grounds, and the birds are relatively silent at other seasons.

Suggested Readings: Dunker, 1974; Peterson, 1976, 1979.

Common Loon (*Gavia immer*)

Identification: This is the only regularly encountered loon of the area, and in breeding plumage adults of both sexes have uniformly blackish heads except for a large "necklace" of white striping on the sides of the neck near its base, and a much smaller area of similar white striping at the back of the throat. White speckling is evident over much of the back; the spots are smaller than those of the arctic loon and less well isolated from the area of black around them. This is the largest of the three loon species of the area, and for birds in non-breeding (winter or immature) plumage this size (at least 29 inches in length) provides a useful fieldmark. This species is also less strongly and sharply patterned on the head and neck in non-breeding plumage; instead there is a gradual gradation in color from black to white. Strong wailing to laughing calls are commonly heard on the breeding grounds.

Habitats and Ecology: Breeding typically occurs on clear and sometimes deep mountain lakes, where fish are abundant, human disturbance is at a minimum, and where small islands provide nest sites. In some areas muskrat houses or similar artificial islands may also be used.

LATILONG STATUS

s	S	S	M
M	S	S	M
S	M	s	M

M	M	S	M
M	S	M	M
M	S	M	M

s	M	M	M
	M	M	M
M	M	M	M

Seasonality: Spring migration in Montana is from March 25 to May 26, with a peak in late April. Wyoming records show a spring peak in May. There is a Montana egg record for June 6, and a Jasper Park record for May 23. A newly hatched brood was observed in Grand Teton Park (Jackson Lake) on July 12, 1983. Fall migration in Montana is from September 7 to November 18; Wyoming records extend to November 28.

Suggested Reading: Olson and Marshall, 1952; Palmer, 1962.

58

Pied-billed Grebe (*Podilymbus podiceps*)

Identification: This is a relatively common and widespread but inconspicuous species, usually found only in areas of overgrown aquatic vegetation, where it is more often heard than seen. It is the only grebe with a short, chicken-like bill that is ringed with black during the breeding season, and the dark eye is ringed with white. There are no head plumes or bright head colors; the entire bird is brownish to grayish, and the size of a small duck. In summer the loud, often whinnying calls or repeated *cow* notes coming from heavy marsh vegetation provide easy identification.

LATILONG STATUS

S	S	s	S
	S	s	s
S	S	s	M

	S	M	M
S	S	S	M

M	S	s	
M	S	S	S
S	S	S	S

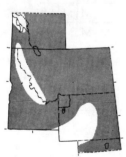

Status: A summer resident on overgrown marshes nearly throughout the area, except in high mountain and semidesert areas. Uncommon to rare in the mountain parks; not yet reported from Yoho and Kootenay parks.

Habitats and Ecology: Primarily found on heavily overgrown ponds or marshes, especially during the breeding season.

Seasonality: Recorded in Montana from the last week of April until the third week of December, and in Wyoming from April 11 to October 25. Nest or egg records for these states range from April 29 to July, and newly hatched young have been observed from June 5 to August 14.

Comments: The unusual bill shape of this species sets it apart from the other North American grebes, and its diet consists of a wide variety of invertebrates including crayfish, leeches, and insects, as well as some fishes. Thus, it often lives in shallow, fish-free ponds, although during migration the birds often are also seen on deeper lakes. Pied-bills tend to leave the area earlier in the fall than other grebes.

Suggested Reading: Glover, 1953; McCallister, 1964; Chabreck, 1963; Kirby, 1976.

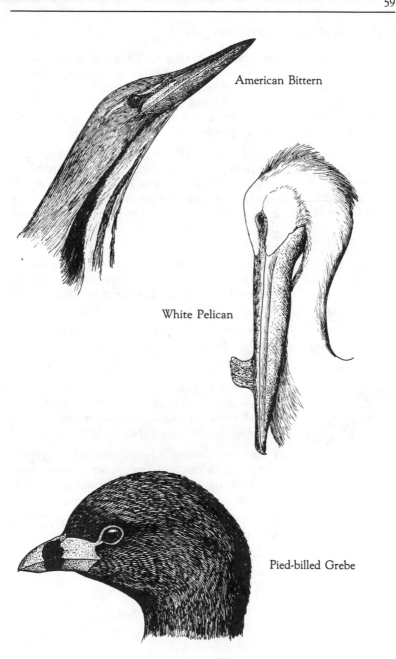

American Bittern

White Pelican

Pied-billed Grebe

Horned Grebe (*Podiceps auritus*)

Identification: In breeding plumage both sexes have a narrow band of golden-yellow feathers extending from the eye to the nape, and a reddish-orange neck and breast. Non-breeding birds have a sharply bicolored head pattern, with a distinct black "cap" well separated from the white cheek area. A wide variety of squeals, croaks and other calls are uttered on the breeding grounds.

LATILONG STATUS

s	S	S	s
M	S	?	S
S	M	M	

M	M	M	
M	S	S	M
M	M		M

M		M	M
		M	M
M		M	M

Status: Breeds in varying abundance in the northern parts of the region, mainly on marshes, ponds and shallow lakes. Has bred on Swiftcurrent Lake, Glacier National Park; may once have bred at Yellowstone but no recent evidence for this. Bred in 1978 east of Yellowstone, at Beck Lake.

Habitats and Ecology: Breeding typically occurs on fairly small ponds (under 2.5 acres), with the pairs scattered and nesting in clumps of rather sparse emergent vegetation.

Seasonality: Spring records in Montana are from April 21 to May 22, and fall migration peaks in late September. Wyoming records are from May 5 to November 11. The earliest egg record is for June 12 (Montana), but young have been reported as early as June 22 in southern Alberta. In Alberta the fall migration usually occurs in October, with a few birds staying on into early November.

Comments: Nesting is largely limited to prairies and foothill areas, and summer occurrences in the mountain parks often appear to be of non-breeding individuals.

Suggested Reading: Fjeldsa, 1973; Faaborg, 1976; McAllister, 1964.

Red-necked Grebe (*Podiceps grisegena*)

Identification: In breeding plumage the black "cap," white cheeks, and rust-colored neck provide a unique combination for grebes. This species is the largest of the three generally dark-colored grebes, and is easily separated from the similar-sized western grebe by color. In non-breeding plumage the large size and rather heavy, tapered and yellowish bill is quite distinctive. Like the other grebes, a variety of rather loud screaming and wailing notes are typical during the breeding season.

Status: Limited as a breeder to the northwestern portion of our area, with park breeding records so far available only for Banff, where it is also regular on the Bow River in spring. It has also been found nesting near Jasper and Watertown lakes. On the lakes of central Alberta it is probably the commonest diving bird, even in areas of rather heavy human use.

LATILONG STATUS

S	S	s	
M	S	S	S
s	M	S	s

M	M		
?	V	M	
	M	S	

			M
			M

Habitats and Ecology: Associated with larger ponds and lakes during the breeding season, and nesting either as solitary pairs or in loose colonies.

Seasonality: Montana migration records are from April 22 to May 22 in spring, and from August 13 to October 17 in fall. Nests in that state have been reported from June 4 until mid-July. That time period also appears typical for southern Saskatchewan. Fall migration in Alberta is usually over by early November. Alberta egg records are from May 24 to June 21.

Comments: Like other grebes, this species apparently migrates at night, and is rarely observed in any kind of flight behavior. Also in common with other grebes, the young are fed for a prolonged period by their parents, and ride on their backs for extended periods prior to fledging.

Suggested Reading: Palmer, 1962; Chamberlain, 1967.

Eared Grebe (*Podiceps nigricallis*)

Identification: This small grebe is the only one that has a black neck and chest in breeding plumage, and a head that is also black except for a fan of golden feathers extending back from the eye. The bill is short and sharply pointed; in non-breeding plumage this short bill, white cheeks that are smudged with black, and a generally dark gray body coloration help to identify the species. On the breeding grounds a frog-like call is often uttered.

LATILONG STATUS

s	S	s	S
M	s	S	S
S	S	s	M

s	M	M	M
S	S	S	S
S	s	s	S

S	S	S	S
	S	S	S
S	M	S	s

Status: A generally widespread breeder in the area, except in the montane lakes. The only park for which there are breeding records is Yellowstone, where breeding has occurred on many lakes and ponds. Occasional breeding in other parks is probable, at least in the foothill areas.

Habitats and Ecology: Associated in the breeding season with rather shallow marshes and lakes having extensive reedbeds and submerged aquatic plants. Generally found in larger and more open ponds than either pied-billed grebes or horned grebes and, unlike these species, typically nesting in large colonies.

Seasonality: Montana migration records are from April 12 to May 14 in spring, and from September 2 to November 23 during fall. Fall records in Wyoming extend to December 4. Egg records for Wyoming are from May 10 to August 14. Colorado egg records are from May 10 to July 20.

Comments: This is the most gregarious of the smaller grebes; only it and the western grebe typically nest in colonies. It also is inclined to migrate in flocks, and often winters in flocks. In North America the species breeds largely in areas of grasslands and semi-arid habitats; it also breeds in Europe and Africa, where it is called the black-necked grebe.

Suggested Reading: McAllister, 1955, 1958; Palmer, 1962.

Western Grebe (*Aechmophorus occidentalis*)

Identification: In all adult plumages this species is easily identified by its strongly black-and-white plumage pattern (black crown and hind-neck, white cheeks and foreneck), and its long, thin and swan-like neck. It is the largest of the grebes in overall length, and has a long, tapering, and stilleto-like bill. During the summer it utters a clear, two-noted whistle, *kreet, kreet,* that in light-phase individuals (which might be a separate species) is only single-noted. These birds also show white extending above the eye, and the bill is more bright orange-yellow rather than greenish yellow.

Status: A local summer breeder over much of the area, with breeding records in the parks apparently confined to Yellowstone (historic records only; no current evidence of breeding) and Glacier (historic record for North Fork of Flathead River).

Habitats and Ecology: Breeding typically occurs on permanent ponds and shallow lakes that are often slightly brackish and have large areas of open water as well as semiopen growths of emergent vegetation.

Seasonality: Montana spring migration dates are from April 20 to May 22, and fall migration is from late August to November 11. Wyoming dates are from April 12 to November 11. Young have been seen in Montana from June 11 to August 5.

Comments: After this book went to press the American Ornithologists' Union taxonomically separated the light-phase form of the western grebe as a separate species, to be called the Clark's grebe (*Aechmophorus clarkii*). In this form the white on the sides of the face extends above the eye, separating the red eye from the black crown, and the bill is more orange-tinted and less yellowish green. Both types occur in our region, but the more southerly Clark's grebe is distinctly rarer here than are the dark-phased birds. Some intermediate plumage types may also rarely occur.

Suggested Reading: Lawrence, 1950; Palmer, 1962; Nuechterlein, 1975.

LATILONG STATUS

s	S	M	S
	S	M	S
S	M	M	M

S	M		M
S	S	S	S
S	S	M	S

M	S	M	M
M	M	S	S
S	M	S	S

American White Pelican (*Pelecanus erythrorhynchus*)

Identification: The huge size, white plumage (except for black wingtips), and enormous yellow bill all make for certain identification of this species. In flight the birds appear ponderous, often gliding between wing-beats, and usually fly in line formation with the head back on the shoulders.

LATILONG STATUS

		M	S
	M	M	S
M	M		s

M	M	s	M
M	S	M	M
s	M	M	S

M	M	M	M
		M	M
M		M	M

Status: Only Yellowstone Park supports a breeding colony of this species, but it is regularly seen in the Teton area as well, especially on Jackson Lake and the adjoining Snake River.

Habitats and Ecology: Associated with lakes and rivers having large fish populations that can be reached by surface-feeding. Gregarious, typically foraging and nesting in groups, and sometimes foraging well away from the nesting grounds, which are typically low islands. The Molly Islands on the southern part of Yellowstone Lake are small and low islands, with extremely limited nesting habitat that is often subject to high wave effects.

Seasonality: Montana migration records are from April 18 to May 27, and from August 15 to October 18. Wyoming records are from April 28, with a peak in May, to October 7, peaking in September. Egg records for these two states are from May 17 to July 10, with young observed as early as June 15.

Comments: The Molly Islands of Yellowstone Lake support an average of more than 180 nests per year (285 in 1980 maximum), and have produced an average of 155 fledglings in 16 of 18 recent years. This is one of the major western breeding areas (*American Birds* 36:250–4); Medicine Lake is another important breeding site in Montana. There are two Idaho breeding records (*Murrelet* 62:19–20). The Montana populations have been in a state of decline since the early 1960s, when over 10,000 breeding birds were present in the state.

Suggested Reading: Palmer, 1962; Schaller, 1964.

Double-crested Cormorant (*Phalacrocorax auritus*)

Identification: The only cormorant of the region; the uniformly glossy black color of adults and orange-yellow throat-pouch are distinctive. In flight the birds resemble geese, but have longer tails, fly with a slight crook in the neck, and appear uniformly black. Immature birds are more brownish, with varying amounts of white on the underparts.

Status: Generally a migrant only in the mountain parks; only Yellowstone has a few breeding birds (on the Molly Islands of Yellowstone Lake).

Habitats and Ecology: Associated with lakes and rivers with good fish populations, often nesting on islands or on cliffs; sometimes also in trees.

Seasonality: Montana migration records are from April 22 to mid-May, and from mid-August to late October. Wyoming records extend from May 7 to November 23. In southern Alberta the birds are present from late April until about mid-September. Egg records for Wyoming are from early June to July 28; in Montana nests have been found as early as late May.

LATILONG STATUS

	M	M	S
		S	S
V			S

S	M	s	S
M	S	M	S
s	s	M	S

M		S	S
	M	S	S
M	V	M	S

Comments: In most areas white pelicans and double-crested cormorants share their breeding areas, and although both species feed on fish they catch them by different means, with the cormorants chasing their prey in underwater pursuit like grebes, loons and mergansers. Like pelicans, cormorants are highly sensitive to disturbance during the breeding season, and colonies should not be visited. Like pelicans too, their foods are primarily non-game fish, and cormorants do little or no damage to sport fisheries.

Suggested Reading: Palmer, 1962; Mitchell, 1967.

Double-crested Cormorant

American Bittern (*Botaurus lentiginosus*)

Identification: This heron is an inconspicuous striped brownish bird of moderate size, usually seen standing among emergent vegetation, often with its bill raised in a concealing posture. In flight it appears uniformly brown, and flies in the usual heron manner with its head back on the shoulders. On the breeding grounds it is more often heard than seen, uttering a low and loud pumping sound associated with territoriality.

LATILONG STATUS

s	S	s	
	s	S	M
S	s	s	s

s	M	M	M
S	S		
S	S	M	S

M		M	
M	M	s	M
S	S	s	s

Status: A widespread but rather inconspicuous breeder, with nesting records for Banff, Yellowstone and the Tetons. Probably breeds locally elsewhere, especially in overgrown edges of beaver ponds or marshes.

Habitats and Ecology: Associated with reedbeds and other emergent marsh vegetation, and rarely observed feeding in open water in the manner of other herons. Foods include frogs, snakes, and other animal life in addition to fish, and thus the species is not limited to areas where fish occur.

Seasonality: Montana records extend from late April to late October, rarely to December. Wyoming records are from April 15 to October 19. Egg records for the region extend from June 1 to June 27.

Comments: I observed courtship and copulation of American bitterns at Christian Pond, Grand Teton National Park, and later found an active nest there (*Auk* 97:868–9). Unlike most herons, bitterns nest solitarily, and their nests are well hidden among marsh vegetation, or sometimes are located on dry land. The nest is fiercely defended by the female.

Suggested Reading: Palmer, 1962; Mousley, 1939; Hancock and Elliott, 1978.

Least Bittern (*Ixobrychus exilis*)

Identification: This tiny bittern is more likely to be seen in flight than while standing, when its large, buffy wing patches are evident and it somewhat resembles a gigantic moth. In shape it resembles a miniature American bittern, but lacks the striping on the breast and neck, and instead is yellowish buff in those areas.

Status: Rare or accidental throughout the area, except in the extreme southeast, where it has bred in the Fort Collins latilong.

Habitats and Ecology: Associated with freshwater or slightly brackish marshes and lakes that have extensive stands of emergent vegetation. Those with scattered woody growth are especially favored habitats.

Seasonality: There are no good migration records for the area. In North Dakota egg records exist for the last half of June, suggesting a seasonality similar to that of the American bittern.

Comments: The nests of this species are usually built above shallow water, and consist of a distinctive structure of leafy materials and twigs that are arranged in a spoke-like manner rising a foot or more above water, and arched over with living vegetation. Nests are often placed fairly close together, so territories are probably fairly small. The male's advertising call is a soft and cuckoo-like series of cooing notes, usually in groups of three to five notes. These are most often heard at dawn and again at dusk.

Suggested Reading: Weller, 1961; Palmer, 1962; Hancock and Elliott, 1978.

LATILONG STATUS

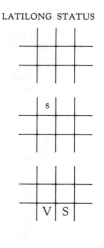

LOCAL MIGRANT

Great Blue Heron (*Ardea herodias*)

Identification: This is the largest of the common herons of the area, and is mainly bluish gray, with a black crown stripe and crest, and a long, yellowish bill. It flies ponderously, with its long legs trailing and its head held back on the shoulders. During the summer it may be seen perching in nesting trees; otherwise it is usually found standing in shallow water.

LATILONG STATUS

S	S	S	s
s	S	S	S
S	S	S	S

S	S	S	S
S	S	M	S
s	S	S	S

S	S	S	S
S	S	S	S
S	S	S	R

Status: Occurs throughout the area, nesting locally wherever conditions permit, but absent from high montane lakes. It nests regularly in Grand Teton, Yellowstone, and Glacier parks.

Habitats and Ecology: This species occurs in a variety of habitats supporting fish life, but usually breeds where there are trees. However, it rarely nests on the ground, on rock ledges, or among bulrushes. Large cottonwoods are a favored location for nesting colonies in the Tetons, such as in the Oxbow area.

Seasonality: Montana migration records extend from March 25 to late September; overwintering occurs occasionally from about Yellowstone Park southward, where open water is present all year. Nests with eggs have been reported in Montana and Wyoming from April 18 to June 13; nestlings have been observed from May 15 to July 28.

Comments: Herons are monogamous and long-lived, and return to the same nesting areas year after year. Old nests are used to establish breeding territories, and the nests are supplemented each year with new materials. Heronries are sensitive to human disturbance, and ravens often use such opportunities to steal eggs or young from the unprotected nests.

Suggested Reading: Pratt, 1970; Mock, 1976; Krebs, 1974.

Great Blue Heron

Great (Common) Egret (*Casmerodius albus*)

Identification: This very large, entirely white bird resembles a great blue heron but is entirely white, with black legs and feet, and lacks head plumes. It is likely to be seen in the same habitats as the great blue heron, namely along the edges of lakes and ponds, fishing in shallow and clear water. In flight it has the same profile as a great blue heron, but is entirely white.

LATILONG STATUS

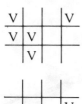

Status: A rare or accidental species over nearly all the area, but breeding has been reported in the extreme southeastern corner (Fort Collins latilong).

Habitats and Ecology: This species occurs in freshwater and brackish habitats, usually foraging in relatively open situations along streams, swamps and lake borders.

Seasonality: There are few migration records for the area, but it is likely to be seen over essentially the same period as the great blue heron, and especially in late summer, when vagrant birds seem to wander freely.

Comments: This species breeds to the east of the region, from Minnesota south to Kansas and Oklahoma, and may gradually be moving northward. Early in this century the species was nearly exterminated in North America by hunters who killed the birds for their plumes, but with protection their range and numbers have slowly increased.

Suggested Reading: Wiese, 1976; Tomlinson, 1976; Hancock and Elliott, 1978.

RARE MIGRANT

Snowy Egret (*Egretta thula*)

Identification: This small egret resembles the great egret, but has a blackish bill and black legs with yellow feet. The head is also distinctly plumed, at least in the breeding season. In flight the yellow-tipped legs are usually evident.

Status: A regular but rare migrant over much of the area, but more common to the south. There are no park breeding records, although summering birds have been seen in Yellowstone and Grand Teton parks. Breeding occurred in the Choteau latilong of Montana in 1979, and has also occurred in the Laramie, Wyoming, latilong.

Habitats and Ecology: These birds occur in a wide range of aquatic habitats, but seem to prefer somewhat sheltered locations for breeding, and often occur in company with other larger heron species. When foraging the birds are fairly active, and sometimes rush about in shallow water in an apparent attempt to flush out their prey.

Seasonality: Montana records extend from May 16 to September 12, and Wyoming records are from April 11 to September 5. Nesting in southern Colorado has been reported from June 1 to June 28, and newly hatched young seen on June 28.

Comments: Like the great egret, this species suffered greatly as a result of plume-hunters, and it has taken a long period for the birds to again become fairly numerous. They nest regularly in southern Colorado, and perhaps the birds are slowly moving northward from this center.

Suggested Reading: Meyeriecks, 1960; Jenni, 1969; Hancock and Elliott, 1978.

LATILONG STATUS

	M		S
	M		M

M			M
M	M		
		M	M	M

M		M	M
M	M	M	S
M	s	S	s

72

Cattle Egret (*Bubulcus ibis*)

Identification: This is a fairly small, dryland foraging egret, usually seen foraging in association with livestock, feeding around their feet. Like the snowy egret it is entirely white, but has yellow legs and a yellow bill; in the breeding season the head and breast area are also tinged with buffy yellow.

LATILONG STATUS

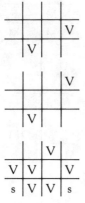

Status: A rare vagrant throughout the region, with no definite nesting records for the parks. There are two breeding records for southern Idaho (*Murrelet* 63:88), and a record for the Greeley latilong of Colorado. Breeding has occurred elsewhere in Colorado, and the species is gradually expanding its range in North America.

Habitats and Ecology: Besides various aquatic habitats, this species also is regularly observed on agricultural lands, especially where there are cattle present. The birds forage on grasshoppers and other insects that are stirred up by the movements of the livestock. The species is highly social, and often nests among colonies of other herons.

Seasonality: There are few records for Montana or Wyoming, but regional records extend from September 12 to December 7.

Comments: This species became self-introduced into eastern North America in 1952, and since that time has gradually spread westward. It has become established as far west as southern California, and in the Rocky Mountain area breeds north to Colorado, with post-breeding wanderers sometimes reaching British Columbia and Alberta.

Suggested Reading: Jenni, 1969; Lancaster, 1970; Hancock and Elliott, 1978.

Green-backed (Northern Green) Heron *(Butorides striatus)*

Identification: This small and stocky heron is generally bluish gray on the upperparts, and has varying amounts of chestnut brown on the head and sides, with little white evident except for one (adults) or several (immatures) streaks down the breast. The neck is rarely extended, and thus the bird appears rather "neckless," especially in flight. A loud, sometimes repeated *skowp* note is often uttered upon flushing.

Status: Generally a rare migrant or vagrant throughout the area, but becoming commoner to the southeast, and breeding locally in Colorado east of the Front Range, just outside the limits of this book.

LATILONG STATUS

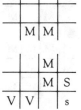

Habitats and Ecology: A wide variety of habitats are used by this adaptable species, which is usually found near trees but also sometimes breeds well away from tree cover. The birds are not very gregarious, and generally are seen as single individuals or territorial pairs. Foraging is done in shallow water; actual baiting of the water to attract prey has been observed in this species.

Seasonality: There are few records for Montana or Wyoming, but in Colorado the species has been observed from April 27 to November 13. In the Great Plains area to the east, nesting records extend from late April or early May to about the middle of June.

Comments: Pairs of this species typically nest in trees, and like other herons the area defended by the male is concentrated on the nest site itself. Various flight displays and displays from perches serve to attract females, and after a pair-bond is formed the female completes the nest. As in other herons, both sexes incubate and care for the young during the nestling period. In some areas two broods are raised in fairly rapid succession.

Suggested Reading: Palmer, 1962; Meyeriecks, 1960; Hancock and Elliott, 1978.

Black-crowned Night Heron (*Nycticorax nycticorax*)

Identification: A small, rather stocky and short-legged heron, with adults having a black crown and back, grayish to white sides and underparts, and bright red eyes. Juveniles also have reddish eyes, but are brownish throughout, with extensive spotting and streaking. A loud, squacking call is usually uttered by birds as they flush.

LATILONG STATUS

	M	
	M	S
V	M	

M	M		M
S	S		M
s	M		M

M	M	M	M
M	M	M	S
S	S	S	S

Status: Generally rare to accidental in the montane parks, but much more common and widespread in prairie and semiarid areas of the region, wherever suitable habitat occurs. The only park where breeding has been reported is Yellowstone, but the species is rare even there.

Habitats and Ecology: This is a highly adaptable species that can use a wide variety of habitats, but in our region it is likely to be associated with shallow bulrush or cattail marshes, often well away from woodlands. The species has very large eyes, and as its name implies, often forages in dim light when it is too dark for most herons to see their prey.

Seasonality: Montana records extend from April 23 to September 29, and Wyoming records are from April 10 to October 21. There are relatively few breeding records, but eggs and young were found in 25 Wyoming nests on June 7, and in northern Utah eggs in an advanced state of incubation were seen on June 3. Colorado nesting records are from May (no date) to July 8.

Comments: As with other herons, males of these species establish a territory that centers around an existing nest or a nest site, and gradually reduce their area of defense to the immediate vicinity of the nest. After finding a mate, the female completes the nest and both sexes participate in the incubation and rearing of the young.

Suggested Reading: Noble, Wurm, and Schmidt, 1938; Allen and Mangels, 1940; Nickell, 1966.

White-faced Ibis (*Plegadis chihi*)

Identification: This unusual wading bird resembles a heron, but has a long, decurved bill and a narrow band of white feathers surrounding a reddish patch of facial skin around the eyes. Otherwise the birds are almost entirely brown to blackish throughout; in flight the neck is stretched out rather than retracted.

Status: A relatively rare migrant or vagrant in southern portions of the area, becoming rarer northwardly. Rarely seen north of Yellowstone Park. Has bred at Brown's Park National Wildlife Refuge adjoining Dinosaur National Monument.

LATILONG STATUS

M		
M		S
V	M	

M		M	M
M	V		
s	M	M	M

M	M	M	
M	M	M	S
S	M	M	s

Habitats and Ecology: Generally associated with freshwater or brackish marshes having an abundance of cattails, bulrushes or phragmites.

Seasonality: A limited number of migration records from Montana are from April 18 to mid-August, while those from Wyoming are from May 6 to September 17. Four nests believed to be of this species were found at Pakowki Lake, Alberta, on July 3, 1975. Colorado records are from May 22 (eggs) to June 20 (nestlings).

Comments: This is an erratic and perhaps eruptive species at the northern part of its range, often appearing and breeding for 'a year or two, and then disappearing again. The birds form monogamous pair bonds, and both sexes help construct the nest, which is often on the ground in dense vegetation, but also may be located in bushes or in trees surrounded by water. As with herons, the young are fed by regurgitation, and by the time they are about seven weeks old they are able to fly with their parents to foraging grounds.

Suggested Reading: Burger and Miller, 1977; Ryder, 1967.

Tundra (Whistling) Swan (*Cygnus columbianus*)

Identification: This is one of two swans native to the area, and the only one likely to be observed in large migratory flocks. It is best told from the very similar but larger trumpeter swan by its higher-pitched, more yelping voice, which lacks the resonant tones of the trumpeter. At close range a yellow spot may be visible just in front of the eye.

LATILONG STATUS

M	M	M	M
M	M	M	M
V	M	M	M

M	M		M
M	M	M	M
M	M	M	M

	M	M	M
		M	M
M		M	M

MIGRANT THROUGHOUT

Status: A migrant throughout, ranging in abundance from rare to fairly common. Probably more common in northern areas, and especially in prairie marshes rather than in forested areas, where trumpeters are more likely to occur.

Habitats and Ecology: On migration, tundra swans frequent favored stopover points that are used every year on their way to and from arctic nesting grounds. These are usually shallow marshes rich in submerged vegetation, which is the major food source for these birds. Field-feeding on dry land has also been observed rarely in migrating birds.

Seasonality: Montana records are from March 27 to May 3, and from September 26 to December 1. Wyoming records are from March 20, peaking in April, and with fall migrants usually gone by the end of November, but with birds rarely overwintering. The nearest breeding grounds are along the west coast of Hudson Bay, in northern Manitoba.

Comments: The swans are among the earliest of the spring migrant waterfowl, often following the breakup of ice by only a few days in order to arrive on their tundra breeding areas as soon as possible after they are snowfree. This allows them the longest possible breeding season, which is required because of their long incubation and fledging periods—often longer than the frost-free periods of the region.

Suggested Reading: Johnsgard, 1975.

Trumpeter Swan (*Cygnus buccinator*)

Identification: This is the largest of all swans, and the one with the most sonorous and penetrating calls, which provide for easy identification. At close range the entirely black bill, without any yellow near the eye, aids in identification. Young swans are much more difficult to identify to species, but are virtually always accompanied by adults.

Status: An uncommon to rare permanent resident in the vicinity of Yellowstone and Grand Teton parks, generally rare or accidental elsewhere. In the late 1970s Yellowstone Park supported about 20 nesting pairs in or very near the park, while Red Rock Lakes Refuge to the west of Yellowstone contained about 27 nesting pairs. Up to a half dozen pairs nest in Grand Teton Park or the general vicinity (including the National Elk Refuge), usually on isolated lakes or beaver ponds. Only local nestings occur outside this region.

LATILONG STATUS

	s	M	M
	M	s	S
V	M	s	

S	M	S	?
S	R	R	
S	R	R	

			V
		s	

Habitats and Ecology: In the Rocky Mountain area this species is mostly limited to fairly large (usually over 30-acre) ponds having considerable aquatic vegetation and relative seclusion from disturbance by humans. Beaver ponds are most often used in the Jackson Hole area, and nests are sometimes built on their lodges.

Seasonality: In the Red Rocks–Yellowstone–Teton area the birds are relatively sedentary, moving off breeding ponds as they freeze over and onto ice-free areas of lakes or rivers. Egg-laying usually begins in late May, and hatching typically occurs before the first week in July. Fledging requires about 100 days, resulting in initial flights by about the end of September.

Suggested Reading: Banko, 1960; Shea, 1979.

Greater White-fronted Goose (*Anser albifrons*)

Identification: This is the more uniformly brown goose in North America, lacking the black head and neck of the Canada goose, and the white head of the blue-phase snow goose. Adults and young have orange-pink bills, and adults have black belly-spotting. The distinctive call, sounding like high-pitched laughter, is perhaps the best fieldmark, and often allows recognition before the birds are even seen.

LATILONG STATUS

LOCAL MIGRANT

Status: A local and relatively rare migrant in the region, becoming more common eastwardly on the Great Plains. Only vagrant birds appear over the montane parks on migration.

Habitats and Ecology: While on migration these birds sometimes fly in company with Canada geese, especially in the case of birds separated from their own flocks. Like Canada geese, they field-forage in grainfields and other croplands, and usually roost in large, open marshes. Breeding is done on open arctic tundra, and in general the birds avoid forested areas.

Seasonality: Migration through the region occurs in March and April, and again in October and November. In some areas of eastern Alberta, such as at Beaverhill Lake, large numbers of these geese stage every year on migration in company with Canada geese and snow geese. The nearest breeding area is only about 50 miles north of the Manitoba border.

Comments: Like other arctic geese, this species varies greatly in its reproductive success from year to year, and the proportion of young birds in the fall and winter flocks provides a valuable index to the success of the previous year's breeding season. Such "age ratios" are important tools in managing goose populations, such as in setting harvest limits and estimating population trends.

Suggested Reading: Johnsgard, 1975.

Snow Goose (*Chen caerulescens*)

Identification: This small goose is found in two color phases; the "snow" phase with a white plumage except for black wingtips, and a "blue" phase, in which only the head and sometimes the neck are white, while most of the body is brownish to dark grayish. In the Rocky Mountain region the vast majority of the birds are of the white plumage type. Interbreeding sometimes occurs, producing birds of intermediate appearance. In all cases, the calls are the same, a dog-like barking or yelping that is easily recognized.

Status: An uncommon to rare migrant throughout the region, becoming more common east of the Rockies, especially to the north.

Habitats and Ecology: Generally associated with large marsh and wetland habitats; feeding in dry fields is done less frequently than in Canada geese, and rootstalks and tubers of marshland plants are more regularly eaten.

Seasonality: In Montana the spring records are from late March to early May, and the fall records are from mid-September to late November. Spring migration in Wyoming peaks in April, and fall migration records are from October 25 to November 11. The nearest breeding grounds are along the west coast of Hudson Bay, in northeastern Manitoba.

Comments: The snow goose is one of the few North American species that exhibits plumage dimorphism unrelated to sex in adults. The plumage type is genetically determined, but these differences do not influence reproductive fertility. Nonetheless, individuals do tend to mate with others of their plumage type, apparently as a result of early imprinting on the plumage type of their parents.

Suggested Reading: Johnsgard, 1975; Cooch, 1958.

LATILONG STATUS

M	M	M	M
M	M	M	M
M	M	M	M

M	M		M
M	M	M	M
M	M		M

M		M	M
M			M
M	M	M	W

MIGRANT THROUGHOUT

Ross' Goose (*Chen rossii*)

Identification: This goose looks like a miniature white-phase snow goose, but additionally has a much higher-pitched call, and a shorter bill, that at very close range can be observed to have a bluish base. However, general size differences of Ross' and snow geese are probably the most useful fieldmark, especially where other geese or ducks are nearby to provide a size comparison.

LATILONG STATUS

M	M	M
M	?	M
M	M	?

M		M	
?			
			M

LOCAL MIGRANT

Status: A highly localized and generally uncommon to rare migrant in the area, with Glacier being the only montane park having any records for the species. However, large flocks of snow geese often have one or more Ross' geese among them, especially in more northerly areas.

Habitats and Ecology: Migrating birds have similar habitat needs to those of snow geese, and often mingle with them, especially in spring. During that time both species often field in grain stubble, where the shorter bill of the Ross' goose presumably allows for closer cropping of grasses or other vegetation.

Seasonality: Montana records are for March 3 to April 25, and again for October. In Alberta the birds usually peak during the last two weeks of September, with some lingering on into October, or somewhat earlier than the main flights of snow geese. They are also regular in late April and early May in such areas as Beaverhill Lake, to the east of Edmonton.

Comments: The Ross' goose populations have apparently increased greatly in recent years, although census data for this species are difficult to obtain accurately, because of its great similarlity to the snow goose. Hybridization between Ross' goose and the snow goose has also been documented in recent years, and adds additional problems in recognizing and managing these birds.

Suggested Reading: Johnsgard, 1975.

Canada Goose (*Branta canadensis*)

Identification: This is the most familiar of all geese, and is easily recognized by the black head and neck, with a white throat and cheek patch. The larger races of Canada geese utter low-pitched honking calls, but the smaller arctic-breeding forms have high-pitched yelping sounds not very different from those of snow geese or Ross' geese.

Status: A widespread, virtually pandemic breeder throughout the area, nesting in nearly all of the montane parks.

Habitats and Ecology: This extremely adaptable goose sometimes nests within the city limits of large cities, but also occurs on prairie marshes, beaver ponds, and forest-edged mountain lakes. Beaver lodges or muskrat houses provide safe and favored nest sites in many areas.

Seasonality: Essentially sedentary over much of the region concerned, moving to open water of large lakes or rivers as breeding areas freeze over, and back again to their breeding territories as soon as weather permits in spring. However, migratory movements do occur in the region, with spring and fall peaks usually in late March and November in the Montana–Wyoming area. Egg records for these two states extend from late March to June 29, while in Jasper Park there are egg records from the first week of May until July 12.

Comments: Populations of Canada geese have increased greatly in recent decades, and the species has now returned to many previously occupied areas from which it was long absent. Canada geese of the Rocky Mountain area represent one of the largest races of this species, while those migrating through the region to and from arctic breeding areas are among the smallest of the races.

Suggested Reading: Johnsgard, 1975; Brakhage, 1965.

LATILONG STATUS

R	R	R	r
R	R	R	R
R	R	R	R

R	R	R	R
R	R	R	R
R	R	R	R

R	R	R	R
R	R	R	R
R	R	S	R

Wood Duck (*Aix sponsa*)

Identification: In breeding plumage the male is unmistakable; the full crest, white throat, and yellowish flank pattern are among the most obvious features. Females and immatures or late-summer males are more difficult, but have a large white eye-ring and generally dark upperparts. In flight the female often utters a squealing alarm whistle, and males also have a whistling note that increases in pitch. Neither sex utters the usual quacking calls associated with ducks.

LATILONG STATUS

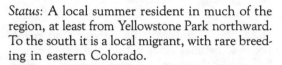

s	S	S	
?	S	S	M
S	S		M

S	s	S	S
?	S		S
s	M	M	M

M		M	M
M	M	s	M
V	M	V	M

Status: A local summer resident in much of the region, at least from Yellowstone Park northward. To the south it is a local migrant, with rare breeding in eastern Colorado.

Habitats and Ecology: During the breeding season these birds are found among woodlands having fairly large trees offering nesting holes, and frequently also those having acorns or similar nut-like foods in abundance. Even outside the breeding season the birds are usually associated with flooded woodlands rather than open marshes.

Seasonality: Montana records extend from March 29 to early May in spring, and from September 14 to November 17 in fall. Most nesting in that state apparently occurs between late May and early July.

Comments: This is the only species of perching duck native to North America, and its relatives are mostly tropical to subtropical species. As in most ducks, pair bonds are reestablished each year during a prolonged period of courtship activity, which in this species is highly developed and spectacular. Females nest in hollow trees, where the inner cavities are at least 8 inches in diameter and with an opening at least 3.5 inches wide.

Suggested Reading: Johnsgard, 1975; Grice and Rogers, 1965.

Green-winged Teal (*Anas crecca*)

Identification: The smallest North American duck; males in breeding plumage have conspicuous yellow triangles at the base of the tail and a white vertical stripe up the side of the breast in front of the wing. Females and late-summer males are best identified by their very small size and a green and black wing speculum pattern (best seen in flight). In spring males often utter sharp *krik-et* whistles; the female has a soft quacking note.

Status: A summer resident over nearly the entire region, probably breeding in all the montane parks except perhaps Rocky Mountain, where it is rare in summer. Also very common on the prairie marshes and foothill areas.

Seasonality: Nearly a year-round resident over much of the region, although the vast majority of birds leave in winter. Montana records are from March 23 to October 30. Spring migration in Wyoming peaks in March, and fall migration in November, with some birds overwintering. Montana nest records extend from May 7 to July 20, and in Wyoming newly hatched ducklings have been observed as late as August 8.

Comments: This small species is able to nest on tiny water areas, usually placing its nest under dense grasses or shrubs that often are some distance from water. The nest is always extremely well concealed from above, and is incubated by the female only. The tiny young grow very rapidly, and in some northern areas may fledge in as few as about 35 days after hatching.

Suggested Reading: Johnsgard, 1975; McKinney, 1965.

LATILONG STATUS

S	R	s	s
s	R	S	S
S	R	R	S

R	r	R	S
S	R	R	R
R	R	R	R

R	R	R	R
R	R	R	R
R	R	S	R

American Black Duck (*Anas rubripes*)

Identification: This mallard-like duck closely resembles a female mallard, but is much darker throughout, and the wing speculum lacks white borders. The only white areas are the underwing coverts, which flash conspicuously in flight. The calls are like those of mallards, and hybrids between the species are relatively common.

LATILONG STATUS

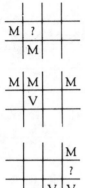

Status: A rare migrant or vagrant over much of the region, but breeding very locally in eastern Alberta. There, nesting has been reported in the vicinity of Kelsey and Hanna, and hybridization has been reported near Calgary.

Habitats and Ecology: This species is similar to the mallard in its foods and general behavior, but is more closely associated with the forests of eastern and northeastern North America. The western edge of its normal breeding range is in eastern Manitoba, with very little nesting in the Canadian prairies.

Seasonality: Migration data are few for this region, but generally black ducks migrate at the same time as mallards, and often with them. Nesting also occurs at about the same time.

RARE MIGRANT

Comments: This species has shown a marked population decline in recent years, apparently as a result of habitat losses, hybridization and competition with mallards, and possible overhunting. Only in extreme northeastern parts of its range is it still fairly secure from hybridization and competition effects, which may eventually spell the end of this form's existence.

Suggested Reading: Johnsgard, 1975; Coulter and Miller, 1968.

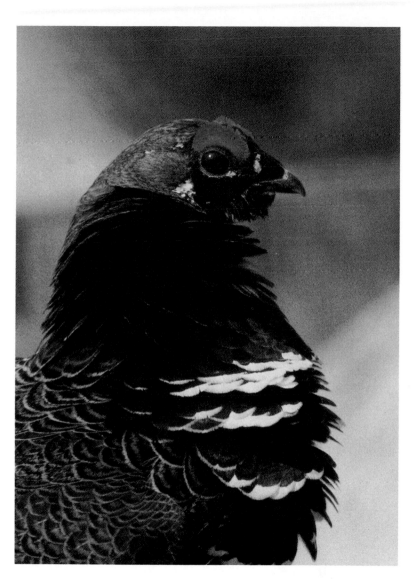

Plate 1. Spruce grouse, displaying male. Photo by author.

Plate 2. American white pelicans and Caspian terns, breeding colony on the Yellowstone Lake. Photo by author.

Plate 3. Double-crested cormorant and California gulls, breeding colony on Yellowstone Lake. Photo by author.

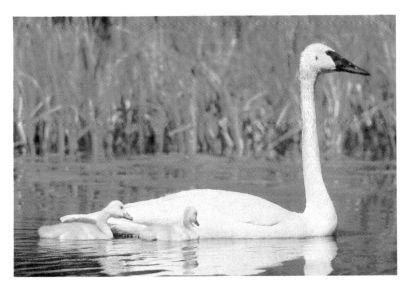

Plate 4. Trumpeter swan, adult and cygnets. Photo by author.

Plate 5. Harlequin duck, male. Photo by author.

Plate 6. Barrow's goldeneye, female and brood. Photo by author.

Plate 7. Common merganser, female and brood. Photo by author.

Plate 8. Osprey, adult. Photo by author.

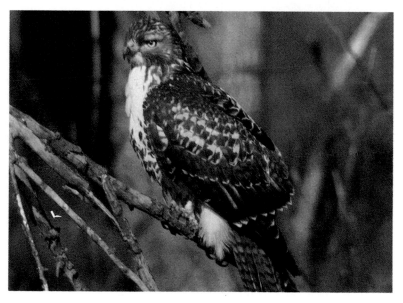

Plate 9. Red-tailed hawk, adult. Photo by author.

Plate 10. Golden eagle, adult. Photo by author.

Plate 11. Prairie falcon, female perched above nest site. Photo by author.

Plate 12. Blue grouse, adult male. Photo by author.

Plate 13. White-tailed ptarmigan, pair in spring plumage. Photo by author.

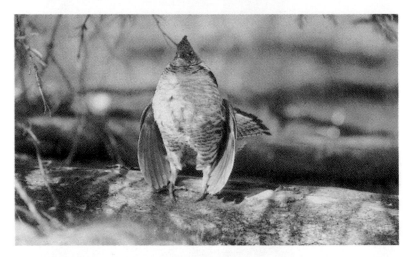

Plate 14. Ruffed grouse, displaying male. Photo by author.

Plate 15. Greater sandhill crane, pair with chick. Photo by author.

Mallard (*Anas platyrhynchos*)

Identification: This familiar "green-head" drake hardly needs description, but females and late-summer males are easily confused with other species, unless the bluish speculum, bordered in front and behind with white, can be seen. Hybrids with black ducks are sometimes present in wild flocks, and are often confusing.

Status: An abundant permanent resident throughout the region, breeding nearly everywhere ponds or marshes occur, and in all of the montane parks.

LATILONG STATUS

R	R	R	R
r	R	R	R
R	R	R	R

R	R	R	R
R	R	R	R
R	R	R	R

R	R	R	R
R	R	R	R
R	R	S	R

Habitats and Ecology: This highly adaptable species nests on nearly aquatic habitats, but prefers non-forested areas over forested ones, and shallow waters over deeper ones. Mallards quickly locate and utilize protected areas, even when close to human activities, and thus remain common in spite of intensive hunting pressures on them.

Seasonality: Mallards are virtually resident in this region, although most birds may leave for the winter. Migration peaks in Montana and Wyoming are in early April and again in October. Nests in Montana and Wyoming have been reported from April 15 to July 21, and in Banff and Jasper parks downy young have been seen between June 8 and July 9.

Comments: This is one of the commonest breeding ducks of the region, and certainly one of the most abundant migrants. During much of the fall and winter mallards are actively forming pairs, and courtship continues on in spring right up to the nesting period. Males abandon their mates as soon as incubation begins, and thereafter the female tends the eggs and young alone. Females often attempt to renest if the first effort is not successful, which accounts for the long spread of nesting records indicated above.

Suggested Reading: Johnsgard, 1975; Girard, 1941.

Northern Pintail (*Anas acuta*)

Identification: Males in breeding plumage have brown heads, a white foreneck with a narrow stripe extending up the side of the nape, and a black rump and elongated black "pin" feathers. Females are almost mallard-sized, but are slimmer, have a narrow blackish bill, and a more uniformly brown head. In flight pintails show grayish underwing linings, and have white only on the trailing edge of the speculum. In spring, males utter fluty whistles both on the water and in the air, during courtship.

LATILONG STATUS

S	R	s	S
s	R	s	S
R	R	r	s

R	M	r	S
R	R	R	R
R	R	R	R

R	R	R	R
R	R	R	R
R	s	S	R

Status: A year-round or summer resident throughout the region, breeding in suitable habitats in most and perhaps all the montane parks.

Habitats and Ecology: This is a tundra- and prairie-adapted breeding species, and it is rarely found in heavily wooded wetlands. It can breed on small and temporary ponds as well as permanent marshes, and frequently nests on dry land in extremely exposed situations well away from water.

Seasonality: Together with the mallard, this is a very early migrant that often overwinters, especially in mild winters. Spring migration in Wyoming and Montana peaks in late March or early April, and is over by early May. Fall migration begins in September or October, peaks in November, with some birds usually remaining until December or wintering on open waters. Egg records in Wyoming are from April 24 onward, with newly hatched young seen as late as July 21. This three-month breeding period is a reflection of early nesting and frequent renesting attempts following initial nest failure.

Suggested Reading: Sowls, 1955; Fuller, 1953; Smith, 1968.

Blue-winged Teal (*Anas discors*)

Identification: Males in breeding plumage have a distinctive white crescent in front of the eye, and an otherwise slate-gray head. Females and late-summer or fall males have a more uniformly grayish brown plumage; females resemble female cinnamon teals but have a pale whitish spot just behind the bill. Both sexes exhibit pale blue upper wing coverts in flight, with a darker speculum behind. Identification of the two teal species during late summer and fall is extremely difficult, and is perhaps best achieved by the somewhat shorter and less spoonlike bill of the blue-winged teal as compared to the cinnamon teal.

Status: A summer resident throughout the area, breeding in most if not all the montane parks. Also common on the grasslands and foothills, especially in the prairie pothole country.

LATILONG STATUS

S	S	s	s
M	S	S	S
S	S	S	S

S	M	S	S
S	S	S	S
S	S	M	S

S	S	S	S
s	S	S	S
S	S	S	S

Habitats and Ecology: This species favors relatively small, shallow marshes over larger and deeper ones, especially those that are surrounded by grass or sedge meadows. Migration in spring is fairly late, as is pair formation, but nonetheless renesting efforts are fairly common following nest failure.

Seasonality: Montana migration records are from April 23 to May 20; Wyoming records extend from April 15 and peak in late April. The fall migration peak in these states is in late September or early October, with few birds remaining beyond the end of October. Wyoming egg records are from June 1 to July 15; Montana records extend from mid-May to August 20. In southern Alberta, newly hatched young have been observed as late as August 1.

Comments: This is a species that is largely eastern in geographic orientation, while the cinnamon teal is more western in distribution. Both are late spring and early fall migrants, with long migration routes that in this species sometimes reach South America.

Suggested Reading: Dane, 1966; Johnsgard, 1975.

Cinnamon Teal (*Anas cyanoptera*)

Identification: Males in breeding plumage are mostly copper-colored, with reddish eyes. Females closely resemble female blue-winged teal, but lack the whitish spots just behind the bill typical of that species. Both sexes resemble blue-winged teal in having pale bluish upper wing coverts that are visible in flight.

LATILONG STATUS

s	S	M	
M	S	s	S
S	S	S	s

S	M	M	S
S	S	S	S
S	S	M	S

S	S	M	M
s	S	S	S
S	S	S	S

Status: A summer resident virtually throughout the entire region, except in the northernmost areas. In Alberta, nesting occurs north to the Tofield area, but apparently not to Banff.

Habitats and Ecology: Associated with small, shallow and often somewhat alkaline marshes of western North America, overlapping with but generally replacing the blue-winged teal in drier regions.

Seasonality: Spring records for Montana are from April 6 to May 31, and fall records are from August 11 to September 11. Wyoming records are from March 27, peaking in April, to October 9, peaking in late September. Wyoming egg records are from June 5 to 15, with young seen as early as June 11, and 15 of 22 Montana nest records are for July. Colorado egg records are from May 3 to June 27.

Comments: Cinnamon teal overlap greatly with blue-winged teal in their nesting ranges, and the two species probably compete with one another. However, they hybridize only very rarely in the wild, in spite of rather similar courtship postures. Male calls and breeding plumages are quite different, however, and this presumably provides an important means of species separation. Where they occur together both often nest on the same marsh areas, and pairs sometimes share adjoining territories.

Suggested Reading: Spencer, 1953; Johnsgard, 1975.

Page 89

Northern Shoveler (Anas clypeata)

Identification: The heavy, spoon-like bill provides a good fieldmark in all plumages, as do the pale blue upper wing coverts of both sexes (also in blue-winged and cinnamon teals). In spring, the male's green head and white breast provide easy identification.

Status: A summer resident essentially throughout the entire region, although uncommon to rare in the montane parks, and common only on wetlands of the plains and foothills.

Habitats and Ecology: The specialized bill of this species allows for filter-feeding of surface organisms, and submerged plants sometimes also provide a supply of organisms that can be reached from the surface.

Seasonality: Spring records in Montana extend from March 24 to May 3, with a peak in late April, while fall records range from late August to late September, with a peak in early September. Wyoming records are from February 26 to November 26. Montana breeding records are from April 18 to the end of July, and Wyoming egg records are from May 25 to June 8.

Comments: This interesting species has a worldwide distribution pattern almost as large as that of the mallard or common pintail, and is replaced by very similar species in South America, Africa, and Australia. All forms forage in essentially the same manner, probably on very similar foods. In spite of their specialized bill shapes, they are all very close relatives of the blue-winged and cinnamon teals.

Suggested Reading: Poston, 1975; March, 1967; Johnsgard, 1975.

LATILONG STATUS

s	S	M	S
M	S	M	S
S	s	M	s

S	M	s	S
S	S	S	S
S	S	M	S

S	S	S	S
s	S	S	S
R	M	S	R

Gadwall (*Anas strepera*)

Identification: This rather inconspicuous duck is often overlooked or misidentified, but in flight can be recognized easily by the white wing patch or speculum on the inner wing feathers. Males in breeding plumage appear mostly gray, except for a black rump, and females are very much like female mallards but have an unspotted orange bill that grades to blackish on the upper ridge.

LATILONG STATUS

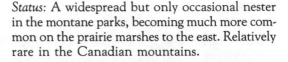

s	M	S	S
M	S	s	S
s	R	s	s

R	M	R	R
R	R	R	R
R	R	M	R

R	R	r	R
r	R	R	R
R	s	S	R

Status: A widespread but only occasional nester in the montane parks, becoming much more common on the prairie marshes to the east. Relatively rare in the Canadian mountains.

Habitats and Ecology: This prairie-adapted dabbling duck prefers shallow marshes with grassy or weedy nesting cover, especially where islands are present.

Seasonality: Spring migration records for Montana are from April 1 to May 10, and fall records are from September 1 to November 18. Wyoming records are from March 14 to November 30, with peaks in April and November. Nesting in Montana extends from late April to the third week of August, and Wyoming egg records are from June 10 to August 1.

Comments: Gadwalls are inconspicuous but extremely interesting ducks that are widespread through the northern hemisphere in grassland and other open-country habitats. Although males lack brilliant coloration, they have a complex courtship display repertoire, and their distinctive throaty calls and whistles add great interest to the spring sounds of the prairie marshes.

Suggested Reading: Oring, 1969; Johnsgard, 1975; Gates, 1962.

Eurasian Wigeon (*Anas penelope*)

Identification: Only males of this species should be identified in the field, since females are extremely similar to those of the next species. Breeding males have a rusty brown head except for a cream-colored forehead and crown, and are grayish rather than reddish brown on the flanks.

Status: This accidental Old World species has occurred rarely in the region, with the largest number of records from Montana, where it has been seen in at least 13 latilongs, while in Wyoming and Colorado it has been reported in 4 and 7 latilongs, respectively.

LATILONG STATUS

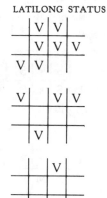

Habitats and Ecology: This species is most likely to be observed in flocks of American Wigeon during spring migration. Its habitats and behavior are virtually the same as in that species.

Seasonality: Too few records are available for estimating migration, but it is only when males are in full breeding plumage from winter until late spring that records for this species are likely to be obtained.

Comments: Although there are still no breeding records for North America, this species is seen every year here, especially in coastal areas. It seems likely that eventually proof of North American breeding will be obtained.

Suggested Reading: Johnsgard, 1975.

ACCIDENTAL VAGRANT

92

American Wigeon (*Anas americana*)

Identification: Breeding-plumage males of this species have white foreheads and a white patch in front of the black rump, but at any distance otherwise appear to be mostly gray to brown. In flight, both sexes show brilliant white upper wing coverts, which easily set them apart from other North American dabbling ducks. Males often utter sharp whistling calls in spring during courtship, which are three-noted, with the middle note loudest.

LATILONG STATUS

S	S	s	S
M	S	s	S
S	R	R	s

R	r	s	S
R	R	R	R
S	R	M	M

R	R	R	R
R	R	R	R
R	S	S	R

Status: A common to rare summer visitor throughout the region, but less common in the montane parks than on the plains, and relatively rare in the Canadian parks, at least in summer.

Habitats and Ecology: Associated with relatively open marshes and lakes having abundant aquatic vegetation at or near the surface, and in the breeding season favoring areas with sedge meadows or with shrubby or partially wooded habitats nearby. Wigeon are strongly vegetarian, and spend more time grazing grassy vegetation than do most ducks.

Seasonality: Spring Montana records are from March 14 to May 14, with a peak in mid-April, and fall records are from early September to November 23, with a peak in late September. Birds sometimes overwinter in Wyoming, but most records are from March 22 to late November, with peaks in April and November. Montana nest records are from mid-May to mid-August, with a peak in early July.

Comments: Social courtship in wigeon is marked by loud calling by males and frequent aerial chases during winter and spring migration. Nests are well hidden, often in sedge cover, and as with other ducks the eggs are tended only by the female.

Suggested Reading: Sowls, 1955; Johnsgard, 1975.

Canvasback (*Aythya valisineria*)

Identification: The long, sloping forehead profile is typical of both sexes, and in breeding males the whitish (not grayish, as in redheads) flanks and red eye color are distinctive. Females are considerably grayer than female redheads, and have a sharper contrast between breast and flank color.

Status: A local and uncommon to rare summer resident over much of the area, being relatively rare in the montane parks (and not known to breed in any), and most abundant in prairie potholes and marshes.

LATILONG STATUS

S	S	M	S
M	s	M	S
S	S	S	M

M	M	M	s
S	M	M	M
S	S	S	M

M	M	S	M
M	S	M	S
S	M	S	s

Habitats and Ecology: In the breeding season, canvasbacks are found on shallow prairie marshes with abundant growths of emergent vegetation and also open water areas that frequently are rich in aquatic plants such as pondweeds.

Seasonality: Spring migration records in Montana are from March 30 to May 9, and fall records are from September 24 to November 24. Wyoming records are from March 9 to November 30, with peaks in April and November. Breeding in Montana occurs from early May to the end of July.

Comments: This fine diving duck has lost much of its breeding range in recent decades, and the prairie pothole country of eastern Alberta is one of its last major nesting grounds. Added to this are heavy hunting pressures on this highly sought after species, which is a major trophy duck among sportsmen.

Suggested Reading: Erickson, 1948; Olson, 1964; Hochbaum, 1944.

Redhead (*Aythya americana*)

Identification: Easily confused with canvasbacks, male redheads in breeding plumage have much grayer flanks and more yellowish eyes, as well as a rounded head profile. Females are more uniformly brown, without the marked contrast between breast and flank color typical of female canvasbacks.

LATILONG STATUS

S	R	s	s
	R	s	S
S	r	S	s

S	M	S	M
S	S	S	S
S	S	S	S

M	S	S	M
s	S	s	S
R	s	S	R

Status: A summer resident over most of the region, and a locally uncommon to rare breeder; rare in the montane parks but fairly common in prairie marshes.

Habitats and Ecology: Breeding habitats consist of nonforested country with water areas sufficiently deep to provide permanent, fairly dense emergent vegetation as nesting cover. Water areas at least an acre in size are preferred for nesting, with substantial areas of open water for taking off and landing.

Seasonality: Spring migration records in Montana are from March 27 to May 9, and fall records are from September 8 to November 29. Some birds overwinter in Wyoming, but there are spring and fall migration peaks in March or April and October or November. Nesting in Montana extends from early May to early August, and in Wyoming flightless young have been observed as late as September 4.

Comments: Like the canvasback, this species is largely associated with shallow prairie marshes, but also occurs in more alkaline wetlands of semidesert regions. Both species often nest in the same marshes, and sometimes females lay their eggs in the nests of the other species.

Suggested Reading: Low, 1945; Olson, 1964; Johnsgard, 1975.

Ring-necked Duck (*Aythya collaris*)

Identification: Males in breeding plumage resemble scaups, but in addition to having blackish heads, breasts and rumps, their backs are also blackish. Females closely resemble female redheads, but have more definite pale eye-rings, more grayish heads, and more grayish and less brownish flank and back coloration.

Status: Relatively common in woodland ponds from the Tetons northward, but rare or absent from prairie marshes during the breeding period.

Habitats and Ecology: Unlike any of its near relatives, the ring-necked duck is strongly associated with beaver ponds and other forest wetlands, where it is often among the commonest of breeding ducks. Sedge-meadow marshes and boggy areas are preferred for nesting, and the presence of water lilies and associated heather cover seem to be an important part of breeding habitats.

Seasonality: Spring migration records in Montana are from April 10 to May 12, and fall records are from September 15 to November 17. Wyoming records extend from April 14 to December 31. Brood records for these states are from June 9 to July 28.

Comments: These birds frequently nest on boggy islands, or hummocks of vegetation in marshy areas, rather than on dry land in the manner of scaups. Ponds that are surrounded by shrubby vegetation seem especially favored.

Suggested Reading: Mendall, 1958; Johnsgard, 1975.

LATILONG STATUS

S	S	s	s
	s	S	M
S	M	M	M

s	M		M
S	S		M
s	S	M	M

M	s	M	M
M	M	M	M
R	s	S	R

Greater Scaup (*Aythya marila*)

Identification: Males in breeding plumage closely resemble lesser scaup, but have a flatter, less peaked head profile, paler gray back color, and are slightly larger than lesser scaup. Females have more white evident around the base of the bill, and both sexes show more white in the primary feathers during flight. While on the water males of both species of scaup tend to appear white in the middle and black on both ends, in contrast to male ring-necked ducks, which have whitish flanks but are relatively dark on the back.

LATILONG STATUS

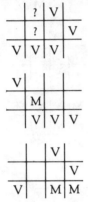

Status: A relatively rare wintering migrant or vagrant throughout the area.

Habitats and Ecology: To be expected on rather large water areas, such as lakes and deeper marshes. The nearest breeding areas are in the Mackenzie District of the Northwest Territories. There, breeding occurs on arctic tundra or in low forests closely adjacent to tundra.

Seasonality: Wyoming records are from October 25 to March 8; Montana records are few, and Colorado records are apparently mostly for October, November, and April.

Comments: Probably this species regularly migrates through the region in small numbers, but is easily mistaken for the lesser scaup, since the birds are rarely seen close to shoreline, and at any distance the two species can be very easily confused.

Suggested Reading: Johnsgard, 1975.

RARE MIGRANT

Lesser Scaup (*Aythya affinis*)

Identification: Male "bluebills" in breeding condition have a rather peaked head profile (reflecting purple in good light), a distinctly barred grayish back pattern, and slightly smaller body size than typical of greater scaups. Females have less white evident behind the bill, and in flight both sexes lack white speculum patterning extending out onto the primaries (which may be pale, however).

Status: A regular migrant and local summer resident through the region, except in the most montane or driest areas. In the montane parks, common only in Yellowstone as a breeder.

Habitats and Ecology: This is largely a prairie-adapted breeder, and is associated also with ponds in the foothill woodlands, especially those supporting good populations of amphipods and other aquatic invertebrates.

Seasonality: Montana migration records are from late March to the end of May, and from September 6 to November 29. Wyoming records are from March 17 to December 11, with peaks in April and November. Montana breeding records are from late May to August 20, and Wyoming egg records are from July 21 to August 7.

Comments: This species, usually called "bluebill" by hunters, is much more adapted to invertebrate animal life than are the redhead, canvasback, and ring-necked duck, and thus has a quite different habitat preference.

Suggested Reading: Gerhman, 1951; Rogers, 1962; Trauger, 1971; Hines, 1977.

LATILONG STATUS

S	M	s	S
M	S	S	S
S	M	S	S

s	M	s	M
S	S	S	R
S	S	M	M

M	S	S	M
s	s	S	S
S	M	S	R

Harlequin Duck (*Histrionicus histrionicus*)

Identification: Males in breeding plumage are unmistakable; they have a gaudy array of white and slate-blue head and body markings, and chestnut flanks. Females are more uniformly grayish brown, with a white spot behind the eye and less clear white markings between the eye and bill. Usually found on torrential mountain streams, where few other ducks occur.

LATILONG STATUS

s	S	S	s
	M	S	M
S	s	S	

s	s		
S	S	M	
	S	M	S

Status: A local summer resident on mountain streams from the Wind River Range northward, becoming commoner to the north. Infrequent and an apparently rare breeder in the Tetons, but regular at LeHardy Rapids in Yellowstone Park. Common at Glacier (Avalanche and Two Medicine Lakes, Roes Creek) and Watertown (Watertown River) and in the rapid streams of the other Canadian parks.

Habitats and Ecology: Associated with clear, rapidly flowing streams, where aquatic insects such as caddis larvae abound; often found where dippers also occur.

Seasonality: There is no good information on migration; the species may be a permanent resident, at least in Wyoming. A few Montana breeding records are for June; young have been seen in July and August. Broods in Jasper Park have been seen from July 19 to August 30.

Comments: This is one of the most beautiful and elusive of all ducks; in spite of the male's bright color he is often hard to see in turbulent water, and the female is even harder to find.

Suggested Reading: Kuchel, 1977; Bengston, 1966a, 1966b.

Oldsquaw *(Clangula hyemalis)*

Identification: Males in fall or winter plumage are mostly white, with elongated tail feathers and a black breast band. Toward summer the entire neck and much of the head turns dark brown. Females are quite variable too, but in winter have mostly white heads with a broad dark patch on the cheeks.

Status: A rare migrant or winter vagrant over much of the area; probably more common to the north. The nearest breeding areas are in northeastern Manitoba, near Hudson Bay.

LATILONG STATUS

Habitats and Ecology: Likely to be observed on reservoirs, lakes, or large rivers, usually far from shore, while on migration. During the breeding season the birds occur on arctic tundra in the vicinity of lakes, ponds, coastlines, or islands. Most wintering occurs in coastal areas, although some deep and large lakes are also used by wintering birds.

Seasonality: Montana records are from mid-October to late April; in Wyoming the birds are most often seen during winter, and in Colorado they have been seen from October 16 to March 7.

Comments: This species is a regular migrant across northern Alberta, but is rather infrequently encountered in more southern areas of that province.

Suggested Reading: Alison, 1975; Johnsgard, 1975.

RARE MIGRANT

Black Scoter (*Melanitta nigra*)

Identification: Adult males have a uniformly black plumage with a bright orange bill-knob. Females and immatures (which are usually seen in this region) are rather brownish and somewhat resemble a female redhead, but the top half of the head is considerably darker than the bottom half, producing a two-toned appearance.

LATILONG STATUS

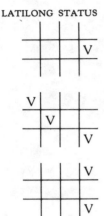

ACCIDENTAL VAGRANT

Status: An accidental vagrant throughout the region. Reported from two Montana latilongs, from 5 in Montana, and 4 in Colorado.

Habitats and Ecology: Likely to be seen on large rivers, lakes or reservoirs in this region. The nearest breeding areas are probably in the Northwest Territories, although most of the North American population breeds in Alaska.

Seasonality: Montana and Wyoming records are for October, November and May; in Colorado there are also October and November records, and probably some birds overwinter on large lakes or reservoirs.

Comments: This is probably the rarest of the scoters in the region, and females may be the most easily overlooked of the three species. Like the other two scoters, this species winters primarily on coastal waters, where it forages for mollusks in tidal areas. In such regions the birds are sometimes called "coots" by hunters, but are quite different from the true coots.

Suggested Reading: Johnsgard, 1975.

Surf Scoter (*Melanitta perspicillata*)

Identification: Adult males have distinctive white nape and forehead patches, and white eyes. Females and immature males, which are likely to occur in this region, have a dark brown body and head, with whitish spotting similar to that of a female harlequin duck, but the spot behind the eye is much larger and more diffuse.

Status: A rare migrant or winter vagrant in the region, more frequent in northern areas. The nearest breeding areas are in northern Alberta, near Lesser Slave Lake. Nonbreeders sometimes summer on montane lakes of Canada.

LATILONG STATUS

Habitats and Ecology: Like the other scoters, this species is most likely to be encountered on large lakes or reservoirs while on migration, and full-plumaged males are almost never seen in this region.

Seasonality: There are no good migration records for Montana or Wyoming; Colorado records are from October to December.

Comments: This species, like other sea ducks, requires two years to attain sexual maturity, and first-year birds may summer south of the breeding grounds. Probably young birds are responsible for most of the U.S. records as well.

Suggested Reading: Johnsgard, 1975.

RARE MIGRANT

White-winged Scoter (*Melanitta fusca*)

Identification: This is the largest of the scoters; males in breeding plumage appear nearly entirely blackish except for a white eye and eye-ring, and a white wing-patch or speculum. Females also have a white wing-patch, but otherwise are rather uniformly brown, with white head markings similar to those of the surf scoter. In flight the scoters are all rather ponderous, and this is especially true of this species.

LATILONG STATUS

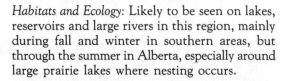

Habitats and Ecology: Likely to be seen on lakes, reservoirs and large rivers in this region, mainly during fall and winter in southern areas, but through the summer in Alberta, especially around large prairie lakes where nesting occurs.

Seasonality: Wyoming records are for the latter half of May and the first half of November; Montana records are from late May to October. Summering birds are also regular in southern Alberta, with some breeding south almost to the Montana border. There are no Montana breeding records.

Comments: This is the most southerly nesting of the three scoters, and it often nests on islands of large lakes in forested as well as grassland habitats.

Suggested Reading: Brown and Brown, 1981; Rawls, 1949; Johnsgard, 1975.

LOCAL MIGRANT

Common Goldeneye (*Bucephala clangula*)

Identification: Males in breeding plumage are mostly white, with a large greenish black head having an oval patch between the eye and bill. Females are almost uniformly grayish, with a dark brown head and a yellow-tipped bill. Immature and non-breeding males closely resemble females.

Status: A local summer resident from approximately Yellowstone Park northward, becoming common in Alberta, and a widespread migrant and wintering species throughout.

LATILONG STATUS

R	R	R	
M	R	R	s
R	R	r	r

r	r	M	r
r	R	R	M
M	R	r	M

M		M	M
M		M	M
W	W	W	W

Habitats and Ecology: During the breeding season goldeneyes of both species are usually found in forested wetland habitats, where large trees offer nesting sites. At other times they occur on deeper and larger bodies of water such as lakes. The breeding status in Grand Teton and Yellowstone parks needs confirmation, as do breeding reports from the Wapati and Dubois latilongs.

Seasonality: This species is resident in Wyoming and Montana, but migration peaks are evident in April and November. Breeding in Montana is from early May to mid-July.

Comments: The common goldeneye has a much broader North American range than does the Barrow's goldeneye, which is largely limited to the western states, plus a very limited Atlantic coast population. The two species overlap greatly, but rarely hybridize.

Suggested Reading: Gibbs, 1961; Carter, 1958; Johnsgard, 1975.

104

Barrow's Goldeneye (*Bucephala islandica*)

Identification: Males in breeding plumage resemble common goldeneyes, but have crescent-shaped facial patches, black extending farther down on the flanks, and exhibit small white wing patches in flight. In this region females of Barrow's goldeneye are best separated from common goldeneye females by their all-yellow bill color (except for a black nail) and slightly flatter head profile.

LATILONG STATUS

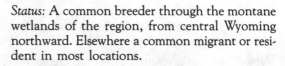

R	R	R	s
	r	S	M
R	r	r	M

r	r	s	r
R	R	M	
R	R	R	R

		M	M
	M		M
V	V	M	W

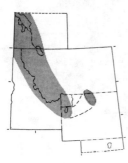

Status: A common breeder through the montane wetlands of the region, from central Wyoming northward. Elsewhere a common migrant or resident in most locations.

Habitats and Ecology: Breeding birds are associated with forested montane lakes, beaver ponds, and slowly flowing rivers in this area; in some regions nesting in cliff or rock crevices also occurs.

Seasonality: Resident in Wyoming and Montana, with migration peaks in April or early May and November. Some birds overwinter as far north as southern Alberta. Broods in Montana and Wyoming from June 28 to late August; in Alberta and Saskatchewan brood records are from June 13 to August 10.

Comments: This is probably the most common breeding duck species in Grand Teton National Park, and also is very common in Yellowstone and Glacier parks. Good data on the relative abundance of Barrow's and common goldeneyes are still lacking for these areas.

Suggested Reading: Johnsgard, 1975; Munro, 1939.

Bufflehead (*Bucephala albeola*)

Identification: This tiny diving duck is easily identified; breeding males are mostly white except for a large dark head with a wedge-shaped white patch behind the eye. Females are very small and dark grayish brown, with a diffuse whitish patch behind and below the eye. Immature males closely resemble females.

Status: A local breeding summer resident from the Tetons northward, mainly in wooded wetlands where tree cavities (especially woodpecker holes) offer nesting sites.

Habitats and Ecology: This species is so small that females can use the old nest holes of flickers (which are also used by bluebirds, starlings, and similar-sized birds) for nesting. Otherwise the birds are generally found on larger and deeper waters.

Seasonality: Montana records are from March 18 to May 31, and from August 27 to November 24, with peaks in late April and October. Wyoming records are mostly from mid-March to late November, with peaks in April or May and October or November, and birds sometimes wintering. There are a few brood records, but a brood on Christian Pond (Grand Teton N.P.) was observed from July 13 until late that month.

Comments: This is the smallest of the North American diving ducks, and one of the most beautiful. In spite of its small size, two years are required to attain sexual maturity, just as in the larger species of sea ducks.

Suggested Reading: Erskine, 1972; Johnsgard, 1975.

LATILONG STATUS

S	S	S	S
	M	M	M
S	s	M	M

M	M	M	M
S	R	M	M
M	R	M	M

M	M	M	M
M	M	M	r
W	M	M	W

106

Hooded Merganser (*Lophodytes cucullatus*)

Identification: Males in breeding plumage have a large, erectile crest that is white, bordered with black, and an otherwise black head. They slightly resemble male buffleheads, but have reddish brown flanks. Females have a full, rusty brown crest that is also somewhat erectile, but are otherwise mostly grayish to brownish. Both sexes have rather slim, pointed bills.

LATILONG STATUS

S	R	S	
	S	S	M
S	R	M	M

M	M		
M	M	M	M
	M	M	M

		M	M
M			M
W		S	M

Status: A local summer resident in wooded areas from western Montana northward; southern breeding limits uncertain. Elsewhere a generally rare migrant, although breeding has been reported for the Powell latilong of Colorado.

Habitats and Ecology: Generally found in river areas bounded by woods and supporting good fish populations associated with clear water.

Seasonality: Montana records are from late February to early April, and again from September to October. In Alberta the birds usually arrive in early May, and sometimes remain on into early November. Breeding in Montana occurs from late June to the end of July. There is a reputed record of a brood near Green River, Wyoming, on June 4, 1929 (McCreary, 1939).

Comments: Hooded mergansers are the smallest North American mergansers, and ecologically approach the goldeneyes, to which they are closely related.

Suggested Reading: Kitchen, 1968; Morse, Jakabosky and McCrow, 1969.

Common Merganser (*Mergus merganser*)

Identification: The largest merganser; males in breeding plumage appear mostly white except for a massive greenish black head and a bright red bill. Females are smaller and mostly brownish, with a contrasting white throat and lower neck. Immatures closely resemble females.

Status: A common resident in most montane rivers during the breeding season, and extending out into nonforested rivers, lakes and reservoirs at other times.

Habitats and Ecology: This fish-eating species occurs in areas of clear water supporting large fish populations, and is much the commonest merganser of the region. Nesting occurs in tree cavities, rock crevices, and sometimes under boulders or dense shrubbery.

Seasonality: A year-round resident throughout the area except perhaps the most northerly regions, but with migration peaks in April and October. Nesting in Montana occurs from late March to late May; Wyoming brood records are from June 12 to July 24, while brood records for Jasper Park are from June 29 to August 8.

Comments: This is one of the commonest ducks of the montane parks, and one of the most beautiful and elegant of all North American waterfowl. Yet it is often wantonly shot by hunters, who usually consider it a "trash duck."

Suggested Reading: White, 1957; Johnsgard, 1975.

LATILONG STATUS

S	R	S	
s	R	s	S
R	R	R	S

R	R	S	R
S	R	R	s
M	R	R	R

R	R	R	R
r	R	R	R
S	R	R	R

Common Merganser

Red-breasted Merganser (*Mergus serrator*)

Identification: Males in breeding plumage have a rather shaggy greenish crest, a brownish breast, and black in front of the flanks. Females closely resemble female common mergansers but are smaller, have a thinner bill, and their head, neck, and breast colors grade into one another gradually, rather than being sharply demarcated.

LATILONG STATUS

	M	?	M
	M	M	M
V	M	M	M

M	M	M	M
M	?	M	
M	M	M	M

		M	
		M	M
			M

Status: A rather uncommon to rare migrant over most of the area, with a few possible scattered breeding records.

Habitats and Ecology: Generally found in similar habitats as the common merganser, but with a more northerly breeding distribution and a more coastal wintering distribution.

Seasonality: Wyoming records are mostly from February 4 to late April, peaking in April, and from October 4 to December 22, while Montana records are from late March to June, peaking in April, and from August to November, peaking in late October.

Comments: Similarities of this species and common mergansers make breeding records suspect, but in addition to reported breeding in Glacier Park (seemingly undocumented), there was an apparent breeding near Dubois, Wyoming (*American Birds* 35:964) and an early but undocumented report of breeding at Yellowstone Park (Meagher, 1963). There have also been other unverified reports of breeding in western Montana and the Green River area of Wyoming.

Suggested Reading: Hilden, 1964; Johnsgard, 1975.

Ruddy Duck (*Oxyura jamaicensis*)

Identification: Males in breeding plumage have a unique ruddy body color, with a bicolored black-and-white head, and a bluish bill. Females are much more uniformly brownish, but have a bicolored brown and white, vaguely streaked head, and a long tail, which is sometimes partially cocked. Non-breeding males resemble females, and lack the bright blue bill coloration.

Status: An occasional to rare summer resident over much of the region, mainly in grassland marshy habitats, becoming rarer in montane areas, and a migrant more or less throughout. Breeding in the montane parks is known only for Grand Teton (Christian Pond).

LATILONG STATUS

S	S	s	
M	S	s	S
S	S	s	s

S	M	S	S
S	S	S	S
S	S	M	S

S	S	S	S
M	M	M	s
S	M	S	S

Habitats and Ecology: Nonbreeding birds are found on larger and generally deeper waters that have silty or muddy bottoms; breeding is on overgrown shallow marshes with abundant emergent vegetation and some open water.

Seasonality: Montana migration records are from April 9 to May 18, and from September 12 to November 3. Wyoming records are from April 6 to November 30, with peaks in April and November. Nest records in both states range from early June to mid-August, with newly hatched young seen from July 15 to August 6.

Comments: This "stiff-tailed" duck is fascinating and amusing; females lay remarkably large eggs for their body size, and are prone to drop them in the nests of others of their species as well as those of other diving ducks.

Suggested Reading: Joyner, 1975; Johnsgard, 1975.

Turkey Vulture (*Cathartes aura*)

Identification: Usually seen in flight, this species soars for long periods on wings that are slightly uptilted and that are two-toned, with black in front and gray behind, and with a blackish body. The head is reddish, and appears small relative to the size of the wings and body.

LATILONG STATUS

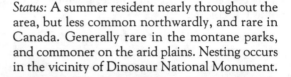

M	M	M	
s	S	s	S
s	S	s	s

s	s	s	s
s	M	s	M
s	M	M	s

M	s	S	S
M	s	s	M
S	S	S	S

Status: A summer resident nearly throughout the area, but less common northwardly, and rare in Canada. Generally rare in the montane parks, and commoner on the arid plains. Nesting occurs in the vicinity of Dinosaur National Monument.

Habitats and Ecology: A scavenger species that consumes only dead remains of large animals such as livestock, which it finds visually. Generally found below 8000 feet, and lower farther north.

Seasonality: Montana records are from March 4 to November 29, and Wyoming records are from April 13 to October 19. There are few nesting records, but Utah and Colorado egg records are from May 20 to June 3, and downy young have been seen as late as July 3.

Comments: Nesting in the Rocky Mountains is usually done on cliff ledges under overhangs, or in rock crevices, often in river valleys. The sites are typically well concealed and shaded, and following a fairly long incubation period of about 40 days the young have a long 70- to 80-day fledging period, which limits nesting to more temperate latitudes.

Suggested Reading: Coles, 1938, 1944; Brown and Amadon, 1968.

Osprey (*Pandion haliaetus*)

Identification: This hawk is unique in its fish-catching adaptations, and the species is limited to areas where fish are found. Its white underside color, with black "wrist-marks" on the wings and a blackish upper surface are distinctive, as are the long wings, which are usually held at a bent angle rather than straight out.

Status: A common summer resident in montane areas near lakes or streams, probably breeding in all the montane parks. Elsewhere mostly a migrant, except around some plains reservoirs, such as Fort Peck, which supports several breeding pairs.

Habitats and Ecology: Commonly seen along clear rivers and lakes, these birds sometimes nest on rock pinnacles (as in Yellowstone Canyon), but more often nest in tall trees near water, in large nests resembling eagle nests.

Seasonality: Montana and Wyoming records are from early April to late October, peaking in late April and September. Nesting in Montana is from late April to late July with eggs usually hatching by mid-June and fledging by mid-August.

Comments: In the late 1970s about 20 osprey pairs nested in Grand Teton N.P., about a third of which hatched young, and brood size averaged 1.4 young. Various lakes supported about twice as many nests as were located along rivers and streams, and the average heights of the nests were 17.2 meters above the ground.

Suggested Reading: Alt, 1980; Swenson, 1975; Green, 1976; Richardson, 1980.

LATILONG STATUS

S	S	S	
S	S	S	s
S	S	S	s

S	s	S	M
S	S	M	M
s	S	S	M

M	M	M	M
		M	M
S	s	S	S

Osprey

Bald Eagle (*Haliaeetus leucocephalus*)

Identification: Adult birds, with their white heads and tails, are unmistakable, but immatures are mostly brown and best told from golden eagles by their relatively heavy bills and underwing linings that are paler than their flight feathers.

LATILONG STATUS

r	R	R	
M	R	R	r
R	R	R	R

R	r	M	r
S	R	R	R
W	R	r	M

M	M	M	M
M	M	R	M
W	R	s	W

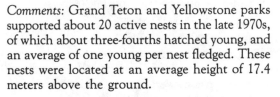

Status: A resident throughout the forested montane areas, breeding in nearly all the montane parks. Otherwise a migrant, and in late fall (especially early November) up to several hundred birds gather in Glacier Park to forage on salmon, producing a unique spectacle.

Habitats and Ecology: This species feeds almost exclusively on nongame fish, such as squawfish, during the breeding season, and also on dying salmon following their spawning period.

Seasonality: Present in much of the area year around, but with major seasonal changes in abundance, with migration peaks in March and November. Egg dates in Wyoming are few, but hatching typically occurs about May 1, with fledging about July 10.

Comments: Grand Teton and Yellowstone parks supported about 20 active nests in the late 1970s, of which about three-fourths hatched young, and an average of one young per nest fledged. These nests were located at an average height of 17.4 meters above the ground.

Suggested Reading: Sherrod et al., 1976; McClelland et al., 1982; Dunstan et al., 1975.

Bald Eagle

Northern Harrier (*Circus cyaneus*)

Identification: This is the only North American hawk with long, narrow wings and a white rump patch; males are mostly silvery gray, while females and immatures are dark brown except for the rump patch. The birds often course low about the ground and marshy areas searching for rodents.

Status: Occurs throughout the region, nesting locally, especially in non-forested habitats such as grasslands, croplands, and meadows.

Habitats and Ecology: Grassy areas, especially those near water, are favored by these birds, which nest on the ground rather than in trees, as with most hawks.

Seasonality: Montana and Wyoming records generally extend from late March to late November, with peaks in April and November, although overwintering often occurs in mild winters. Wyoming egg records are from May 2 to June 6, with nestlings observed as late as August 8.

Comments: This species is called the "hen harrier" in Britain, and has previously been known generally as the "marsh hawk."

Suggested Reading: Snow, 1973b; Hamerstrom, 1969; Hammond & Henry, 1949; Watson, 1977.

LATILONG STATUS

M	r	s	s
s	S	s	S
R	R	S	S

R	r	r	S
s	S	s	S
R	S	s	S

S	S	S	s
S	S	S	S
R	R	S	R

114

Sharp-shinned Hawk (*Accipiter striatus*)

Identification: This is a pigeon-sized hawk with short, rounded wings, a long and square-tipped tail, and usually found in woodland habitats. It flies swiftly, with little gliding, typically catching birds in flight.

LATILONG STATUS

S	S	R	
S	s	s	S
R	R	R	S

S	S	M	r
s	R	M	S
M	S	S	S

s	s	S	M
s	S	S	S
R	s	R	R

Status: A common to rare summer resident or year-round resident of montane woodlands of the region; breeding in most and perhaps all the montane parks. Rarely found far from woodland, even on migration.

Habitats and Ecology: Fairly dense forests, either mixed or coniferous, are the preferred habitats of this species, which is swift and elusive, and usually nests in dense groves of trees. Aspens, riparian woodlands, and coniferous forests are all used for breeding.

Seasonality: Although locally resident in Wyoming, most birds appear by April, and are gone by November. Wyoming egg records are for the first half of June; egg records from Montana are from June 6 to July 3, and nestlings have been seen as late as August 25.

Comments: Perhaps the commonest of the "accipiter" hawks of the area; this species is a major predator on small birds.

Suggested Reading: Wattle, 1973; Reynolds, 1978; Platt, 1976.

Cooper's Hawk (*Accipiter velox*)

Identification: This accipiter hawk is larger than the sharp-shinned, and has a more rounded tail. Both species are swift woodland hawks, and are rarely seen for extended periods in open country.

Status: Found in essentially the same wooded montane and foothills habitats as the previous species, and perhaps slightly less common in most areas.

LATILONG STATUS

S	s	R	s
s	S	S	s
R	S	S	R

s	R	M	s
s	S	S	S
M	S	S	S

s	S	M	M
M	S	s	M
S	S	S	R

Habitats and Ecology: Associated with mature forests, especially deciduous or mixed, and less often in pure coniferous stands. Aspen groves are favored breeding locations; nonbreeders use riparian woodlands, scrub oaks, and mountain meadows.

Seasonality: Occasionally resident in southern areas, but in Wyoming and Montana most birds are present between April 1 and September, with migrants commonly seen in April and May. There are few regional nesting records, but in Colorado egg records extend from May 18 to July 1, and nestlings have been seen as early as July 1. Nestlings in northern Utah have been found by June 19, and fledged young by July 30.

Comments: Like the sharp-shinned hawk, this is an effective bird predator, probably taking slightly larger prey on average than does that species.

Suggested Reading: Craighead & Craighead, 1956; Henny & White, 1972; Meng, 1951, 1952.

116

Northern Goshawk (*Accipiter gentilis*)

Identification: This is the largest accipiter, and the palest underneath, at least in adults. Immatures are more brownish and streaked, but are still large and have conspicuous white eyebrows as in adults.

LATILONG STATUS

R	R	R	
s	r	r	s
R	R	S	R

R	R	M	M
s	R	R	M
r	R	R	R

R	r	r	R
R	R	R	R
R	R	R	R

Status: An uncommon to rare permanent resident in woodland and montane forests of the region.

Habitats and Ecology: This species is found in many habitats from aspen groves to timberline, but favors dense conifers or aspens near water for breeding, and ranges into low woodlands, riparian woods, and sage areas at other times.

Seasonality: Although largely resident, there is some seasonal migration, with migrants evident during March and November. Wyoming egg records are from May 10 to June 17, with nestlings reported from June 17 onwards. Egg records in Colorado are from May 30 to July 27, and nestlings have been seen from mid-July onward.

Comments: This magnificent hawk is the largest of the North American accipiters, and sometimes takes prey as large as grouse or pheasants. It also regularly hunts mammals such as hares and ground squirrels in this general region.

Suggested Reading: Schnell, 1958; Brown & Amadon, 1968.

Red-shouldered Hawk (*Buteo lineatus*)

Identification: This rare hawk is best identified by its strongly banded tail and pale "windows" at the base of the primaries when in flight. Adults are also strongly rufous on the breast and underparts. It is similar to the equally rare broad-winged hawk, but that species is less slim, and has shorter wings, as well as whitish rather than dark brown wing linings.

Status: An accidental vagrant in our region; reported once at Glacier National Park and reported as a rare migrant in the Fort Collins latilong of Colorado.

Habitats and Ecology: This is a hawk of the moist eastern deciduous woodlands, and is unlikely to be encountered in most of the region.

Seasonality: There are two (April and September) records for Montana; Colorado records extend from March through July.

Comments: Although a common species in eastern North America, red-tailed hawks occupy at least part of this species' niche in western forest habitats.

Suggested Reading: Wiley, 1975; Portnoy & Dodge, 1979; Stewart, 1949.

LATILONG STATUS

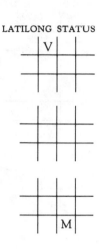

ACCIDENTAL VAGRANT

Broad-winged Hawk (*Buteo platypterus*)

Identification: Another rare eastern hawk which, like the red-shouldered hawk, is unlikely to be seen in the region. Like that species, it also has a strongly banded tail and pale areas at the base of the primaries, but the rest of the underwing is also whitish except for a posterior black border.

LATILONG STATUS

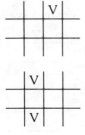

Status: A rare vagrant or migrant in most of the region, but breeding in the Cypress Hills of Alberta, as well as north-central Alberta; and has bred once in the Fort Collins latilong of Colorado. A rare migrant east of the Front Range in Colorado.

Habitats and Ecology: Associated with deciduous woodlands, including riparian woods and (in Alberta) aspen grovelands.

Seasonality: Migration in Alberta occurs during April and September; in Colorado the birds have been seen from April 15 to December 1.

Comments: In Alberta these birds are found in deciduous or mixed woods, where they hunt within the confines of the forest rather than along forest edges or open fields, as do red-tailed hawks. They migrate south early, sometimes in rather large flocks.

Suggested Reading: Matray, 1974; Brown & Amadon, 1968.

Swainson's Hawk (*Buteo swainsoni*)

Identification: A typical prairie and grassland hawk, easily identified by the combination of whitish wing linings and a dark grayish primary and secondary color. In most adult color phases the breast color is darker than the abdomen as well, but in dark-phase birds this may not be evident.

Status: A common to rare summer resident over most of the region; least common in the heavily forested montane parks, especially to the north.

Habitats and Ecology: Associated with open grasslands, sagebrush, agricultural lands, and rarely with riparian areas, typically nesting in isolated trees, but sometimes in bushes, on man-made structures, or on cliffs.

Seasonality: Reported in Montana from mid-March to early November, with migration peaks in late April and early September. Montana and Wyoming egg records are from May 3 to mid-July, with nestlings observed as late as September 12.

Comments: Like the red-tail, this is a highly beneficial hawk that primarily consumes rodents while on the breeding grounds (insects in wintering areas).

Suggested Reading: Bowles & Decker, 1934; Smith & Murphy, 1973; Dunkle, 1977.

LATILONG STATUS

s	S	S	s
s	s	S	S
s	M	S	s

S	s	s	s
s	S	M	s
s	S	M	s

s	S	S	M
S	S	S	S
S	S	S	S

120

Red-tailed Hawk (*Buteo jamaicensis*)

Identification: This is the commonest buteo of the region, especially in wooded areas. The rusty tail of adults (visible from above only) is diagnostic, but immatures lack this feature. Both age groups have blackish leading edges to their wings from the base to the "wrist"; the underwing surface is otherwise quite whitish, except in rare dark-phase individuals, which are extremely hard to identify.

LATILONG STATUS

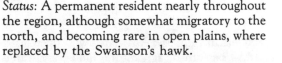

S	S	S	s
S	R	S	S
R	R	S	S

S	R	S	S
s	R	R	R
R	R	R	R

R	R	R	R
R	R	R	R
R	R	R	R

Status: A permanent resident nearly throughout the region, although somewhat migratory to the north, and becoming rare in open plains, where replaced by the Swainson's hawk.

Habitats and Ecology: A tree-nesting buteo that also extends to open woodlands and even treeless areas, where nesting may occur on cliffs. However, trees, especially large cottonwoods and pines, are favored nest sites.

Seasonality: Present year around in most areas, but with migration evident in April and again in September and October, especially in some areas. Montana and Wyoming egg records are from April 20 to late June; in northern Utah hatched young have been seen as early as April 14; in Jasper Park newly hatched young have been seen in late May.

Comments: A highly beneficial and adaptable hawk, which concentrates on rodents and rabbits as its prey, but also utilizes a variety of other vertebrate foods.

Suggested Reading: Fitch et al., 1946; Austin, 1964.

Ferruginous Hawk (*Buteo regalis*)

Identification: A large open-country hawk with mostly white underparts except for brownish thighs with silvery white bases to the primaries, often forming a "window" effect from above. Dark-phase birds may have entirely blackish undersides and wing linings, but their primaries still show white bases.

Status: A relatively uncommon to rare summer or year-round resident in the region; primarily in open-country habitats, and rare in the montane parks, with only Yellowstone reporting breeding.

Habitats and Ecology: Found during the breeding season in grasslands, sagebrush, and sometimes also mountain meadows, and nesting in pygmy conifers, cliff ledges, rock outcrops, and sometimes on man-made structures.

Seasonality: Usually present in Montana and Wyoming from March to late September, but often overwintering, especially from southern Wyoming south. Nesting in the two states is from late April to mid-July, with nestlings seen as early as June 23.

Comments: A large and spectacular prairie hawk, now relatively rare because of shooting, poisoning, and illegal killing, in spite of the species' great value as a rodent hunter.

Suggested Reading: Smith & Murphy, 1978; Snow, 1974a; Watson, 1969; Angell, 1969.

LATILONG STATUS

	M	s	s
	M		S
V	M	M	s

S	M	s	S
s	R	R	R
M	R	r	M

R	R	R	R
R	R	R	R
R	R	R	r

Rough-legged Hawk (*Buteo lagopus*)

Identification: Similar to the ferruginous hawk, but usually much darker underneath, especially across the abdomen, and with more conspicuous dark "wrist" patches underneath. From above, the primaries show the same kind of pale "window" as the ferruginous hawk, but the tail is usually black-tipped or (in immatures) somewhat banded.

LATILONG STATUS

W	W	W	W
	W	W	W
	W	W	W

W	W	W	W
W	W	W	W
W	W	W	W

W	W	W	W
W	W	W	W
W	W	W	W

WINTERING MIGRANT

Status: A regular winter visitor throughout the region, especially in open habitats.

Habitats and Ecology: Usually found in grasslands, sagebrush, or sometimes over marshes or mountain meadows.

Seasonality: Present in Montana and Wyoming from late August or September until April or May, depending on weather conditions, but usually migrating north by March. In Alberta they may not always overwinter, but usually appear by March and leave before the end of April, appearing again in October and November. The nearest breeding areas are in the tundra areas of northeastern Manitoba.

Comments: This is one of the common winter hawks of the region, replacing the Swainson's to a large degree.

Suggested Reading: Springer, 1975; Schnell, 1968; Brown & Amadon, 1968.

Golden Eagle (*Aquila chrysaetos*)

Identification: Unlike the bald eagle, this species never has an entirely white tail or a white head. Further, it typically has white or pale color near the base of the dark tail, and pale to whitish "windows" at the base of the flight feathers when viewed from below, although in adults the undersurface is rather uniformly dark. Unlike the bald eagle, it is often found far from water.

Status: A permanent resident throughout the region, but most common in montane or rimrock country that is relatively open.

Habitats and Ecology: This is a mountain- and plains-adapted species, that often occurs in grasslands, semidesert areas, pinyon–juniper woodlands, the ponderosa pine zone of coniferous forests, and sometimes forages above mountain meadows or alpine tundra. It nests over a broad altitudinal range, usually on cliffs or in trees, rarely on the ground.

Seasonality: A permanent resident in the southern parts of the region, but somewhat migratory in northern areas, especially in the case of young birds. In Montana egg records are from mid-April to June; in Wyoming egg records are as early as March 15, and in Rocky Mountain N.P. nestlings have been seen as early as April 20. Nestlings have been reported to late July in Montana and Alberta.

Comments: One of the grandest of American birds, this species has been greatly harassed by humans, especially ranchers, who have wrongly condemned it of sheep-killing. It is now largely confined to remote or protected nesting areas.

Suggested Reading: Snow, 1973a; McGahan, 1968; Brown & Amadon, 1968

LATILONG STATUS

R	R	S	?
s	R	r	S
R	R	R	S
R	R	R	S
r	R	R	R
r	R	R	R
R	R	R	R
R	R	R	R
R	R	R	R

American Kestrel (*Falco sparverius*)

Identification: This tiny falcon is commonly observed on telephone wires or on the wing, where it hovers frequently. The rusty brownish upperparts, the conspicuous "mustache," and (in males) the bright rusty tail color are all useful fieldmarks. Females are slightly less colorful, and have a brownish barred tail.

LATILONG STATUS

S	S	S	S
S	R	S	S
R	R	S	s

R	S	s	R
S	S	S	S
R	S	S	S

S	S	S	S
S	S	S	S
R	R	S	R

Status: A relatively common summer or permanent resident over nearly all the area, becoming more seasonal in occurrence farther north, and probably breeding in all the montane parks.

Habitats and Ecology: This is an open-country falcon, occurring in agricultural areas, grasslands, sagebrush, desert scrub, and nesting in tree cavities, rock or building crevices or cavities, and rarely in earthen holes. It avoids forests, but sometimes forages as high as mountain meadows.

Seasonality: In Alberta these birds are present primarily between April and late October, while in Montana and Wyoming they usually arrive in March and leave in November, with some birds overwintering in mild winters. Wyoming egg records are from May 15 to June 20, and eggs have been found as late as July 3 in Rocky Mountain N.P. In Montana fledged young have been seen in July, but small young have been observed as late as August 18. Alberta and Saskatchewan egg records are from May 22 to June 17.

Comments: This species, previously known as the "sparrow hawk," is the most insectivorous of the falcons, often catching grasshoppers or other large insects.

Suggested Reading: Cade, 1955, 1982; Enderson, 1960; Balgooyen, 1976; Willoughby & Cade, 1964.

Merlin (*Falco columbarius*)

Identification: This medium-sized falcon is much more elusive than the American kestrel, and is somewhat larger, with streaked to spotted underparts, and much less conspicuous "mustache" markings. It is brown to bluish gray above, with a somewhat barred tail.

Status: A relatively rare migrant or summer resident in the region, mainly in montane wooded areas, and breeding locally.

Habitats and Ecology: This is a forest and woodland-adapted falcon, usually breeding in tree clumps of open woodlands, often in bottomlands or valleys. During the non-breeding season it also appears over grasslands, agricultural lands, desert scrub, and marshes or shorelines.

Seasonality: Migration records in Wyoming are from March 13 to May 24, and again from September 2 to November 27, with some overwintering. Breeding records are few, but egg records in Colorado are from May 26 to July 3. Alberta and Saskatchewan egg records are from May 7 to June 6.

Comments: Like most falcons, this species has declined in recent years, and sightings have become ever rarer. It hunts like the other falcons, catching birds in shallow dives, or sometimes from stooping in steep dives. Commonly also known as the "pigeon hawk."

Suggested Reading: Cade, 1982; Craighead & Craighead, 1940.

LATILONG STATUS

s	s	s	
	M	S	S
R	r	r	M

s	r	M	s
M	S	r	M
	r	M	r

M		R	R
r	M		R
S		M	S

Peregrine Falcon (*Falco peregrinus*)

Identification: This is a large falcon, the same size as the more common prairie falcon, but darker throughout, with more conspicuous "mustache" markings, and lacking the dark underwing linings of that species.

LATILONG STATUS

s	r	s	
s	S	S	s
	R		s

S	r	M	S
S	R	r	r
M	R	R	M

M	M	M	M
R	M	M	R
S		s	M

Status: Now largely extirpated as a breeding species from the region; once a nester throughout the montane areas and nesting in at least Rocky Mountain and Yellowstone parks. Efforts are currently underway to re-establish the species in some of these areas by releasing hand-reared birds. There are no recent Alberta nesting records.

Habitats and Ecology: This species is largely a cliff-nesting species, typically in woodland habitats. Non-breeders occur over a wide habitat range, from mountain meadows to grasslands, marshes, and riparian habitats.

Seasonality: Wyoming records are from April 2 to November 30; wintering birds sometimes occur as far north as southern Alberta. There are Montana records of nestlings from July 10–17. Alberta and Saskatchewan egg records are from May 6 to June 13.

Comments: Widespread use of pesticides in the post-war era nearly destroyed this species in North America, and it is still gone from most of its historic breeding range. Between 1976 and 1982 a total of 330 peregrines were released in the Rocky Mountains. Golden eagles and great horned owls have been major predators of the released birds.

Suggested Reading: Snow, 1972; Cade, 1960, 1982; Porter et al., 1973; Hickey, 1969.

Gyrfalcon (*Falco rusticolus*)

Identification: This large, arctic-breeding falcon is usually seen in its gray phase, which resembles a very large prairie falcon, but lacks dark underwing linings, and is more grayish throughout. A rarer white phase has also been reported from the region.

Status: A rare winter migrant or accidental vagrant, more likely to occur in the northern portions of the region, but sometimes wandering south to Colorado.

Habitats and Ecology: An arctic tundra species, in our region the birds are likely to be found in open habitats such as grasslands and plains, where large prey such as rabbits or waterfowl may be found.

Seasonality: All the regional records are for the winter, from December to early March.

Comments: Probably most regional records are of first-year birds that have not yet established breeding territories, after which they are more likely to remain in northern areas throughout the year.

Suggested Reading: Cade, 1960, 1972; Langvatn & Moksnes, 1979.

LATILONG STATUS

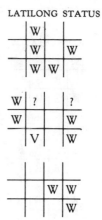

RARE MIGRANT

Prairie Falcon (*Falco mexicanus*)

Identification: This is the most common of the large falcons of the region, and is usually seen in flight, when its distinctive blackish underwing linings are conspicuous. The underparts are spotted with brown, the upper surface is more uniformly brown, and the facial "mustache" markings are only moderately developed.

LATILONG STATUS

M	r	S	s
S	R	S	S
R	S	R	S

R	R	r	R
S	R	R	R
s	R	R	R

R	R	R	R
R	R	R	R
R	R	R	R

Status: A widespread summer resident or year-round resident throughout the region, mainly in mountain or rimrock areas offering open country for hunting.

Habitats and Ecology: Breeding birds are largely associated with plains, sagebrush, or desert scrub habitats with steep cliffs nearby for nesting; sometimes tundra areas also support breeders, and foraging may be done on mountain meadows or similar alpine habitats.

Seasonality: In Wyoming the birds are essentially present year around, but some migration occurs in March and November, as they move to more favored wintering areas such as deep canyons. Montana egg records are from mid-April to late June, and Wyoming records are from May 1 to June 8. In Rocky Mountain N.P. nestlings have been observed in late June. Alberta and Saskatchewan egg records are from April 22 to June 14.

Comments: Unlike the peregrine, this species has not suffered great losses of breeding range, although its numbers have sharply declined. It feeds to a considerable degree on rodents and rabbits in the Rocky Mountain region, and thus perhaps has not been so seriously exposed to pesticide poisoning.

Suggested Reading: Enderson, 1964; Snow, 1974b; Cade, 1982; Hickey, 1969.

Prairie Falcon

Gray Partridge (*Perdix perdix*)

Identification: This quail-sized bird is mostly grayish, with a more brownish head, and with a dark brown blotch on the belly. When flushed, a short, rusty tail is apparent. The birds form quail-like coveys in winter, and in spring males utter a hoarse *kee-uk*, with the first note louder.

LATILONG STATUS

r	R	r	r
	R	?	r
V	R	R	R

R	R	r	R
R	V	R	r
R	R	r	R

R	V	R	R
		V	

Status: Widespread in the prairies and agricultural lands of the region, but rarely entering the montane areas. Only a vagrant in the montane parks except for the Jackson Hole area, where apparently a rare resident.

Habitats and Ecology: Generally associated with grainfields and nearby edge habitats, such as shelterbelts, but also sometimes extending into sagebrush areas, especially where local water supplies are present. Nesting usually occurs in grainfields or hayfields, under grassy or herb cover.

Seasonality: A permanent resident, but limited movements occur in winter as food supplies demand. Nesting occurs over a fairly long period, with frequent renesting; from about mid-May to mid-August, with most hatching (in North Dakota) in late June and early July. Washington egg records are from May 25 to June 10.

Comments: This is an introduced species, which has adapted well to life in the northern plains of central North America, and is an important gamebird there. The largest populations occur now on the central plains of southern Canada, where small grain cultivation and native grasslands or prairie hayfields provide a combination of food and nesting cover.

Suggested Reading: Johnsgard, 1973; McKinney, 1966.

Chukar (*Alectoris chukar*)

Identification: This species is the size of a large quail, but has strongly barred black and white flanks and a white throat and face, surrounded by a black band. The repeated *chuck* or *chuck-or* call is a good field-mark. Normally limited to rocky slopes and canyons.

Status: A local permanent resident in arid lands of southern Idaho and Wyoming; reported in the montane parks only as a vagrant; reportedly once casual in Rocky Mountain N.P., but apparently not reported since 1957. Probably a rare resident in the Jackson Hole area.

LATILONG STATUS

	r		
	R		

r		?	R
		R	R
	r	R	R

R	r	R	R
			R
R			r

Habitats and Ecology: Primarily associated with sagebrush habitats during the breeding season, extending into grassland and sometimes also riparian habitats at other times. Nearly always found in hilly, rocky areas.

Seasonality: A permanent resident in the region. Nesting records are few for the region, but broods have been seen as late as August 15. In Washington the nesting begins in early April, and hatching begins in late May and June, with the latest nests not hatching until mid-August.

Comments: This species competes little if at all with any North American grouse or quails, and has thus seemingly occupied a vacant ecological niche. In Nevada the populations are especially high, and over most of its best Nevada range the chukar is associated with sagebrush vegetation. The species' common and scientific name reflects its most typical call, which serves to reassemble scattered coveys and may also help to disperse breeding pairs.

Suggested Reading: Johnsgard, 1973; Christensen, 1952, 1970; Galbraith & Moreland, 1953.

Ring-necked Pheasant (*Phasianus colchicus*)

Identification: The familiar "ring-necked" male hardly needs description; females may be confused with sharp-tailed grouse if their long tails and more generally mottled brownish plumage isn't noted. Males utter a distinct crowing call, a double-noted *caw-cawk* during late winter and spring, that can be heard for more than a half-mile.

LATILONG STATUS

r	R	r	R
	R		R
R	R	R	R

R	R	R	R
		R	R
		R	R

R		R	R
			r
R	R	s	R

Status: A widespread resident throughout the general region, but rare or accidental in the montane parks, and mostly limited to low-altitude grasslands, croplands, and similar non-wooded environments.

Habitats and Ecology: Breeding occurs mainly in native grasslands and grain croplands or their edges, but sometimes also in marsh edges, hayfields, or shelterbelts, as well as roadside ditches.

Seasonality: Present throughout the year, with almost no movements. The breeding season is fairly long, with eggs (in North Dakota) from late April to the end of July; the hatching peak is usually in early June.

Comments: The most widespread and successful of the introduced gamebirds, and harvested in the millions every year. During winter, males gather small groups of females, or "harems," within their home ranges, and advertise their location by wing-whirring and crowing calls. Like grouse, males do not participate in nesting or brood-rearing activities, although females typically nest within their mate's home range.

Suggested Reading: Baxter & Wolfe, 1973; Allen, 1956; Baskett, 1947.

Spruce Grouse (*Dendragapus canadensis*)

Identification: This species is associated with coniferous forests, and usually is very tame. Males are mostly gray to black, with black and white patterning on the breast, and with white-tipped tail coverts. Females are heavily barred on the underparts, unlike the otherwise similar blue grouse females.

Status: A permanent resident from southwestern Montana northward, mainly in montane forests. Although reported earlier for Yellowstone Park, there have been no records for at least a decade, and any sightings are likely to be of vagrants.

LATILONG STATUS

R	R	R	?
R	R	R	R
R	R	R	R

Habitats and Ecology: Largely limited to coniferous forests throughout the year. Conifer needles are their primary food source, supplemented by insects and berries in late summer.

Seasonality: A year-round resident, with no significant seasonal movements. In Montana chicks have been reported during July, with fledged broods typical by the end of that month, and in Banff N.P. young have been seen from mid-July to early August. Egg records from Alberta, Saskatchewan, and Montana are from May 18 to July 19.

Comments: The tameness of this species, resulting in its common name "fool hen," along with destruction of coniferous forest habitats, has been responsible for its decline in many areas. During late winter and spring males perform "strutting" displays, and utter extremely low-pitched hooting calls, which apparently serve to attract females to them. The males are polygynous, attracting as many females to them as they can, and play no role in breeding after fertilization has occurred.

Suggested Reading: Johnsgard, 1983b; Robinson, 1980; Stoneberg, 1967.

134

Blue Grouse (*Dendragapus obscurus*)

Identification: Very similar to the spruce grouse, but males are more uniformly grayish throughout, without white on the breast or tail coverts, while females are more uniformly brownish below, lacking the strong breast and underpart barring found on the spruce grouse. In spring, males produce "hooting" calls that carry for some distance.

LATILONG STATUS

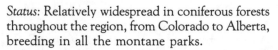

R	R	R	?
R	R	R	R
R	R	R	R

R	R	R	R
R	R	R	R
r	R	R	R

R	R	R	R
R	R	R	R
R	R	R	R

Status: Relatively widespread in coniferous forests throughout the region, from Colorado to Alberta, breeding in all the montane parks.

Habitats and Ecology: Like the spruce grouse, closely associated with coniferous forests, but also reaching alpine timberline during the breeding season, and as low as the ponderosa pine zone. Extends in late spring and summer into the sagebrush zone as nonbreeders.

Seasonality: A permanent resident throughout its range. Hatching of 120 broods in Montana ranged from May 25 to July 14, with a peak the third week of June. Broods in Alberta have been reported from July 7 to August 10. Egg records from British Columbia, Alberta, and Montana are from May 19 to July 29.

Comments: Hooting behavior by males begins n late winter, and extends to early June. While hooting, the males expose bare reddish neck skin surrounded by a white rosette of feathers, and engorge yellow to reddish eye-combs. Like the other forest grouse, males are polygynous, and attempt to fertilize several females.

Suggested Reading: Johnsgard, 1983b; Blackford, 1958; Mussehl, 1960; Bendell & Elliott, 1967.

Willow Ptarmigan (*Lagopus lagopus*)

Identification: In any plumage, this is the only ptarmigan of the region having a brownish to black tail. It is also larger than the white-tailed ptarmigan, and in spring or summer plumage males are much more rusty brown.

Status: An accidental vagrant in the region, with a single winter record for Glacier Park (Skaar, 1980). The nearest breeding area is in northern Jasper Park, south to the Tonquin River and Cairn Pass.

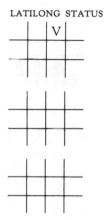

LATILONG STATUS

		V	

Habitats and Ecology: Associated with low arctic and alpine tundra, especially that rich in shrubby willow vegetation. Widespread through the arctic tundra zones of northern Canada, south along the west coast of Hudson Bay.

Seasonality: A permanent resident, but with some winter movement southward in some areas, accounting for vagrant birds appearing south of the breeding range. Nesting in Jasper Park has been observed in July, and broods have been observed as late as early September.

Comments: This is the largest of the ptarmigans, and is the same species known as the "red grouse" in Britain, where the birds do not turn white in winter. However, in North America all the races turn completely white in winter, except for the black tail feathers. Like many other grouse, snow-burrowing is common in these birds; temperatures beneath the snow are usually much warmer than at the surface.

ACCIDENTAL VAGRANT

Suggested Reading: Weeden, 1960; Johnsgard, 1983b.

White-tailed Ptarmigan (*Lagopus leucurus*)

Identification: This alpine grouse is totally white in winter, including the tail, and in summer is variably grayish to brownish, but retaining a white tail. It is about the size of a pigeon, and is usually overlooked by the casual observer, because of its remarkably concealing coloration.

LATILONG STATUS

	R	R	
		R	R
R	?	?	

	?	?	
	V	?	
	?	?	

		r	
	R	R	R

Status: Occurs in alpine areas from Alberta southward more or less continuously to western Montana, and on scattered alpine areas farther south throughout the region.

Habitats and Ecology: Limited to the alpine and timberline zones, moving slightly lower during winter, especially where willows remain exposed above the snow. In Alberta the birds sometimes descend to mountain valleys in hard winters, where they may feed on seeds and waste grain.

Seasonality: A permanent resident where found, usually breeding in June. Hatching in Montana has been reported from July 18 to 28, with fledging occurring between July 27 and August 8. Broods in Jasper N.P. have been reported between mid-July and mid-August. Nesting in Rocky Mountain N.P. occurs from early June until late July.

Comments: This is the smallest and most southerly ranging of the ptarmigan, and the only one breeding in this region. Unlike most grouse, the birds are normally monogamous, although at times males have two mates. Male ptarmigans of all species are highly territorial, and advertise by loud calls and by making long display flights over their territories.

Suggested Reading: Choate, 1963; Braun & Rogers, 1971; Scott, 1982

Ruffed Grouse (*Bonasa umbellus*)

Identification: Ruffed grouse have relatively long, fanlike tails, with a black terminal band, and which (in this region) are mostly grayish. The flanks tend to be barred, but unlike the spruce grouse the underparts are mostly white. Drumming by males, sounding like a muffled drum, is performed from late winter until early June.

Status: Widespread and relatively common in wooded areas from south-central Wyoming northward; absent from Colorado.

Habitats and Ecology: Especially associated with aspen woodlands, the buds and catkins of which provide a major food source. However, also up to the spruce–fir zone of coniferous forest. Nesting is often in or near aspen clumps.

Seasonality: A permanent resident in the area. In Wyoming hatching occurs from mid-June to mid-July, and in Jasper N.P. broods have been observed from June 13 to September 6.

Comments: This widespread species of grouse is best known for the drumming behavior of males in spring, which typically is done on dead logs in fairly heavy forest. The sound is produced by changes in air pressure rather than direct striking of the wings against one another, and serves as territorial advertisement signals by males. Males usually display from a primary log, but often have secondary display sites within their territory as well. Ruffed grouse of this region are all "gray-phase" birds, but farther east in their range "red-phase" types, with a rusty brown tail, are fairly common.

Suggested Reading: Johnsgard, 1983b.

LATILONG STATUS

R	R	R	?
R	R	R	R
R	R	R	R

R	R	R	R
R	R	R	
R	R	R	r

R		R	

Sage Grouse (*Centrocercus urophasianus*)

Identification: The largest of the grouse, and the only one restricted to the sagebrush habitat. Both sexes have rather long, pointed tails, and blackish underparts, while being rather uniformly grayish brown above.

LATILONG STATUS

	?		?

R	R	R	R
R	R	R	R
r	R	R	R

R	R	R	R
R	R	R	R
R	R	R	R

Status: Relatively common on sage habitats in the plains and foothills; rare or absent in the montane parks except for the Jackson Hole area of Grand Teton N.P.

Habitats and Ecology: Closely associated with sagebrush, which is the primary food and which also is used for nesting cover. Occurs locally in sage to 9000 feet elevation.

Seasonality: A permanent resident, but exhibiting considerable seasonal movements associated with snow cover and food availability. Wyoming egg records extend from April 18 to July 27.

Comments: The range of this species is slowly decreasing, as sagebrush areas are being cleared and converted to irrigated cultivation. However, Wyoming still supports the nation's largest sage grouse population. Sage grouse display socially in spring, with as many as 50 or more males "strutting" in local display grounds, or "leks." Grand Teton Park has one such display ground, near the Jackson Airport, where display can be watched during spring.

Suggested Reading: Johnsgard, 1983b; Patterson, 1955.

Sage Grouse

Sharp-tailed Grouse (*Tympanuchus phasianellus*)

Identification: This "prairie grouse" is found on grasslands and other open habitats, and is mostly buffy white below, with a short, pointed tail. It is lighter in color than the sage grouse or any of the forest grouse, and has a shorter tail than do female ring-necked pheasants.

Status: Widespread on plains and foothills, but rare or absent in the montane parks except at Watertown Lakes N.P., where local on grassland areas.

Habitats and Ecology: Associated with grasslands and grassy sagebrush areas, and sometimes also mountain meadows during the breeding season, and extending into cultivated fields during fall and winter. Brushy foothills and similar edge habitats are often used in Alberta.

Seasonality: A permanent resident, with only limited seasonal movements. Nesting in Montana occurs from mid-May to mid-June; in southern Alberta broods are usually observed in early July.

Comments: In spring, male sharp-tails "dance" on traditional display areas in groups of from a few to 20 or more males, during which dominance is determined and the relative access of males to females for fertilization is established. These activities begin in late winter, and may continue until May. In this species the displays consist of active "dancing," cooing sounds, and actual or ritualized fighting activities among the competing males.

Suggested Reading: Johnsgard, 1983b.

LATILONG STATUS

r	r	r	R
	r	r	R
R	r	r	r

R	R	R	R
r	r		r
r	r		

		r	r
	R		r
	R	R	

Wild Turkey (*Meleagris gallopavo*)

Identification: Virtually identical to the "bronze-type" domestic turkey, with iridescent bronze plumage, a large fan-like tail, and naked head skin. In spring the gobbling call of males is a good fieldmark.

LATILONG STATUS

Status: Locally present in the region as a result of fairly recent introduction efforts, mainly in open forests of ponderosa pines or mixed woods, especially those with oaks or other mast-bearing trees. Virtually absent from the montane parks, but there was a Yellowstone sighting in October of 1954, probably from birds planted near Cody in the early 1950s.

Habitats and Ecology: In this area, wild turkeys are typically associated with ponderosa pines and red cedars, running water, and fairly rugged topography. The birds nest on the ground in forested areas, often under a log or at the base of a large tree.

Seasonality: Permanent resident wherever found, with only limited movements. There are no regional nesting records, but in most areas the nesting season is quite extended.

Comments: Turkeys spend much of the year in small flocks typically dominated by one or more adult males. Males are polygynous, and display singly or gathered in small groups on "gobbling grounds," where they call and attempt to attract females for fertilization.

Suggested Reading: Watts & Stokes, 1971; Lewis, 1973; Sanderson & Schulz, 1973.

Bobwhite (*Colinus virginianus*)

Identification: This widespread and familiar quail is similar to a gray partridge, but has a black and white head pattern (in males) and a grayish rather than rust-colored tail. Females have a dark brown and tawny head pattern, but otherwise are similar to males. The "bob-white" call of males in spring is a useful fieldmark.

LATILONG STATUS

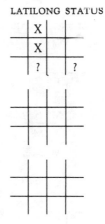

Status: Highly local in the region, as a result of introductions, and absent from the montane parks at present. Once introduced in the Flathead Valley, and reported from Swan Lake and the North Fork of the Flathead River (Bailey, 1918).

Habitats and Ecology: Generally associated with brushy edge areas, where a combination of grassy, shrubby, and woody cover all occur in close proximity, and where water can usually be found nearby. Eastern Colorado and Wyoming represent the nearest natural range of this species.

Seasonality: Permanent residents where found, with only very limited seasonal movements. Nesting seasons are usually greatly prolonged through spring and summer, as a result of persistent renesting by unsuccessful pairs.

Comments: Unlike the grouse, pheasants, and turkeys, quail are strongly monogamous, with the male participating in brood care, although not normally assisting in incubation. Sometimes the male will take over the care of the brood entirely, freeing the female to begin a second clutch, at least in captivity.

Suggested Reading: Johnsgard, 1973.

Virginia Rail (*Rallus limicola*)

Identification: This small marsh bird is elusive, and is usually seen when flushed from marshy vegetation, when its fairly long and reddish bill is evident, and its rusty to brownish plumage can be seen. It is more often heard than seen; a metallic ticking and a descending laughing call are its most common vocalizations.

LATILONG STATUS

M		
S		
V	S		

s	M	S	M
s		M	
s	M	M	M

	M	S	M
		M	S
S	S	S	R

Status: A local summer resident and breeder in the general region, especially at lower elevations; generally rare or lacking in the montane parks except for Rocky Mountain N.P., where a rare breeder in the Colorado River valley (in 1954).

Habitats and Ecology: Inhabits marshes with dense stands of emergent vegetation, nesting on wet ground or over shallow water in such stands.

Seasonality: A summer resident in the region, with rather few actual records, but in Montana and Wyoming the birds apparently arrive in May, and remain until October or November. There are several nest records for Wyoming in mid-May.

Comments: This little-known species is often detected in late spring, when its distinctive "ticket, ticket" notes are uttered, during territorial establishment by males. Besides the actual nests, these rails also construct several "dummy" nests, some of which later serve as brooding sites. Both sexes participate in incubation, and the blackish chicks are cared for until they are about two months old, after which they are chased from their parents' territory.

Suggested Reading: Sanderson, 1977; Pospichal & Marshall, 1954; Horak, 1970; Kaufmann, 1970.

Sora (*Porzana carolina*)

Identification: Similar to the Virginia rail, but with a much shorter bill that is more yellowish and (in adults) with a black facial mask. The calls are similar, but sora rails utter a distinctive descending whinny-like call, and a plaintive *kerwee* note.

Status: An uncommon to occasional summer resident in the region; infrequent in the montane parks, but known to nest in several of them.

Habitats and Ecology: Found in essentially the same marshy habitats as the Virginia rail, and apparently having very similar niche adaptations, but perhaps somewhat more vegetarian in its diet. More surface-feeding and less probing for food is also done by soras than Virginia rails.

Seasonality: In Montana and Wyoming these birds usually arrive in April and leave in late October or November, with probable migration peaks in May and September. Egg records in Wyoming extend from May 14 to June 7; nest records for Montana and Rocky Mountain N.P. are for the latter half of June.

Comments: Sora rails are highly aggressive, and tend to evict Virginia rails from their territories as well as males of their own species. As in Virginia rails, the birds construct several extra nests that may be used by the broods for resting after hatching. The birds fledge in about five weeks, and in some areas the birds may possibly rear two broods in a single season.

Suggested Reading: Sanderson, 1977; Horak, 1970; Pospichal & Marshall, 1954; Tanner & Hendrickson, 1956.

LATILONG STATUS

s	S	S	?
	S	s	S
S	S	S	S
S	s	s	S
S	S	M	S
S	S	S	S
S	s	S	M
	s	S	S
S	S	S	S

144

Common Moorhen (*Gallinula chloropus*)

Identification: This marsh-dweller is rather elusive, but if seen its bright red and yellow bill is distinctive (immatures have duller yellow-tipped bills), and it shows a triangular patch of white below the tail. The birds swim in a coot-like manner, and also walk through the marshes somewhat like rails. Their calls are also coot-like, but higher pitched.

LATILONG STATUS

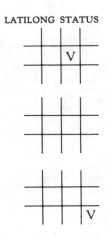

Status: An accidental vagrant in the region, with few records. Recorded in Rocky Mountain N.P. on two occasions.

Habitats and Ecology: During the breeding season this species is associated with fresh waters, ranging in size from lakes to small ponds, and with small and slowly flowing streams to large rivers. It forages primarily on the seeds and fruits of weeds and grasses, and to some degree on various invertebrate foods.

Seasonality: The few Colorado records are inadequate to estimate seasonal occurrence; they extend from May to August. In Kansas, eggs are laid in May and June, which represents the nearest regular breeding area.

Comments: This species, previously known as the "common gallinule," is coot-like in that the birds are highly territorial, and both sexes vigorously defend their territories. In this species two broods are sometimes raised in a single season; in such cases the young of the first brood often help feed the chicks of the second brood.

Suggested Reading: Sanderson, 1977; Frederickson, 1971; Wood, 1974.

ACCIDENTAL VAGRANT

American Coot (*Fulica americana*)

Identification: This duck-like species of the rail family is mostly dull gray except for a white, chicken-like bill, and a triangular white patch under the tail similar to that of a common moorhen. The two species' behavior and calls are also similar, but coots are more often seen swimming in open water, and their calls tend to be lower and more grating.

Status: Widespread and a summer resident on wetlands throughout the region, especially at lower elevations. Present and probably breeding in all the montane parks.

Habitats and Ecology: Associated with ponds and marshes having a combination of open water and emergent reedbeds, in which nesting occurs. Besides foraging on aquatic plants, the birds sometimes also graze on nearby shorelines and meadows.

Seasonality: In Montana the birds are present from late March to late November, with migration peaks in late April and late September. Wyoming nest records extend from May 11 to August 8, and Montana nest records are from May 27 to July 8, with hatched young observed as early as June 28.

Comments: Coots are highly successful members of the rail family, which have adopted a quasi-ducklike niche, and more often mingle with ducks than with other species of rails. They are monogamous, and like the other rails are highly territorial, often fighting fiercely for territorial space. The young are cared for by both parents, and fledge at about 75 days of age.

Suggested Reading: Sanderson, 1977; Fredrickson, 1970; Gullion, 1954.

LATILONG STATUS

S	R	s	S
s	R	R	S
R	R	S	s

S	R	s	R
S	S	S	S
R	S	S	S

S	S	S	S
s	S	S	S
S	S	S	R

Sandhill Crane (*Grus canadensis*)

Identification: Sandhill cranes are hard to mistake for any other species except perhaps the larger herons; their grayish to rust-brown plumages and bare red crowns are distinctive, as are their loud, penetrating and rattling calls, and their goose-like manner of flying with the neck fully stretched out.

LATILONG STATUS

M	s	S	
	M	M	M
V	s	S	s

S	s	S	M
S	S	M	M
S	S	S	S

S	M	M	M
M	S	M	M
M	S	S	M

Status: A local summer resident in the more remote wetlands of the region, north almost to Glacier N.P. The range is probably now expanding, and may soon include the Glacier area, where the last reported breeding occurred in 1899.

Habitats and Ecology: In the Rockies, sandhill cranes are especially associated with beaver impoundments, where the birds nest along shorelines or sometimes on beaver lodges, often in dense willow thickets. The birds are highly territorial and usually are well scattered. Their loud calls serve to advertise territories and communicate over long distances.

Seasonality: Records in Montana and Wyoming are mostly from mid-April to mid-October, with migration peaks in late April and late September. Nests in both states are from mid-May to late June.

Comments: After a prolonged period of range retraction, this species began to recover under full protection, and now the Rocky Mountain population of this species is thriving. The densest breeding population occurs at Gray's Lake N.W.R. in southern Idaho, but perhaps a half-dozen pairs nest yearly in the Jackson Hole region as well. Coyotes and human disturbance represent the major sources of losses during nesting; incubating cranes should never be disturbed from their nests.

Suggested Reading: Johnsgard, 1983c; Drewein & Bizeau, 1974; Sanderson, 1977.

Sandhill Cranes

Whooping Crane (*Grus americana*)

Identification: This is the largest marsh bird in North America; its white plumage and black wingtips are unmistakable. Immatures are more rust-colored, but not as brown as young sandhill cranes. The loud, resonant calls of adults may be heard for up to a mile or so.

LATILONG STATUS

Status: Still a rare and local migrant, mainly in southern Idaho, where a flock of whooping cranes is being established by rearing young under sandhill crane foster parents.

Habitats and Ecology: Once widespread in central North America, the whooping crane currently breeds only in a limited area of northwestern Canada. It nests in remote wetlands there, in a muskeg wilderness area. On migration the birds utilize a variety of wetland types, but they are usually shallow and broad, with safe roosting sites and nearby foraging opportunities.

Seasonality: There are still few regional records, but probably migration times are similar to those of sandhill cranes. There are still no nesting records for this population.

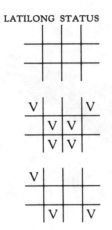

Comments: The "cross-fostering" of whooping crane eggs under sandhill crane parents represents a daring attempt to establish a new flock of this endangered species that began in 1976. The new flock is slowly growing (about 30 birds in 1983), and the results should soon be apparent as to whether the young birds will be able to breed successfully in their new habitats.

Suggested Reading: Johnsgard, 1983c.

Black-bellied Plover (*Pluvialis squatarola*)

Identification: In spring, this large plover is easily identified by its pale white and marbled upperparts, contrasting with the black underparts. Immature and winter-plumaged birds are much more difficult to identify, but in flight they exhibit a patch of black feathers in the "armpit" area, as do breeding adults. Flying birds also often utter a plaintive whistled *pee-a-wee* note.

Status: A rare migrant or vagrant throughout the region, more prevalent on the plains than in montane regions.

Habitats and Ecology: Migrant birds are likely to be found along lakes or reservoir areas, but sometimes also are seen on plowed fields or forage on short meadows or pasturelands. The birds breed on high arctic tundra areas of northern Canada and Alaska.

Seasonality: Wyoming migration records are from May 11 to 21, and from August 21 to November 9. Montana records are also for May and from August through October. In Colorado the records extend from April 11 to December 28.

Comments: This species, called the "grey plover" in Great Britain, is adapted to breeding on dry stony tundra, usually on rocky slopes, where its pale dorsal plumage is hard to detect. On their wintering grounds the birds occupy tidal flats and ocean shorelines, and sometimes also inland habitats.

Suggested Reading: Johnsgard, 1981; Stout, 1967.

LATILONG STATUS

M	M		
M		M	
M			

M		M	?
M	M	M	M
M	M		M

		M	M
	M	M	M
V		M	M

RARE MIGRANT

Lesser Golden Plover (*Pluvialis dominica*)

Identification: A smaller plover than the black-bellied, this species in spring has a dorsal plumage beautifully dappled with golden spots, and black underparts like those of the black-bellied plover. In flight, it lacks black "armpits," and this fieldmark is useful in recognizing winter-plumage and immature birds, which closely resemble those of the black-bellied plover. The flight call is shorter and harsher than that of the black-bellied plover.

LATILONG STATUS

LOCAL MIGRANT

Status: A local migrant in the region, especially on the plains and prairie areas, but only a vagrant in the montane parks. The nearest breeding area is along the western coast of Hudson Bay in northern Manitoba.

Habitats and Ecology: Migrant birds are usually found along lakes or reservoirs, or on agricultural lands, during migration. They often occur on plowed or recently burned fields on migration, where surface-foraging opportunities are available.

Seasonality: The records for Montana are mainly for the fall period, from August through September. Similarly, in Colorado, the birds are most often seen in summer and fall, although the spread of records is from March 30 to November 5.

Comments: Like the black-bellied plover, this is an arctic tundra breeder, but the birds nest on both dry upland tundra and on knolls of lowland tundra, especially those having many boulders with golden lichens, which closely match the dorsal coloration of the breeding plumage. Most adults migrate south in the fall along the Atlantic coast, leaving immatures to move south through the Great Plains region.

Suggested Reading: Johnsgard, 1981; Stout, 1967.

Semipalmated Plover (*Charadrius semipalmatus*)

Identification: This small "ringed" plover has a broad breastband and a black-tipped orange bill in breeding plumage, as well as orange legs and a narrow orange eye-ring. Immatures and winter adults are less colorful, but the general patterning remains the same. A clear, rising whistle is often uttered in flight.

Status: A local migrant throughout the region, mainly on the plains, and a rare migrant or vagrant in the montane parks. The nearest breeding areas are in northern Saskatchewa, or possibly extreme northeastern Alberta.

LATILONG STATUS

M		
M	M	M
	M	M

Habitats and Ecology: Migrating birds are usually observed on open sandy or gravelly habitats along rivers or beaches, where they feed by running about and picking up morsels from the surface, rather than probing for foods.

M		M
M		M
	M	M	M

Seasonality: In Montana these birds are primarily fall migrants, from July to October. Wyoming records are mainly for May and August, and in Colorado they have been reported between April 21 and December 7, but mainly occur during spring and fall. In Alberta they are most common in mid- to late-May, and again from mid-August to late September

	M	M
	M	M	M
		M	M

Comments: This small tundra-breeding plover is widely distributed in North America, and is replaced by the very similar ringed plover in the Old World. The semipalmated plover winters from the southern United States all the way south to Argentina.

Suggested Reading: Johnsgard, 1981; Stout, 1967.

LOCAL MIGRANT

Killdeer (*Charadrius vociferus*)

Identification: The most widespread and common of the North American plovers, the killdeer is easily recognized by its rusty brown tail and its double breast band, together with its incessant *kill-dee* calling, especially during the breeding season. Often found well away from water.

LATILONG STATUS

R	R	S	S
S	S	S	S
R	R	S	R

R	R	S	R
S	R	S	S
S	S	S	S

S	S	S	s
S	S	S	S
R	R	S	R

Status: A common summer or permanent resident throughout the region, both on the plains and montane areas, but more common at lower elevations and not reaching alpine areas.

Habitats and Ecology: Widely distributed in openland habitats, including pastures, roadsides, gravel pits, golf courses, airports, and sometimes suburban lawns. Gravelly areas are favored, and gravelled rooftops are sometimes used for nesting in urban areas. Migrating and wintering birds are more closely associated with water, but also use mud flats and open fields.

Seasonality: Locally resident in much of the area, but in Montana most records are from late March to October, with stragglers remaining until December. In Colorado most migrants arrive in April and leave by mid-November, but many birds overwinter. Nesting records for Montana are from March 15 to June 30, and in Banff and Jasper young have been observed from May 29 to July 23.

Comments: This familiar shorebird is able to exploit a wide variety of human-associated habitats, and is highly protective of its nest and young, performing elaborate injury-feigning or "broken-wing" displays when they are threatened by humans or other mammals. The birds are insectivorous, feeding on surface-dwelling insects such as beetles for the most part.

Suggested Reading: Johnsgard, 1981; Phillips, 1972; Bunni, 1959.

Mountain Plover (*Charadrius montanus*)

Identification: This species has the shape and size of a killdeer, but lacks breast banding and is rather uniformly brownish dorsally, with (in adults) a black forehead patch. Its calls include a variety of whistled notes, and the birds are almost invariably found on grassland habitats.

Status: A local summer resident on the plains east of the mountains; a rare migrant or vagrant in the montane parks.

Habitats and Ecology: This species breeds exclusively in April on arid grasslands, where grasses are usually no more than three inches in height, and sometimes in semidesert areas with cacti and scattered shrubs, often far from water. During the nonbreeding seasons they also are found in relatively dry habitats.

Seasonality: Wyoming records are from March 25 to September 13, and Montana records are from late April to September. In Colorado the records extend from March 25 to October 5. Egg records in Wyoming are from May 22 to July 9, and in Montana from late May to late July.

Comments: This shorebird is one of the distinctive breeding species of the short-grass plains of interior North America, and its range is gradually retracting as these areas have come under irrigation or otherwise have been eliminated as suitable breeding habitat.

Suggested Reading: Johnsgard, 1981; Graul, 1974, 1975; Laun, 1957.

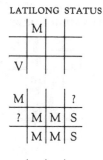

LATILONG STATUS

	M		
V			
M			?
?	M	M	S
	M	M	S
s	S	S	S
s	S	S	S
s		M	M

Black-necked Stilt (*Himantopus mexicanus*)

Identification: This distinctive shorebird is unique in having long red legs and a strongly bicolored plumage, with black above and white below. The bill is black and needle-like, and the birds are usually found wading in shallow waters.

LATILONG STATUS

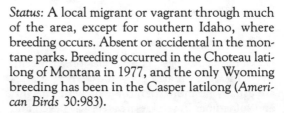

Status: A local migrant or vagrant through much of the area, except for southern Idaho, where breeding occurs. Absent or accidental in the montane parks. Breeding occurred in the Choteau latilong of Montana in 1977, and the only Wyoming breeding has been in the Casper latilong (*American Birds* 30:983).

Habitats and Ecology: Breeding in this species usually occurs in the grassy shoreline areas of shallow freshwater or brackish pools of wetlands having extensive mudflats, or sometimes along the shorelines of salt lakes where vegetation is essentially lacking. Often found in company with American avocets, which use similar habitats.

Seasonality: The few regional records are from April 14 to September 12.

Comments: This species is part of a nearly worldwide complex of stilts that vary greatly in plumage, but probably all represent variations of the same species. Although rather different in appearance from avocets, the two groups are quite closely related and have rather similar behavior patterns.

Suggested Reading: Johnsgard, 1981; Hamilton, 1975; Stout, 1967.

Black-necked Stilt

American Avocet (*Recurvirostra americana*)

Identification: The recurved and long bill makes this species easy to identify; additionally, it has very long legs and a strong black and white wing pattern that flashes in flight. Usually observed wading in shallow waters.

Status: A local summer resident over much of the region, mainly on shallow marshes of the plains. Rare in the montane parks, with breeding reported only for Yellowstone (no recent records).

	M	M	M
M	S		S
V	M	s	M

S	M	S	s
S	S	S	M
S	M	M	M

M	S	S	M
S	S	M	S
S	s	S	S

Habitats and Ecology: During breeding this species favors ponds or shallow lakes with exposed and sparsely vegetated shorelines, and somewhat saline waters that have large populations of aquatic invertebrates, which are gathered by making scythelike movements of the curved bill through the water.

Seasonality: Reported in Wyoming and Montana from mid- or late April through September, rarely as late as early November. Colorado records extend from March 15 to November 30. In Alberta the birds usually do not arrive until early May, and are rarely seen after the end of August. Wyoming egg records are from May 15 to July 15.

Comments: This large and attractive shorebird is a frequent feature of the shallow, alkaline marshes of the arid plains and prairie areas, north to the Alberta parklands. In Alberta many of the prairie marshes support these birds, which often nest in loose colonies, and collectively defend their breeding areas from intruders by loud calling and diving attacks.

Suggested Reading: Johnsgard, 1981; Gibson, 1971; Hamilton, 1975.

Greater Yellowlegs (*Tringa melanoleuca*)

Identification: This medium-sized shorebird has rather bright yellow legs, and a long black bill that is nearly straight. The dorsal plumage is dark and finely spotted, and the undersides are white to slightly flecked with spotting. In flight, the birds usually utter three to five sharp, clear whistled notes.

LATILONG STATUS

M	M	M	M
M	M	?	M
V	M	M	M

M		M	M
M	M	M	M
M	M	M	M

M		M	M
	M	M	M
M	M	M	M

Status: A migrant over most of the region, but a summer resident in the mountains of Alberta, occasionally breeding south to Banff Park. Breeds rather commonly in Jasper Park (Miette and Athabasca valleys, Willow and Blue creeks, etc.).

Habitats and Ecology: In migration these birds occupy the edges of marshes and slow-moving rivers, foraging along the shorelines and sometimes wading out belly-deep to probe in the mud or skim the surface for invertebrates. On the breeding grounds the birds favor muskeg areas, with a mix of ponds, trees, and clearings, and sometimes extend into subalpine scrub near timberline. In Alberta a favored nesting habitat consists of muskeg with spruce and tamarack.

Seasonality: Wyoming and Montana records extend from April to late May, and from August to October 24, with peaks in early May and early September. Colorado records are from March 26 to November 12. In Alberta the birds begin nesting in early May, and have usually left by late September.

Comments: This species and the lesser yellowlegs are very close relatives, and their niches overlap considerably, both on breeding and nonbreeding areas. In the Old World the closely related greenshank and spotted redshank replace them ecologically.

Suggested Reading: Johnsgad, 1981; Stout, 1967.

Lesser Yellowlegs (*Tringa flavipes*)

Identification: Like the previous species, this one has conspicuous yellow legs, but it is generally smaller, with a bill that is only about as long as its head, and it usually utters only one or two whistled notes when alarmed. In flight both species exhibit whitish rumps and tails.

Status: A migrant nearly throughout the entire region, but breeding locally in central Alberta, south to about 53°N. latitude, but not known to breed in either Jasper or Banff parks, where it is a rare migrant or vagrant.

Habitats and Ecology: Breeding typically occurs in habitats that have a combination of rather open and tall woodlands, with low and sparse brushy undergrowth, and fairly close to grassy or marshy ponds. Broken hills, covered with burned or fallen timber, and low poplar second growth, are a favored Alberta nesting habitat. Outside the breeding season the birds occur along mud flats and shallow ponds, often with vegetated shorelines, and sometimes visit flooded fields.

Seasonality: In Montana and Wyoming these birds arrive in late April or early May, leaving by the end of that month, and again occur from mid-July to middle or late October, peaking in September. Colorado records are from April 5 to November 8. Few birds are seen after September in Alberta.

Comments: Breeding areas of these species are often well away from water, on sandy ridges covered by jackpines, or on dried burned-over areas with fallen timber present. The nest is often at the base of a tree or log, and the bird blends well with the browns of the bark or branches of the fallen trees. Usually one bird stands watch, often from the top of a small tree, while the other incubates.

Suggested Reading: Johnsgard, 1981; Stout, 1967.

LATILONG STATUS

M	M	M	
	M	M	M
V	M	M	M

M	M	M	M
M	M	M	M
	M	M	M

M	M	M	M
M	M	M	M
M		M	M

158

Solitary Sandpiper (*Tringa solitaria*)

Identification: This small sandpiper is about the size of a lesser yellow-legs, but with darker legs and a strongly barred tail (evident in flight), as well as a conspicuous white eye-ring. A sharp alarm whistle is often uttered when flushed.

LATILONG STATUS

M	M	M	M
	M	M	M
M	M	M	M

M	M	M	M
M	M	M	M
M	M	M	M

M	M	M	M
M	M	M	M
M	M	M	M

Status: A migrant throughout the region, breeding along the northern edge of this book's boundaries, in central and northern Alberta (south to about the North Saskatchewan River and, in the montane parks, to Kootenay N.P. In Jasper N.P. known breeding areas include Willow, Blue, and Isaac creeks, Topaz and Southesk lakes, and Rocky Forks.

Habitats and Ecology: Breeding is done around muskeg ponds, along woodland lakes, and near forest ponds, where the old nests of tree-nesting birds such as American robins are utilized.

Seasonality: Montana and Wyoming records are from April 20 to May, and from late July to October 4. Colorado records are from April 20 to September 23. In Alberta there are egg records for early June, and observations of young during the latter half of July.

Comments: This is the only North American shorebird that nests in the abandoned nests of various passerine birds, such as blackbirds and robins, usually in rather wet and open terrain, where rusty blackbirds also often breed and provide nesting sites. Old nests of waxwings, king-birds, and jays have at times also been used.

Suggested Reading: Johnsgard, 1981; Stout, 1967.

Willet (*Catoptrophorus semipalmatus*)

Identification: This is a fairly large sandpiper with a straight, blackish bill and a brown to grayish dorsal coloration. The best fieldmark is provided by the wings, which when opened show a broad white wing stripe, bordered in front and behind with black. Loud *willet* calls are often uttered by disturbed birds.

Status: A migrant or summer resident over much of the area, breeding mainly in grassland marshes; generally rare or lacking in the montane parks. Although a reported nester in Yellowstone N.P., current evidence indicates that it is only a spring and fall migrant.

LATILONG STATUS

	M	M	s
M	M		S
V	M	M	

S	M		M
S	s	s	M
S	S	M	M

S	s	S	M
S	S	M	M
s	s	S	s

Habitats and Ecology: Breeding habitats of this species consist of prairie marshes, usually brackish to semialkaline, seasonal ponds, and sometimes also intermittent streams in grassland areas. The birds are effective probers, and spend much of their time feeding in this way, but also peck at objects on the water surface.

Seasonality: Montana and Wyoming records extend from April 25 to October 1, and Colorado records are from April 14 to October 24. They are usually present in Alberta from late April to early September. Nesting in Montana is from mid-May to mid-June, and in the Cypress Hills downy young have been seen from June 13 to 26.

Comments: The willet is a species that breeds both on coastal shorelines and in the continental interior, in rather different habitat types. They often share their breeding habitats with marbled godwits, American avocets, and Wilson's phalaropes. Wintering occurs from the southern states to central South America.

Suggested Reading: Johnsgard, 1981; Tompkins, 1965; Stout, 1967.

Wandering Tattler (*Heteroscelus incanus*)

Identification: This rarity resembles a solitary sandpiper, but is unspotted dorsally, much grayer throughout, and has yellowish legs. Its call is a distinctive series of rapid whistles, all uttered on the same pitch. Its yellow legs might cause confusion with yellowlegs, but the birds are much grayer throughout, and have no noticeable dorsal spotting with buffy white.

LATILONG STATUS

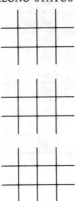

Status: An accidental vagrant in the area, with records for Jasper and Watertown Lakes parks, a few other Alberta records, and few if any other regional records. The nearest known breeding areas are in northwestern British Columbia and the mountains of Yukon Territory.

Habitats and Ecology: Associated with rocky shorelines in wintering areas and on migration, and with mountain streams in breeding areas, usually in the alpine zone. Nesting occurs on gravelly or rocky bars and islands of arctic streams.

Seasonality: The few Alberta records are from May 30 to September 2.

Comments: Very few nests of this elusive species have been found, and little is known of its reproductive biology. Its common name "tattler" derives from the fact that the birds are highly vocal on their breeding grounds whenever any danger threatens, thus alerting all the locally nesting birds to such danger.

Suggested Reading: Johnsgard, 1981; Stout, 1967.

ACCIDENTAL VAGRANT

Spotted Sandpiper (*Actitus macularia*)

Identification: This common small sandpiper is usually best identified by its behavior, a teetering motion that is almost constantly used. In flight the birds have a distinctive vibrating flight and appear to fly with strongly downcurved wings. Breeding-plumage birds have spotted underparts, but these are lacking in immature and non-breeding individuals.

Status: A common summer resident throughout the region, breeding in all the montane parks.

Habitats and Ecology: Associated with forest streams, pools, and rivers, usually at lower elevations, but extending locally to alpine timberline and utilizing a wide array of open terrains with water present, and rarely even in the absence of nearby water. Shaded watercourses are favored, and sometimes the birds are found along rapidly flowing mountain torrents.

Seasonality: Reported in Wyoming from April 26 to October 5, but mostly present between early May and late August. Colorado records are from April 28 to October 12, and in Alberta the birds are usually present from late April to early September. Egg records in Wyoming are from May 23 to August 7, and in Alberta eggs have been reported from June 14 to July 15, with young reported between July 9 and 25.

Comments: The distinctive bobbing motion of this species may make visual recognition easier in its sometimes noisy environment, such as along ocean surfs or on the edges of mountain streams. Similar behavior is performed by American dippers, which are often found in the same habitats.

Suggested Reading: Hays, 1973; Oring & Knudson, 1973; Nelson, 1939; Miller & Miller, 1948.

LATILONG STATUS

M	M	S	s
?	S	s	S
s	M	s	

S	s	s	s
?	S	S	S
S	S	S	M

S	s	S	M
s	S	S	S
S	S	S	S

Upland Sandpiper (*Bartramia longicauda*)

Identification: This prairie-adapted sandpiper is usually seen perched on fenceposts in grassland habitats or flying above the prairies, often calling loudly and showing a bowed-wing flight similar to that of the spotted sandpiper, which is much smaller. The head is small relative to the body, and the bill and legs are yellowish to pale orange in color.

LATILONG STATUS

M	M	S	S
?	S	s	S
V	M	s	

S	s	s	s
?			M
	M		M

		s	
	M	M	

Status: A summer resident in native grassland areas east of the Rockies; a rare migrant or vagrant in the montane parks.

Habitats and Ecology: Generally associated with wet meadows, hayfields, mowed prairies, or mid-length prairies, avoiding shortgrass steppe areas and extremely tall grasses. Often found far from water, and rarely if ever wading for its foods.

Seasonality: Reported in Montana and Wyoming from April 27 to September 16, with probable migration peaks in May and August. Colorado records extend from April 14 to September 2, and eggs have been found from May 20 to June 28. In Montana eggs have been noted from May 25 to mid-July.

Comments: This is an indicator species of native prairie, and as such is one that has been declining in range and abundance. On its breeding grounds it utters a flight song of great beauty while fluttering above the territory like a giant butterfly. On landing, it momentarily holds its wings above the body like a graceful ballet dancer.

Suggested Reading: Johnsgard, 1981; Higgins & Kirsch, 1975; Stout, 1967.

Long-billed Curlew (*Numenius americanus*)

Identification: This very large shorebird is easily identified by its extremely long (at least 5 inches), decurved bill. In flight the birds exhibit cinnamon wing linings underneath (also present on marbled godwits), and frequently utter loud *cur-lee* calls when alarmed. Often seen far from water, in native grasslands.

Status: A summer resident in grassland areas over much of the region, but rare to absent in the montane parks except Grand Teton (where not yet known to breed) and rare but recorded as breeding in Yellowstone.

Habitats and Ecology: On the breeding grounds this species occurs in shortgrass areas, grazed taller grasslands, and overgrazed grasslands with scattered shrubs or cacti. Hilly or rolling areas seem favored over flatlands, and the birds often nest rather far from standing water. However, migrating birds usually are found on beaches or other shoreline habitats.

Seasonality: Montana records are from April 9 to mid-September; peak migration records for Montana and Wyoming are in April and August. Colorado records are from April 10 to October 24, and in Alberta the birds are present from late April until late August. Montana, Idaho, and Wyoming egg records are from April 20 to July 4; in Colorado eggs have been found from May 4 until June.

Comments: This is one of the finest shorebirds of the world, and one whose range continues to decline. In Alberta it still nests north to about Elnora and Castor, but in the western U.S. it has generally lost much of its original breeding range, especially at the eastern edges of its nesting range.

Suggested Reading: Johnsgard, 1981; Bicak, 1977; Fitzner, 1978.

LATILONG STATUS

M	M	s	s
	S	s	S
S	s	S	s

S	s	s	S
S	S	M	S
S	S	S	s

S	M	S	M
s	M	M	M
s	M	s	s

164

Hudsonian Godwit (*Limosa haemastica*)

Identification: Adults in breeding plumage have barred rusty red breasts and underparts, and a black-tipped orange bill, as well as broad white wing stripes (visible in flight). Immatures and non-breeding birds are much more grayish throughout, with less colorful bills, but the white wing stripe still provides an excellent fieldmark.

LATILONG STATUS

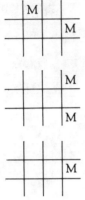

Status: A rare to occasional local migrant in the region, mostly in the plains area, and rare or accidental in montane areas.

Habitats and Ecology: On migration this species is likely to be found along shorelines of prairie marshes, singly or in small numbers, and usually probing dowitcher-like for food. They breed in subarctic areas where woods and scrub tundra intermix, and where wet meadows or ponds are close by. The nearest breeding grounds are along western Hudson Bay, in northeastern Manitoba.

Seasonality: There are not many records, but in southern Alberta the birds are typically present in late April or early May, and again in August and September.

LOCAL MIGRANT

Comments: This beautiful tundra nester becomes, on the breeding grounds, a perching bird, with the non-incubating member of the pair resting on low trees and watching for danger. Its loud cries of alarm warn not only its mate but other nesting birds in the vicinity. Because of this effective alarm system, the species' nests are among the hardest of arctic shorebirds to locate.

Suggested Reading: Johnsgard, 1981; Stout, 1967.

Marbled Godwit (*Limosa fedoa*)

Identification: This species is about the size of a long-billed curlew, and like it exhibits cinnamon-colored underwing linings, but also has a nearly straight, black-tipped orange bill. It is usually seen on native grasslands, or foraging along the shorelines of prairie ponds and marshes.

Status: A local migrant and summer resident in northern parts of the region, including eastern Alberta and northern Montana. In Alberta it breeds from the foothills north to St. Paul and Athabasca, but is absent from the montane parks.

LATILONG STATUS

	M	M	s
M	M		S
V	M	M	

s	M	M	s
M	M	M	
	M	M	M

	M	M	M
M	M	M	M
M		M	M

Habitats and Ecology: On the breeding grounds this species occupies wetlands associated with prairies, including intermittent streams, ponds, and shallow lakes ranging from fresh to strongly alkaline. Semipermanent ponds and lakes are especially preferred, with nesting occurring in grassy flats nearby.

Seasonality: Wyoming records are from April 25 to September 17; peak migration in Wyoming and Montana seems to be in May and early September. Colorado records extend from April 11 to November 25, and in Alberta the birds are usually present from late April to the end of August or early September. There are few regional nest dates, but in North Dakota eggs have been found from April 17 to June 22.

Comments: This species and the long-billed curlew are often seen together in breeding areas, but the curlew tends to be an upland nester, while the godwit remains much closer to water. Both species defend their nests strongly from humans, screaming loudly and diving at the intruder.

Suggested Reading: Nowicki, 1973; Johnsgard, 1981.

Ruddy Turnstone (*Arenaria interpres*)

Identification: In breeding plumage the bright rusty and black dorsal plumage, and a black-and-white head pattern, is easily recognized, but immatures and birds in winter plumage have a much more subdued pattern. The legs are always orange-red, and the bill is short and sharply pointed. Often found in gravelly or rocky shoreline areas, where it pokes about for food and flips small rocks over to expose the invertebrates below.

LATILONG STATUS

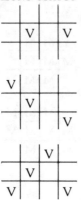

Status: A local migrant east of the mountains; an accidental vagrant in the montane parks.

Habitats and Ecology: On migration, these birds are likely to be found foraging on stubble fields where they sometimes forage with other species of shorebirds such as plovers, or on sandy shorelines of lakes or reservoirs. They are high-arctic breeders, nesting along the shorelines of the Arctic Ocean of northern Canada.

Seasonality: There are few regional records, but in spring they usually are seen during the second half of May. Colorado records are from April 26 to September 28.

Comments: The turnstones are highly specialized shorebirds in that their bills are adapted for probing and flipping over rocks or similar objects. Partly for this reason, their relationships to other shorebirds are rather obscure.

Suggested Reading: Johnsgard, 1981; Stout, 1967.

LOCAL MIGRANT

Black Turnstone (*Arenaria melanocephala*)

Identification: Very similar to the ruddy turnstone in shape and size, but lacking rusty color at all seasons, and without orange-colored legs.

Status: An accidental vagrant in the area; there is a single Montana sighting for Glacier National Park (*Condor* 60:337), but no other apparent record.

Habitats and Ecology: This species is similar in its behavior and ecology to the ruddy turnstone, but is more westerly in distribution, breeding in Alaska.

Seasonality: The only regional record is of a bird reported on August 28, 1957.

Comments: Very little is known of the biology of this species, which breeds in remote areas of western and southwestern Alaska, in lowland and coastal tundra. During the winter the birds inhabit rocky coastlines, foraging in the manner of ruddy turnstones just above the limits of the surf, probing in sand and turning over rocks or patches of seaweed with the bill, sometimes running toward the heavier objects and using their heads as battering rams.

Suggested Reading: Johnsgard, 1981; Stout, 1967.

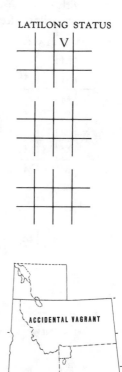

LATILONG STATUS

ACCIDENTAL VAGRANT

Sanderling (*Calidris alba*)

Identification: This small sandpiper is one of the palest of the regional shorebirds; a black bill, black legs, and blackish shoulder areas contrast with an otherwise mostly white to grayish body. The birds feed at the waterline, often following retreating waves to find exposed foods.

LATILONG STATUS

	M		
	M		M
V	M		

M			
M	M	M	M
	M		M

		M	M
M	M	M	M
M			M

MIGRANT THROUGHOUT

Status: A rare migrant or vagrant in the region, more common on the plains, and rare or accidental in the montane parks.

Habitats and Ecology: Migrating birds are usually seen around the larger lakes, especially those with wave-swept sandy beaches. The birds are high-arctic nesters, and the nearest breeding areas are extreme northern Canada.

Seasonality: Wyoming records are from May 3 to 26, and from August 26 to October 30. Most Montana records are for fall, from late August through September. In Alberta the birds arrive rather late in spring, mostly passing through in late May and early June, and occur again in August and September, with stragglers remaining until early November.

Comments: This is one of the most widespread of all shorebirds, and one with one of the longest of all migration routes, with North American birds wintering in extreme southern South America and probably migrating close to 20,000 miles a year.

Suggested Reading: Johnsgard, 1981; Stout, 1967.

Semipalmated Sandpiper (*Calidris pusilla*)

Identification: A small nondescript sandpiper, with a medium-long bill that is straight and tapers gradually. Both the bill and legs are black. Several other small and very similar sandpipers occur regionally, and require very careful identification.

Status: A migrant throughout the region, fairly common in the plains, but only a vagrant or rare migrant in the montane parks. The nearest breeding areas are in northern and northeastern Manitoba, in tundra areas near Hudson Bay.

LATILONG STATUS

M	M		
M	M		M
V	M	M	M

M	M	M	M
M		M	
	M	M	M

	M	M	
	M	M	M
			M

Habitats and Ecology: A very close relative of the western sandpiper, this species is more prone to occur on wet and dry mud, where it often picks up surface organisms, while the western sandpiper is more often found standing in water or in wet mud, where it probes for food. The semipalmated sandpiper is also less prone to move out into grassy flats to forage than is the Baird's sandpiper.

Seasonality: Wyoming records are for mid- to late May, and for the latter half of August. In Alberta the birds pass through in May, peaking near the end of the month, and appear again in early July, peaking in August, with some remaining until late September. Colorado records are from April 15 to October 1.

MIGRANT THROUGHOUT

Comments: This species has a somewhat elliptical migratory pattern, moving north through the Great Plains, but with at least many adults moving south along the Atlantic Coast, in a similar manner as lesser golden plovers. However, some adults move south through Alberta in August, followed by immatures, so perhaps these represent unsuccessful breeders leaving the nesting grounds early.

Suggested Reading: Johnsgard, 1981; Stout, 1967.

Western Sandpiper (*Calidris mauri*)

Identification: Another small nondescript sandpiper, or "peep." In spring it is rather more rusty brown dorsally than is the semipalmated, and in any plumage its slightly longer and very slightly drooping bill aids in identification. Like the semipalmated, both the legs and the feet are blackish.

LATILONG STATUS

	M		
M	M	M	
M	M		M

M			?
M			
	M	M	M

		M	M
M	M	M	M
M	M	M	M

MIGRANT THROUGHOUT

Status: A migrant nearly throughout the region, mainly in the plains, and rare or accidental in the montane parks. The nearest breeding areas are in western Alaska.

Habitats and Ecology: Migrants are likely to be seen in the same areas as semipalmated and least sandpipers, and frequently mingle with both these species, allowing for each comparison. Their breeding areas are farther to the west than those of the other two species, and thus they are more likely to be seen west of the Rockies than to the east.

Seasonality: There are few regional migration records, but they appear to be about the same as for the semipalmated sandpiper. Confusion with that species makes the status of the western sandpiper difficult to determine. Colorado records are from March 23 to October 17, and more northerly records fall within these limits.

Comments: The slightly longer bill of the western sandpiper helps to reduce competition from least and semipalmated sandpipers, and allows it to forage in very slightly deeper waters. Like the semipalmated, its feet have slight webbing, perhaps facilitating swimming. However, they are less adapted to deep probing than such sandpipers as dunlins, with which they also overlap in wintering areas.

Suggested Reading: Johnsgard, 1981; Stout, 1967.

Least Sandpiper (*Calidris minutilla*)

Identification: Very slightly the smallest of the common "peep" sandpipers, and having the shortest bill. The legs and feet are olive-yellow rather than black, and at least in breeding plumage the upperparts are somewhat sooty-toned, without rufous patterning.

Status: A migrant throughout the region; more common on the plains, where it is generally among the commonest of the "peeps," but rare to accidental in the montane parks. The nearest breeding areas are the coastal tundra of northeastern Manitoba.

Habitats and Ecology: While on migration these sandpipers are found on a variety of moist habitats, often in company with semipalmated, Baird's, or western sandpipers, and probably feeding on much the same foods as these species.

Seasonality: Wyoming and Montana records are from late April to June 1, and from July 8 to October 21. In Alberta they are present from May to early June, and again from early August to September. Colorado records range from April 14 to November 7.

Comments: Least sandpipers breed widely through the North American subarctic, and extend somewhat farther south as breeders than do semipalmateds and the other small *Calidris* species. They tend to forage by making pecking rather than probing movements with their short bills, and forage along shorelines or in very shallow water areas.

Suggested Reading: Johnsgard, 1981; Stout, 1967.

LATILONG STATUS

M	M		
M	M		M
V	M		

M	M	M	?
M	M	M	M
	M	M	M

M		M	M
M	M	M	M
M	M	M	M

MIGRANT THROUGHOUT

White-rumped Sandpiper (*Calidris fuscicollis*)

Identification: A "peep" sandpiper that differs from the others in having a white rump that is very conspicuous in flight, but otherwise is not easily seen. On the ground, the birds exhibit distinctly spotted breasts and flank feathers in spring, setting them apart from the other small *Calidris* forms.

LATILONG STATUS

Status: A local migrant through the eastern portions of the region, east of the Rockies; absent from the montane parks.

Habitats and Ecology: Migrants utilize the same kinds of prairie ponds as do the other "peeps," but on the breeding grounds the birds seek out wet tundra around the edges of ponds or lakes. It breeds in very similar habitats as does the least sandpiper, but occurs farther north than that species.

Seasonality: Wyoming spring records are from May 15 to mid-June, with a peak in late May; fall records are lacking. In Colorado the records extend from May 7 to October 21, but are primarily for May and early June. Most Alberta records are for early June.

Comments: The white-rumped sandpiper is notable for its unusual swollen-neck display performed by males on the breeding grounds, which seems related to its non-monogamous mating system, approaching that of the pectoral sandpiper.

Suggested Reading: Johnsgard, 1981; Stout, 1967.

LOCAL MIGRANT

Baird's Sandpiper (*Calidris bairdii*)

Identification: Another "peep" sandpiper; this one has a distinctly buff cast to its plumage, with pale and broad feather edgings, a buffy wash on the breast, and relatively long wings that extend beyond the tip of the tail. Both the bill and the feet are black.

Status: A migrant throughout the region; more common on the plains and rare in the montane parks. A high-arctic nester, breeding in extreme northern Canada.

LATILONG STATUS

M	M	M	M
	M	M	M
V	M		M

M	M	M	M
M	M	M	M
	M		M

		M	M
M	M	M	M
M		M	M

Habitats and Ecology: Migrants are associated with wet meadows and shallow ponds, often feeding in grassy areas somewhat away from water, but also along muddy shorelines, where they tend to peck at food sources rather than to probe for them. On the breeding grounds they seek out dry tundra areas rather than wet coastal habitats.

Seasonality: Wyoming records are from March 22 to June 3, and from July until October 21, with peaks in mid-May and late August. In Alberta the birds occur from mid-May to early June, and again from late July onward, with stragglers reported as late as November. Colorado records are from March 14 to January 2.

Comments: This rather drab species of "peep" is rather inconspicuous but often very common among the migrants using prairie ponds and shallow lakes, perhaps because they often forage somewhat back from the water's edge. They often seem to use somewhat alkaline ponds, to which their rather pale plumage seems appropriately colored.

Suggested Reading: Johnsgard, 1981; Stout, 1967.

MIGRANT THROUGHOUT

Pectoral Sandpiper (*Calidris melanotos*)

Identification: The largest of the "peep" sandpipers, and the one with the most distinctive chest bib, which is sharply separated from the white belly. The bill is black, and the legs and feet are yellowish. Often feeds with the smaller species of "peeps," when its larger size is easily evident.

LATILONG STATUS

M	M	M	
	M	M	M
V	M		

M			M
M	M	M	M
	M		M

		M	M
	M	M	M
		M	M

MIGRANT THROUGHOUT

Status: A migrant throughout the region, more common on the plains; rare or accidental in the montane parks. A high arctic nester, with the nearest breeding grounds in extreme northern Canada.

Habitats and Ecology: Migrants are commonly seen along prairie marshes or potholes, where they wade in shallow water and probe or peck for food. Often found near grassy cover rather than on open mud flats, and on the breeding areas they also select rather wet and grassy tundra. This is the most polygynous of the *Calidris* sandpipers, and males perform display flights above their territories to attract females to them, while swelling their chests and expanding their "bibs" greatly.

Seasonality: Montana and Wyoming records are from late February to April, and from mid-August to late October. Fall records in Alberta are from early August to late October; spring records are fewer and concentrated in May. In Colorado the records range from April 2 to November 17. The birds tend to arrive somewhat earlier in spring than the smaller "peeps," and stay somewhat later in the fall.

Comments: The remarkable male territorial display of this species, with inflated chest and throat area while the bird flies about making a grunting noise, is quite striking, and one of the memorable sights of the high arctic tundra. A close Asian relative, the sharp-tailed sandpiper, has a similar aerial display, and the white-rumped sandpiper seems to represent a rudimentary evolutionary development toward this condition.

Suggested Reading: Johnsgard, 1981; Stout, 1967.

Dunlin (*Calidris alpinis*)

Identification: This small but long-billed sandpiper is easily recognized in spring plumage, when it exhibits a unique black belly and a whitish breast. Juveniles and winter-plumaged birds are much more difficult; but at all times the very long, black bill, which droops slightly, is an excellent fieldmark. Rusty back feathers are also conspicuous in spring-plumaged birds, and aid in identification.

Status: A rare migrant in the region, with few records, and none for the montane parks. The nearest breeding areas are in Manitoba, along the west coast of Hudson Bay.

Habitats and Ecology: Migrant birds are likely to be seen with other small sandpipers such as the "peeps," and usually occur on mud flats or sandy beaches, where they probe for food. In breeding areas they seek out wet coastal tundra for nesting, and sometimes extend into areas of low foothills.

Seasonality: There are few migration records for the area. Alberta records extend from May 1, but most are for the latter part of May. It has also been recorded there in late October. In Colorado it has been seen from April 20 to May 9, and again on December 30.

Comments: Most dunlins migrate to and from their tundra breeding areas along the coasts, with only a small number migrating throughout the Great Plains, in spite of a rather extensive breeding area on the west coast of Hudson Bay.

Suggested Reading: Johnsgard, 1981; Stout, 1967.

LATILONG STATUS

M		
		M
M	M	

M		M
M		
	M	

	M	M	
		M
		V

RARE MIGRANT

176

Stilt Sandpiper (*Calidris himantopus*)

Identification: This long-legged relative of the typical "peeps" has a longer black bill than any of these, and considerably longer legs. In breeding plumage it shows a bright patch of chestnut feathers in the ear region, and pale "eyebrows"; nonbreeding birds and immatures lack the ear-patch but have more conspicuous eyebrow lines.

LATILONG STATUS

M	M	M
	M	
V	M	

M		?
	M	

M	M	M
M	M	M
M		M

LOCAL MIGRANT

Status: A local migrant, mainly east of the mountains; very rare to absent from the montane parks. The nearest breeding areas are in northeastern Manitoba, along the coast of Hudson Bay.

Habitats and Ecology: Migrants are usually found in company with the typical "peeps," but usually are wading in belly-deep water, and thrusting their bills at organisms or probing the bottom with their rather long bills. They are fairly gregarious on migration, and often occur in moderately large flocks. On breeding areas they seek out well drained tundra areas or sedge meadows.

Seasonality: Wyoming records are from May 15 to 25, and from July to September 11. Colorado records are from April 20 to October 8.

Comments: Stilt sandpipers are among the most attractive of the small North American sandpipers; their dainty feeding behavior, trim profiles, and elegant chestnut cheek patches in spring make them a delight to watch. Their foraging at times resembles the somewhat similar dowitcher, which are relatively more bulky birds and which, when they flush, show distinctive white rump markings.

Suggested Reading: Johnsgard, 1981; Stout, 1967.

Short-billed Dowitcher (*Limnodromus griseus*)

Identification: Both species of dowitchers are very similar, but this one is rarer in the region, and has less heavy spotting on the underparts in spring plumage, less heavily barred flanks, and generally less reddish color of the underparts to help identify it. The bill is only slightly shorter in this species, and thus immature and winter-plumaged birds are extremely difficult to identify to species. This species tends to utter three-noted whistles, rather than single notes or a long series of notes, which may help to identify it.

Status: Apparently less common on migration than the long-billed, although in Alberta it is likely that the reverse is true. Breeding occurs in northern Alberta, south to about Edmonton.

Habitats and Ecology: On migrations, dowitchers are found in grassy marshes, where they feed by probing their long bills in belly-deep water. The breeding habitat consists of marshy, boggy, and muskeg areas, with the nests often placed on hummocky sites, and well hidden from above by overhanging grasses.

Seasonality: There are few good migration dates, but Colorado records are for July and August only. There is one August record for Wyoming, and the few Montana records are not definite as to date. The birds arrive in Alberta in May, and dowitchers are seen there well into October, although species identification is uncertain.

Comments: The nesting range of dowitchers in Alberta has retracted in recent years, and they apparently no longer nest near Edmonton, as was the case in the 1950s.

Suggested Reading: Johnsgard, 1981; Stout, 1967.

LATILONG STATUS

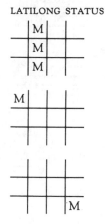

LOCAL MIGRANT

Long-billed Dowitcher (*Limnodromus scolopaceus*)

Identification: In spring plumage, this species is rather brighter red than the short-billed, and tends more toward barring than spotting on the flanks. Its bill length is slightly greater, and its calls tend not to be three-noted. However, the species are extremely similar and very difficult to separate under most field conditions.

LATILONG STATUS

M	M	M	
	M	M	M
M	M		

M		M	M
M		M	M
M	M		M

M		M	M
M	M	M	M
M		M	M

Status: A migrant throughout the region, rarer in montane areas and virtually absent from the montane parks. The nearest breeding areas are in Alaska.

Habitats and Ecology: Migrating birds probably use the same habitats as do short-billed dowitchers. Likewise, their breeding habitats seem to be very similar muskeg and wet tundra habitats.

Seasonality: Migration records for Wyoming are from April 8 to May 18, and from July to November 11. In Montana the spring migration is in May and the fall records are from mid-August to late October, with a peak in early October. Possibly some of these records apply to the preceding species, but both appear to migrate at about the same time.

Comments: Dowitchers of both species are distinctly gregarious, foraging in small groups and flying in compact flocks. In many ways they appear to be intermediate between typical sandpipers and the snipes, and their downy young also have some snipe-like features.

Suggested Reading: Johnsgard, 1981; Stout, 1967.

MIGRANT THROUGHOUT

Common Snipe (*Gallinago gallinago*)

Identification: The very long bill of this species, and the stout body shape, help to identify it and separate it from dowitchers, the only other shorebird with nearly this beak-length and body size. Snipes often utter a grating, scraping call when flushing, and while on territory the eerie "winnowing" sound made by vibration of tail feathers while in flight is a very easily recognized fieldmark.

Status: A summer or year-round resident nearly throughout the region, both in mountains and plains wetlands; probably breeding in all of the montane parks.

Habitats and Ecology: Common snipes nest in marshy areas, often beaver ponds in the Rocky Mountains, and in muskeg ponds or other heavily vegetated marshes elsewhere in their extensive range. Peatland habitats are especially favored, but the birds may also occur along slow-moving rivers, marshy shorelines of lakes, or sometimes even wet hayfields.

Seasonality: Locally resident in Wyoming and Montana, with some migrations evident in April and again in September. In Alberta the birds also occasionally overwinter. Locally resident in Colorado, but with a migration peak in September. Montana and Wyoming egg records are from May 14 to June 26, and in Colorado young have been seen as early as May 8. Alberta egg records are from May 16 to July 20.

Comments: The spectacular aerial displays of territorial common snipes are an unforgettable aspect of the Rocky Mountains and the other northern breeding grounds. The sounds of the snipe are entirely mechanically produced by the narrow outer tail feathers which are held at right angles to the body in flight, and allowed to intercept the periodically interrupted airflow from the wings, producing a wavering sound.

Suggested Reading: Johnsgard, 1981; Sanderson, 1977; Mason & Macdonald, 1976; Tuck, 1972.

LATILONG STATUS

r	R	S	s
s	R	s	S
R	R	S	r

S	R	R	S
S	R	s	S
R	R	S	S

S	s	s	s
s	s	S	S
S	S	S	R

180

Wilson's Phalarope (*Phalaropus tricolor*)

Identification: Phalaropes typically are seen swimming rather than wading, and this species is much the commonest of the species of the region, with a long bill (longer than the head) and a blackish stripe through the eye and extending down the side of the neck, and a whitish breast with chestnut tinting. Males are duller than females, but have the same general plumage pattern.

LATILONG STATUS

s	S	S	s
	S	s	S
S	S	s	s

S	s	S	s
S	S	S	S
S	S	M	S

S	s	S	M
s	S	S	S
s	S	S	S

Status: A summer resident over most of the area, becoming rarer in the mountains, and a rare migrant in most of the montane parks. Breeding has been reported only from Yellowstone N.P., in some of the marshy areas. Common on the prairie marshlands.

Habitats and Ecology: Breeding habitats are typically wet meadows adjoining shallow marshes, which range from fresh to highly saline. Ditches, river edges, and shallow lakes are sometimes also used for breeding. Similar areas are used by migrating birds.

Seasonality: Wyoming records are from late April to September 27, with peaks in May and August. Montana records extend from the end of March to early September, with a spring peak the third week of May. Wyoming egg records are from May 25 to June 21; Colorado records are from May 19 to June 24.

Comments: These shorebirds are unique among North American birds in that the females are larger, more brightly colored, and more aggressive than males, establishing territories, courting males, and leaving the clutch for the male to incubate. In contrast to the other two species of phalaropes, multiple or polyandrous matings with males are still unproven for this species, although such a possibility is present, given the relatively long available breeding season in the Great Plains.

Suggested Reading: Hohn, 1967; Kangarise, 1979; Johns, 1969; Johnsgard, 1981.

Red-necked Phalarope (*Phalaropus lobatus*)

Identification: This phalarope resembles the previous species, but has a bill about the same length as the head, and (in breeding plumage) a mostly grayish black head with a white throat patch and chestnut on the foreneck. Males are less colorful, but have a similar general appearance. Juvenile and non-breeding plumages of phalaropes are quite different, and are much paler throughout. The swimming behavior helps identify the birds as phalaropes, and the relative bill length should assist in determining the species under these conditions.

Status: An uncommon to rare migrant through-out the region, mainly in the plains, with most montane records in the Alberta parks. The nearest breeding region is in northern Saskatchewan or possibly extreme northern Alberta.

LATILONG STATUS

	M	M	M
	M		M
V	M		M

M	M		?
M	M	M	
M	M		M

M		M	M
M	M	M	M
M		M	M

Habitats and Ecology: Breeding habitats of these species are subarctic ponds, marshes, and lagoons having adjacent grassy or sedge vegetation, where nesting occurs. Proximity to lakes or other fairly permanent bodies of water may also be a part of the habitat characteristics. On migration the birds are found in the same areas as Wilson's phalaropes, and often are seen in company with them.

Seasonality: Wyoming records are from May 13 to 29, and from August 29 to October 26. Montana records are for late May and June, and from August 5 to October 4, with peaks in late May and mid-September.

Comments: This is one of two species of phalaropes for which sequential multiple matings by females with males has been observed, although it is by no means common behavior, judging from available data. At times the female will remate with her original mate for a second clutch, if the first one was unsuccessful. Apparently no more than two clutches are laid by females of this species. This species is frequently referred to as the "northern phalarope."

Suggested Reading: Johnsgard, 1981; Stout, 1967; Hohn, 1971.

MIGRANT THROUGHOUT

Red Phalarope (*Phalaropus fulicarius*)

Identification: This phalarope has the shortest and stoutest bill of the three species, and in breeding plumage is the only one that is mostly reddish on the breast, flanks, and underparts. Immature and winter-plumaged birds are very whitish, but the typical phalarope swimming behavior and short bill (shorter than the head) should help to identify the species.

LATILONG STATUS

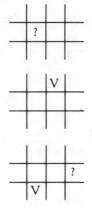

Status: A rare migrant to accidental vagrant in the area, most likely to be seen toward the north; a single record for Banff N.P.

Habitats and Ecology: A high-arctic breeder, nesting in wet tundra areas of extreme northern Canada. The nearest breeding areas are in Yukon Territory. On migration similar habitats to those used by the other two phalarope species are utilized.

Seasonality: There are very few regional records, but in Alberta the species has been observed during June, July, and September.

Comments: Sequential polyandrous breeding has been observed in this species; in one study four of eleven females were found to form second pair bonds, while in another study there were three suspected cases of polyandry among a population of about 100 birds. In Britain this species is generally known as the "grey phalarope."

Suggested Reading: Johnsgard, 1981; Stout, 1967; Hohn, 1971.

RARE MIGRANT

Pomarine Jaeger (*Stercorarius pomarinus*)

Identification: This gull-like bird is dark-colored, with dark underwings and variably dark underparts. In adults the central tail feathers are blunt-tipped and extend a rather short distance beyond the rest of the tail feathers. Immature birds cannot be readily identified in the field.

Status: An accidental vagrant in the region. There is a specimen record for Rocky Mountain N.P. There are at least three Montana records, one Wyoming record, and three other Colorado records.

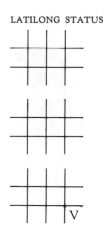

LATILONG STATUS

Habitats and Ecology: Like the other jaegers, this is an arctic-breeding species, which most often winters in coastal areas, and is unlikely to be seen in interior regions. The nearest breeding areas are in extreme northern Canada.

Seasonality: Too few records exist to judge migration times; the record for Rocky Mountain N.P. was for November, and probably many records involve immatures wandering. The Colorado records are from late September to mid-November.

Comments: Jaegers, especially immatures, are often very difficult to identify, and vary greatly in their plumage coloration, typically existing in both light and dark phases. Up to three or four years may be required for the tail feathers to reach their full length and make identification easier.

ACCIDENTAL VAGRANT

Suggested Reading: Anderson, 1973; Pitelka et al., 1955.

Parasitic Jaeger (*Stercorarius parasiticus*)

Identification: Very similar to the preceding and following species, but in adults the central tail feathers are pointed and extend out only a short distance for the other feathers. Field identification of non-adults is almost impossible except for experts.

LATILONG STATUS

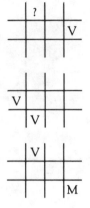

Status: An accidental vagrant throughout the region except in Alberta, where it is very rare. The nearest breeding area is in northeastern Manitoba, along the coastline of Hudson Bay.

Habitats and Ecology: Breeds in arctic and subarctic tundra areas of North America and Eurasia, extending somewhat farther south in Canada than the other two species. Like the other jaegers a predator rather than a scavenger, feeding on small rodents, nestling birds, and the like.

Seasonality: There are relatively few regional records, but they are largely for late summer and fall.

Comments: This is the jaeger most likely to be seen in the region; it has been reported from six latilongs in Montana, three in Wyoming, and six in Colorado.

Suggested Reading: Pitelka et al., 1955; Perdeck, 1963.

ACCIDENTAL VAGRANT

Long-tailed Jaeger (*Stercorarius longicaudus*)

Identification: Very similar to the two preceding species, but adults can be identified by their two long, pointed, and streamer-like central tail feathers. The birds are more graceful and tern-like than the other species, and adults always lack whitish areas on the underwings that the others usually show.

Status: An accidental vagrant throughout the region, more common northwardly, and reported from April 15 to June 17. Nestlings have been seen in Wyoming in June, and fledged young in Montana in early July.

LATILONG STATUS

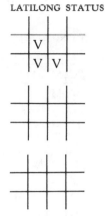

Habitats and Ecology: Similar to the other two jaegers in being an arctic tundra nester, where it has circumpolar breeding distribution. The nearest breeding areas are in the Northwest Territories.

Seasonality: There are few regional records, but they are largely for late summer or fall. Reported from at least four Montana latilongs, and one in Colorado.

Comments: Jaegers are important predators on lemmings in the arctic, and in addition they are effective in stealing waterfowl eggs or taking laggard ducklings or goslings. They have at times been observed catching small birds in flight.

Suggested Reading: Pitelka et al., 1955; Anderson, 1971.

ACCIDENTAL VAGRANT

Franklin's Gull (*Larus pipixcan*)

Identification: In breeding plumage, the black head and dark upper wing surface of this gull make it easily identifiable and unique among gulls of the region. In late summer or fall the head is mostly grayish, but retains black ear patches; immature birds are similar. The rarer Bonaparte's gull is also black-headed, but shows white upper wing patches on the outer flight feathers.

LATILONG STATUS

		S	M
M	M	M	S
V	M		

M	M	S	M
S	S	S	
S	M	M	

M		M	M
M	M	M	M
M	M	M	M

Status: A migrant and local summer resident in the area, primarily on the plains; relatively rare in the montane parks, but there is an undocumented breeding record from Yellowstone N.P., where it is rare in summer.

Habitats and Ecology: Breeding occurs in large, relatively permanent prairie marshes having extensive stands of emergent vegetation, where the birds nest in colonies. Unlike other gulls of the region, the nest is constructed over water, in dense vegetation. On migration they typically feed on dry land, often in fields that are being cultivated prior to planting.

Seasonality: Migration in Wyoming and Montana is from late April to mid-May, and from late August to early October. Extreme Colorado records are April 1 and November 22. In Alberta the birds usually arrive the third week of April, and most have left by the end of September. Egg records are few, but in Montana nesting occurs from early June to early July, with hatched young reported by June 23.

Comments: This gull is a highly beneficial species, eating grasshoppers, cutworms, and many other agricultural pest insects that are exposed by plowing. Unlike the other gulls they rarely if ever scavenge at dumpgrounds, but instead are almost entirely insectivorous.

Suggested Reading: Burger, 1974.

Bonaparte's Gull (*Larus philadelphia*)

Identification: This small and beautiful gull closely resembles the Franklin's gull in having a black head, but it also has primarily white primary feathers that are tipped with black. Immature and fall-plumaged birds lack the black head, but instead have a small back ear-patch, and the white outer wing feathers provide the best fieldmark.

Status: An uncommon to rare migrant throughout the region, mainly in the plains regions, and rare or accidental in the montane parks. The nearest breeding areas are in central Alberta (from Battle Lake and Edmonton northward), in muskeg forests.

LATILONG STATUS

M	M	M	
	M		M
V	M	M	M

M	M		
M	M		
M	M		

	M		
	M	M	M
M	M	M	M

Habitats and Ecology: This gull is unique in its tree-nesting adaptations; it typically nests in small coniferous trees well above ground level, but at times also nests in reedbeds of marshes. Jackpines, spruces, and other conifers are the usual nesting site; typically the mate stands watch in a nearby tree as the other bird incubates. The birds often nest in loose colonies, and outside the breeding season they are highly gregarious, often forming flocks numbering in the hundreds.

Seasonality: Wyoming records are from April 18 to late May, peaking in late April, and from early September to November 18, peaking in November. Montana records are from May 13 to 18, and from July 31 to November 5. Large numbers concentrate in the lakes of central Alberta in mid-September.

Comments: This is among the most beautiful of the North American gulls, and a fairly close relative of the common black-headed gull of Europe, which is starting to colonize eastern Canada, probably from Iceland.

Suggested Reading: Henderson, 1926; Twomey, 1934.

Mew Gull (*Larus canus*)

Identification: This rather large, white-headed gull resembles a California gull, but has a small and short bill that is uniformly yellow in adults. The legs are more pinkish than those of the California or ring-billed gulls. Immature birds are best identified by their short and relatively weak bill.

LATILONG STATUS

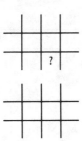

Status: An accidental vagrant in the region; reported most often in Alberta and a rare migrant or vagrant in its montane parks. The nearest breeding areas are in extreme northern Saskatchewan or perhaps Alberta.

Habitats and Ecology: Nesting occurs on the shorelines of lakes in northwestern Canada and Alaska, usually on the ground but occasionally in trees as typical of the Bonaparte's gull. A common breeder in Europe (where it is called the "common gull").

Seasonality: Regional records south of Alberta are few (one October record for Montana; one August record for Wyoming); records for Banff and Jasper parks are from May through September.

Comments: This species used to be called the "short-billed gull," a better name for it than the current one (which does not refer to its voice). Recognition of it in the field is difficult in any plumage unless the bill color and shape are seen clearly.

Suggested Reading: Weidmann, 1955.

ACCIDENTAL VAGRANT

189

Ring-billed Gull (*Larus delawarensis*)

Identification: This abundant and familiar gull is best identified by the black band that surrounds it near the tip (in adult and second-year birds); younger birds have a black-tipped bill and are very difficult to separate from other white-headed gull species.

Status: A summer resident and local colonial breeder over most of the region, primarily on the plains; reported breeding in the montane parks is limited to Yellowstone, where the last known breeding was in 1949. The range is gradually expanding in western North America, and there are now numerous colonies in the region (*Wilson Bulletin* 95:362–83).

Habitats and Ecology: A highly adaptable gull, exploiting new habitats in the form of reservoirs. Breeding usually occurs on isolated and sparsely vegetated islands of lakes and reservoir impoundments, sometimes in colonies of a thousand pairs or more.

Seasonality: Wyoming records extend from March 7 to November 30; most migration there and in Montana occurs from late March to early May, and from mid-August to late November. Some overwintering occurs as far north as Colorado. Egg records in Montana are from April 27 to June 18.

Comments: Probably this is the commonest gull of the region, incorrectly called "seagulls" by the average person. However, many ring-bills do winter coastally, mixing with several other species of white-headed gulls.

Suggested Reading: Tinbergen, 1959; Vermeer, 1970; Johnson & Forster, 1954.

LATILONG STATUS

M	M	s	S
M	S	M	S
s	M	M	s

M	M	S	M
M	S	M	M
M	M	M	M

M	M	M	M
M	M	M	S
M	M	s	r

California Gull (*Larus californicus)*

Identification: Similar to the ring-billed gull, but adults show red and black spots near the tip of the lower mandible, thus resembling a small herring gull. The birds have darker upper surfaces than ring-billed or herring gulls, and are somewhat intermediate in size between them. Immature birds are extremely difficult to identify in the field except by experts.

LATILONG STATUS

	?	s	S
M	S	M	S
s	M	s	M

M	M		M
M	S	M	
r	M	M	S

M	M	S	M
M	M	M	S
M	M	s	r

Status: A summer resident and local breeder over much of the region, mainly on the plains; the only breeding in the montane parks is in Yellowstone, where 200–300 pairs breed yearly on the Molly Islands of Yellowstone Lake. The breeding range of the species in the general region is increasing at present (*Wilson Bulletin* 95:362–83). There are large colonies in Wyoming (Banforth and Ocean lakes) and Montana (Freezeout Lake).

Habitats and Ecology: Like the ring-billed gull, this species usually nests on gravelly islands of large lakes or reservoirs or along their shorelines, and in many areas the two species nest in close proximity. In Alberta the California gulls tend to nest on more elevated and boulder-strewn sites, while ring-bills occupy more level terrain. Ring-bills also tend to cluster their nests more strongly, while California gulls space their nests more randomly.

Seasonality: Wyoming records are from April 18 to November 16, while Montana records extend from March 27 to late October. In Wyoming egg-laying begins as early as April 21 and extends to June 11; fledged young have been seen by the end of June. In Montana nesting records are from April 27 to June 18, with nestlings seen as late as July 19.

Comments: This is the species made famous by the Mormons, when it helped save their crops in Utah by feeding on locusts.

Suggested Reading: Vermeer, 1970; Baird, 1976; Behle, 1958; Diem & Condon, 1967; Green, 1952; Raper, 1976.

Herring Gull (*Larus argentatus*)

Identification: This is the largest of the regional gulls, and probably the most abundant North American gull. In adults, the yellow bill with a red spot near the tip of the lower mandible provides the best field-mark, in addition to its large size; however, some California gulls also have very similar bill markings during the breeding season, complicating identification. Immature birds are highly variable, and require expert identification.

Status: An uncommon to rare migrant or vagrant in much of the area, becoming more common northwardly in the montane parks. There is only a single regional nesting record, for Big Lake (Stillwater County), Montana, in 1918. The nearest breeding areas in northern Alberta south to Namur Lake and Lower Therien Lake.

LATILONG STATUS

	M	M	
		M	M
M	M		

M		?	?
?	M		
		M	

	M		
	M	M	
	M	W	

Habitats and Ecology: This is primarily a coastal gull, but also breeds in small colonies across northern Canada on the islands of larger lakes, sometimes among colonies of ring-billed or California gulls where they often nest as single pairs. They usually winter coastally, but sometimes spend the winter on ice-free lakes or impoundments in the more southerly states.

Seasonality: In southern Alberta these birds are usually seen as spring and fall migrants, generally appearing in April and early May, and again in September and October. In Montana the records are mostly from September to May, including several winter occurrences. In Wyoming they are occasional from summer to winter, mostly in eastern areas, and in Colorado they are largely winter visitors, from October 17 to May 17.

Suggested Reading: Keith, 1966; Burger, 1977.

Sabine's Gull (*Xema sabini*)

Identification: The only black-headed gull (in breeding plumage) that has a large, triangular patch of white feathers formed by the inner primaries and secondaries, and bounded in front by black. The bill is black, with a yellow tip. Younger birds lack the black head, but have the distinctive wing patterning. This is also the only gull with a forked tail, although the tail forking is rather slight and easily overlooked.

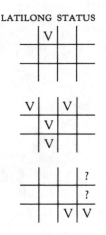

LATILONG STATUS

Status: A rare migrant or vagrant in the region; accidental in the montane parks. A high-arctic breeder, with the nearest nesting areas in extreme northern Canada. Most likely to be encountered in northern portions of the region.

Habitats and Ecology: This arctic nester is most likely to be observed in flocks of migrating Franklin's or Bonaparte's gulls, but it is extremely rare south of Canada. It nests in colonies on the high-arctic tundra, and migrates to a restricted wintering area off the coast of Peru.

Seasonality: In Alberta this gull has been observed between May 18 and July 4, and again during mid- to late September. Colorado records are from September 3 to November 17, as well as a summer (July) record. Most of these are of immature birds.

Comments: Apparently these birds must migrate more or less directly from the Pacific Ocean to their high-arctic breeding grounds, as few observations are made of them on spring migration.

Suggested Reading: Bent, 1921; Brown et al., 1967.

RARE MIGRANT

Caspian Tern (*Sterna caspia*)

Identification: The largest tern of North America, almost as large as a ring-billed gull, and with a massive red bill and a slightly crested black "cap." Immatures have more grayish "caps" and less colorful bills.

Status: An uncommon to rare migrant and very local summer resident in the region; a regular breeder in Yellowstone N.P. where a few pairs nest yearly on the Molly Islands. Has bred once in Montana (*American Birds* 36:991) and has nested in southern Idaho and southeastern Wyoming.

LATILONG STATUS

M	M	M	
	M		M
V			

M		M	M
	S		
	M		M

M		S	
		S	S
			V

Habitats and Ecology: This species usually nests near coastlines, but has also nested interiorly on shorelines or islands of large lakes or reservoirs, usually on sandy or stony beaches. Often in these locations only one or two pairs nest among other terns or gulls, but normally nesting is done in colonies.

Seasonality: In Alberta the birds are regular in summer, and nested irregularly at the west end of Lake Athabaska as recently as 1952; now they are mainly seen as summer visitors in the province. Terns nesting on the Molly Islands of Yellowstone Lake begin incubation in late May, and flightless young have been seen from July 10 to August 24.

Comments: This large and beautiful tern usually nests in colonies of more than 150 pairs in the Great Lakes area, sometimes with as many as 500 pairs. Territories are very small, with nests as close as about two feet apart in such colonies. Caspian terns seem to be gradually extending their range in interior North America, nesting rarely as far inland as North Dakota and Minnesota.

Suggested Reading: Bent, 1921; Ludwig, 1965.

Common Tern (*Sterna hirundo*)

Identification: This species is not as common regionally as the Forster's tern in spite of its name; the two species are easily confused, but the common tern has a less silvery upper wing surface, showing a medium gray in this area. Breeding adults have a more reddish than orange bill color, and their outermost tail feathers are edged with black. Late summer and immatures of the two species are very similar, but common terns tend to have darker heads than Forster's terns. Common terns are also more likely to be seen on lakes than on prairie marshes.

LATILONG STATUS

	M	M
S	M	S
M	M	M

M	M	M	?
s	M		
	M		

M		
		M
		V

Status: A regional migrant and local summer resident from central Montana northward, mainly on plains lakes; rare in the montane parks. Although breeding has reportedly occurred in Yellowstone N.P. this is not documented, and there is no current evidence of this.

Habitats and Ecology: Islands in large lakes are favored breeding grounds in this region; sparsely vegetated areas are used for colonial nesting. Occasionally a pair or two will also build nests in reedy vegetation over water, but this behavior is much more typical of Forster's terns. Sometimes the two species will nest in close proximity, but normally they are well isolated ecologically.

Seasonality: Montana and Wyoming records are from mid- to late May to September; Colorado records extend from May 15 to October 18. In Alberta they usually arrive about the second week of May and rarely remain beyond the end of October, as feeding areas begin to freeze over. There are few regional nesting records, but in North Dakota egg dates range from June 8 to July 28, within which range the available regional records fall.

Suggested Reading: Palmer, 1941.

Forster's Tern (*Sterna forsteri*)

Identification: Differs from the preceding species in having (in breeding plumage) a more orange bill, more silvery gray upper wing surfaces, and pale outer tail feathers. Late-summer and immature birds have little or no black on the nape. The birds are associated with shallow prairie marshes rather than deep lakes.

Status: A summer resident locally in the area, mainly on plains marshes; rare or accidental in the montane parks, with no breeding records.

Habitats and Ecology: Large marshes having extensive reedbeds or muskrat houses for nest sites are the typical breeding habitats of this species, which breeds colonially in such locations, with as many as five nests sometimes situated on a single muskrat house. Such sites that are close to open water areas for foraging are especially favored nesting locations.

Seasonality: Wyoming records are from May 2 to September 30, with migration peaks in May and early September. Montana records are from late April to late September, and extreme Colorado dates are April 10 and October 18. In Wyoming egg records are from July 2 to 30, and in Colorado from May 15 to July 1.

Comments: Apart from their habitat differences, Forster's terns and common terns are very similar in appearance and behavior, including similar courtship displays. Thus, the two species are often confused by all but the keenest observers.

Suggested Reading: Bergman et al., 1970; McNicholl, 1971.

LATILONG STATUS

	M	M
	s		M
V	M	

M	M		M
s	M	
S	M	M	M

		M	M
M	M	M	S
M		S	s

Black Tern (*Chlidonias niger*)

Identification: This is the only slate-colored North American tern; in breeding plumage the body is mostly black, with gray wings and almost no white evident. In late summer and non-adult birds the body is much whiter, but there are usually darker patches of feathers on the breast, at least while the birds are in this region. This is the most insectivorous of the regional terns, and often can be observed catching insects in the air or skimming them from the surface of the water, rather than plunge-diving for fish.

LATILONG STATUS

s	S	S	
	S	S	s
S	M	S	s

M		S	M
s	S	M	
S	M		M

M		M	M
M	M	M	S
M	M	S	S

Status: A summer resident over much of the region, mainly in plains marshlands; rarer in the montane parks but breeding occasionally or regularly in several.

Habitats and Ecology: Typical nesting habitat consists of small to large marshes with extensive stands of emergent vegetation and some areas of open water. Fish populations are not necessary, as the birds feed mostly on insects while on the nesting grounds. Nests are more often placed among emergent vegetation than on muskrat houses, although the latter are sometimes used.

Seasonality: Wyoming records are from May 11 to October 1, with migration peaks in May and September. In Alberta the birds usually arrive the third week of May and are usually gone by September. Wyoming and Montana nest records are from June 12 to early July; nesting in Colorado extends from late May to the end of June.

Comments: In this species, courtship "fish-flights" usually involve the carrying of insects rather than fish, but otherwise the social behavior patterns of the birds resemble those of the other terns. The adaptive significance of black coloration, rather than white as in most terns, is not obvious, but it may have to do with reduced need for camouflage from fish during foraging.

Suggested Reading: Bergman et al., 1970; Bailey, 1977; Goodwin, 1960.

Rock Dove (*Columba livia*)

Identification: This is the familiar barnyard pigeon, well known to everyone, living in the wild state. Wild-type rock doves rather resemble bandtailed pigeons, but lack yellowish bills and a white band, although the plumage patterns sometimes vary greatly. Rarely found far from humans, but at times living on cliffs or other natural sites.

Status: Present virtually throughout the region, although rare or lacking in high montane areas, and declining northwardly.

Habitats and Ecology: Largely associated with cities and farms in North America, and infrequent in forested areas. Buildings that provide narrow nesting ledges are preferred for nesting, but cliff ledges or crevices are sometimes also used.

Seasonality: Resident throughout the area, with a prolonged nesting season that probably extends over most of the year except perhaps in the most northerly areas.

Comments: This species is one of the avian "pests," sometimes causing serious health problems in areas where the birds are abundant and generally protected, as in most cities.

Suggested Reading: Goodwin, 1967; Murton & Clarke, 1968; Murton et al., 1972.

LATILONG STATUS

r	R	r	r
r	r	r	R
R	R	R	R

R	R	r	R
r	r	R	R
	r	R	R

R	r	R	R
R	R	r	R
R	R	R	R

198

Band-tailed Pigeon (*Columba fasciata*)

Identification: This forest-dwelling pigeon has the general shape of a rock dove, but has a pale band across the posterior half of the tail, and a white crescent across the nape. The bill is yellow, with a black tip. The call is a double-noted cooing, repeated several times.

LATILONG STATUS

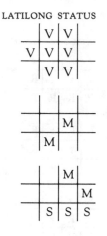

Status: Limited to the southernmost part of the region, breeding in Rocky Mountain N.P. and its vicinity, north to the Wyoming border. A vagrant farther north, rarely to Montana and southern Alberta.

Habitats and Ecology: This species is generally associated with western oak woodlands or mixed oak and pine woodlands, and extending into the ponderosa pine zone locally, especially where Gambel oaks are also present. Available foods in the form of acorns are an important determinant of local distributions. During July and August found up to 10,000 feet in Rocky Mountain N.P., gradually moving to lower altitudes in late summer.

Seasonality: In Colorado these birds are present from the end of March to late October, with a few birds sometimes wintering. The height of the nesting season is from late July to the first week in August, and nests with young have been seen from August 22 to 31.

Comments: Besides acorns, this species also feeds in summer on a variety of cultivated crops, gradually shifting in fall to a mixture of nuts, fruits, and berries. The birds are quite gregarious, and flocks often gather at salt deposits or sources of mineral water. At least in some parts of their range up to three broods (two eggs per clutch) may be reared.

Suggested Reading: Sanderson, 1977; Gutierrez et al., 1975; Neff, 1947; Houston, 1963.

Mourning Dove (*Zenaida macroura*)

Identification: This familiar dove is almost uniformly grayish brown, with a long, pointed, and white-edged tail. The bill is blackish, and there is a narrow bluish eye-ring. The call is a five-noted series, *who-ah, who, who, who.*

Status: A widespread and common breeder in the region, occupying nearly all vegetational zones up to the lower coniferous forest zone. Present and probably breeding in all of the montane parks.

Habitats and Ecology: Breeds from riparian woodlands and cultivated areas through grasslands and sagebrush to woodlands, aspen, and open coniferous forest habitats, as well as in cities and farmsteads. Nests either on the ground or, preferentially, in shrubs or trees.

Seasonality: Resident in Colorado and sometimes also in Wyoming, although there a migration peak in April and October is evident. In Alberta and Montana the birds usually arrive in mid-April and leave in October. The breeding season is prolonged; egg records in Colorado are from May 7 to August 12, in Wyoming from May 6 to September 4, and in Montana from early June to September 11, reflecting a nesting season of at least four months and repeated broods per pair.

Comments: Like other doves of the region, these birds lay only two eggs per cycle, but repeated nesting efforts result in high productivity. Both sexes help feed the young, initially providing them with "pigeon milk," a secretion of the crop lining. By late fall the adults and young are feeding on seeds of weeds, grain, berries, and the like.

Suggested Reading: Sanderson, 1977; Nice, 1922; Cowan, 1952; Hanson & Kossack, 1963.

LATILONG STATUS

s	R	s	s
s	S	s	S
S	S	S	S

R	S	S	S
s	S	S	S
S	S	S	S

S	S	S	S
S	S	S	S
R	R	S	R

Black-billed Cuckoo (*Coccyzus erythropthalmus*)

Identification: The two cuckoo species are similar in appearance, but this species lacks rusty coloration on the wings, has no yellow on the lower beak, and has less white on the tail. Its calls are fast cooing notes, in groups of three and four units.

LATILONG STATUS

?	M	?
s		s
		s

s	s	s	S
M	V	M	S
	M		S

M	s	S	M
	M	M	M
		V	S

Status: A fairly common summer resident east of the mountains in plains woodlands; a rare vagrant in the montane parks. Breeds commonly on the plains of southern Alberta and northern Montana; local elsewhere.

Habitats and Ecology: Associated during the breeding season with somewhat dense woodland cover, such as upland woods with a variety of trees, shrubs, and vines, offering shady hiding places and nest sites.

Seasonality: Wyoming records are from May 21 to September 12, and Montana records extend from late May to early October. Nesting in Montana occurs in June, and in Colorado eggs have been seen as late as July 23. The birds leave soon after nesting, and even in Colorado the latest record is October 19.

Comments: Cuckoos are much more often heard than seen, as they skulk about in shady areas, acting in a generally secretive manner. Unlike the Old World cuckoo they are not obligatory nest parasites, although the two North American cuckoos sometimes drop their eggs in one another's nests in areas where both occur.

Suggested Reading: Spencer, 1943; Bent, 1940.

Yellow-billed Cuckoo (*Coccyzus americanus*)

Identification: Similar to the preceding species, but with a bill that is yellowish below, and with rusty brown on the wings, and also a long tail that has large white tips on the outer features.

Status: A summer resident in the southern and eastern parts of the region, mainly east of the mountains, but breeding west locally at least to southern Idaho, and probably north to southeastern Montana. Rare at higher elevations, and absent from the montane parks.

Habitats and Ecology: Associated with thickety areas near water, second-growth woodlands, deserted farmlands, and brushy orchards. Dense woodlands are avoided.

Seasonality: Reported in Colorado from May 8 to September 22, with egg records from June 18 to August 15, and nestlings reported as late as September 10. The long nesting season suggests double-brooding is present, as is known to occur in some other areas.

Comments: This species utters rather strange-sounding *kaw* or *kawp* notes, usually in long series, and sounding something like the noise made by pounding on a hollow wood drum. In some parts of their range the birds are called "rain crows," as they often call on dark and cloudy days, such as before rainstorms.

Suggested Reading: Poulter, 1980; Preble, 1957; Bent, 1940.

LATILONG STATUS

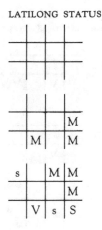

Flammulated Owl (*Otus flammeolus*)

Identification: This small owl closely resembles a screech-owl, but has somewhat shorter "ears" and dark brown eyes, which are surrounded by rufous feathers. More easily recognized by its song, a cadenced series of *hoop* or *hoop-hoop* notes at intervals of three or four seconds.

LATILONG STATUS

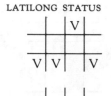

Status: Of uncertain extent in the region, probably more common and widespread than currently known, but known to breed in western Idaho, present during the breeding season in the River of No Return wilderness area of central Idaho, and also reported from Sawtooth and Caribou national forests. Breeds uncommonly in northern Utah, just beyond this book's coverage. At least until the early 1900s it bred in Rocky Mountain N.P., but there are no recent records for the species in the park.

Habitats and Ecology: Associated with aspen and ponderosa pine forests in both breeding and non-breeding periods, particularly ponderosa pine areas. In Colorado nesting up to 10,000 feet has been noted, usually in pines or aspens with woodpecker holes about ten to twenty feet above ground. Sometimes natural cavities are used.

Seasonality: A migratory owl, with Colorado records from April 16 to October 4. Egg records for that state are from May 27 to June 27, and mostly from June 5 to 20; nesting dates for Rocky Mountain N.P. are for mid-June.

Comments: This is a highly insectivorous little owl, often feeding on grasshoppers or beetles rather than vertebrate prey. Probably because of this, the species is quite migratory, usually wintering in Mexico or Central America.

Suggested Reading: Marshall, 1939; Winter, 1974; Richmore et al., 1980; Bergman, 1983.

Western Screech-owl (*Otus kennicottii*)

Identification: Screech-owls are small, yellow-eyed owls with ear-tufts and (at least in the western species) a grayish overall plumage. The eastern species also occurs in a reddish brown phase. The western form is best identified by its call, a "bouncing ball" series of notes that begin slowly but speed up toward the end. The typical eastern song is a whinny-like series of rising and falling notes.

Status: The species limits are still uncertain, but at least in most montane areas of the region the western form is present, and from the Wyoming Bighorns east the eastern species is apparently the resident form. Both species appear to be present in northern Colorado and perhaps southern Wyoming, although past practices of classifying the two forms as a single species make range interpretation difficult. Screech owls probably breed in all of the montane parks south of Canada, but become rarer northwardly, and are apparently absent from the Alberta parks.

Habitats and Ecology: Associated with a variety of wooded habitats, including farmyards, cities, orchards, etc., and from riparian edges through pinon–juniper and oak–mahogany woodlands to aspens and ponderosa pine forests.

Seasonality: A resident in southern areas, usually a summer resident northwardly. In Colorado egg records (both species) are from April 7 to May 6, and nestlings seen as late as June 14. There are April egg records for Wyoming, and dependent young have been observed in Montana by early July.

Comments: Much more information from Wyoming and Colorado is needed before the status of the eastern screech-owl can be determined for these areas. Thus, no separate species account is provided for it here.

Suggested Reading: Marshall, 1967; Earhart & Johnson, 1970.

LATILONG STATUS

s	s	S	
	R	s	s
R	s	R	R

S	s		R
	R		M
	s		R

M		R	M
	M	R	
R	R	s	R

Great Horned Owl (*Bubo virginianus*)

Identification: The largest of the "eared" owls of the region, and one of the commonest. Up to two feet long, with wingspreads of almost four feet, only the rare great gray owl is comparable in size, and it has a larger head that lacks ear-tufts. The usual call is a low hoot, *Who-who-ah-who, who-ah-whoo.*

LATILONG STATUS

R	R	R	
	R	R	R
R	R	R	R

R	R	R	R
R	R	R	R
R	R	R	R

R	R	R	R
R	R	R	R
R	R	R	R

Status: A common resident in wooded habitats throughout the region; probably breeds in all the montane parks.

Habitats and Ecology: A powerful and adaptable owl, this species occurs everywhere from riparian woodlands through the coniferous forest zones, and extends into city parks, farm woodlots, and rocky canyons well away from trees. Nesting is thus highly variable, but often occurs in abandoned bird or squirrel nests, or on tree crotches, rock ledges, or even on the ground.

Seasonality: A permanent resident throughout the region, with Wyoming egg records from March 1 to May 23, and Montana records from mid-March to mid-April. Nestlings are usually evident in late May and June. Alberta egg records are from February 23 to May 29.

Comments: This is one of the most efficient avian predators in North America, taking a wide variety of mammalian and avian prey. They have been found to be serious predators on peregrine falcons that have been released in the Rocky Mountain region, and often take coots and ducks from the water at night.

Suggested Reading: Craighead & Craighead, 1956; Errington et al., 1940; Bent, 1938.

Snowy Owl (*Nyctea scandiaca*)

Identification: This is the only white or nearly white owl of the region, and it is about the same size as a great horned owl. Likely to be seen only in winter.

Status: A wintering migrant over most of the region, becoming rarer farther south. Rarely reported from the montane parks; usually found on open plains.

Habitats and Ecology: An arctic breeding species that periodically is forced south in winter when food supplies on the breeding grounds are limiting. The nearest breeding area is in extreme northern Manitoba, along the coast of Hudson Bay.

Seasonality: Present in Alberta from about mid-November to late March, and in Montana from late November to mid-February. Rare and irregular farther south in the region.

Comments: Snowy owls feed primarily on mice and other rodents while wintering in this region, and thus are beneficial birds. Unlike other owls, they often hunt during the day, using their ears to locate mice under the snow if necessary.

Suggested Reading: Taylor, 1974; Walker, 1974.

LATILONG STATUS

W	W	W	W
	W	W	W
W	W	W	W

W	W	W	W
	W		W
	W		W

		W	W
			W
	V	W	W

WINTERING MIGRANT

Northern Hawk-owl (*Surnia ulula*)

Identification: A medium-sized grayish owl without ear-tufts, a long tail, and underparts barred with brown. Often active in daylight hours, perched on exposed areas, and sometimes pumping its tail.

LATILONG STATUS

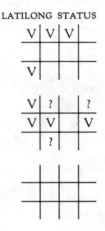

Status: A rare resident in the montane parks of Alberta; farther south a wintering vagrant.

Habitats and Ecology: This rare owl is most likely to be seen in burned over or open muskeg areas, where dry boggy areas offer excellent hunting, and tall trees or stumps provide look-out sites for the birds to perch while hunting. Hunting is done largely by visual means, in a hawk-like manner.

Seasonality: In central or southern Alberta these birds often appear in fall and establish wintering hunting territories. In Montana it has been reported from November to late February. It has been reported from three Wyoming latilongs, but not from Colorado. Alberta egg records are from April 1 to June 4, with a majority between April 13 and May 18.

Comments: This handsome and elusive owl also breeds in Eurasia in similar habitats, and has been better studied in that region.

Suggested Reading: Mikkola, 1983; Pulliainen, 1978.

Northern Pygmy-owl (*Glaucidium gnoma*)

Identification: A tiny owl (about six inches long) with a relatively long tail and no ear-tufts. Its breast is streaked and its back spotted. A mellow, dove-like and three-noted hooting is distinctive, with the notes spaced one or two seconds apart. The notes are higher pitched than those of the saw-whet owl and are repeated more rapidly. Black nape patches, somewhat resembling false eyes, are visible from behind. Like the hawk owl, it sometimes nervously jerks or pumps its tail.

Status: An inconspicuous species that probably occurs through the montane forests south at least to Jackson Hole. Also breeds in the mountains of western Colorado, although apparently rare in Rocky Mountain N.P.

LATILONG STATUS

r	R	r	
r	r	r	M
R	r	r	M

r	R		
	R	M	
	r		

M			
			R

Habitats and Ecology: Found in similar habitats as saw-whet and flammulated owls, but apparently ranging higher, to about 12,000 feet in Colorado, and more active in daylight hours than these species. Nesting is done in woodpecker holes or similar tree cavities.

Seasonality: A resident over most or all of the region. Colorado egg records are from May 17 to June 22, and in Montana young have been observed in June.

Comments: Imitating this owl or playing its song on a tape recorder often provides a means of learning if it is in an area, as it is otherwise only rarely detected. A recording of the call will often also attract many small songbirds, attempting to "mob" the apparent unwelcome owl.

Suggested Reading: Earhart & Johnson, 1970; Walker, 1974.

Burrowing Owl (*Athene cunicularia*)

Identification: Usually seen at or near colonies of prairie dogs or ground squirrels, whose vacant burrows the owls use for nesting. The owls are small, "earless," and rather long-legged, with pale spotting on the back. The call is a soft and dove-like *coo-cooo.*

LATILONG STATUS

M	M		
	s		s
S	M		

s	s	s	S
s	M	s	S
	M	M	S

S	s	S	S
S	S	S	S
S		s	S

Status: A summer resident on the plains over much of the region; rare in most of the montane parks, and not proven to breed in any.

Habitats and Ecology: This is the only North American owl closely associated with plains rodents such as prairie dogs, and as the range and abundance of these mammals have decreased, so too has the status of the burrowing owl. It is largely an insectivorous species, often eating beetles, but also takes many small mice in some areas.

Seasonality: Wyoming records are from March 12 to November 2, with migration peaks in April–May, and again in September–October. In Montana the birds are present from mid-May to mid-September, and in Colorado most records extend from March 26 to November 4, with rare later occurrences. Colorado egg records are from May 9 to June 19, which encompasses available nesting records from farther north.

Comments: This attractive and interesting ground-dwelling owl is a member of the "high-plains" fauna that once included bison, prairie dogs, prairie wolves, and other animals that have now largely disappeared from the American scene. Large areas of grassland are needed to preserve these species, and few such areas still exist.

Suggested Reading: Coulombe, 1971; Thomsen, 1971; Martin, 1973; Grant, 1965.

Barred Owl (*Strix varia*)

Identification: This rather large woodland owl is much more often heard than seen; its distinctive "Who cooks for you; who cooks for you-all" call is unmistakable after hearing it for the first time. If seen, the owl appears "earless," with a large rounded head and a heavily streaked and barred breast and belly. Its eyes are dark brown rather than yellow as in most owls.

Status: A relatively rare resident in the northern part of the region in wooded areas, especially along river bottom forests. Breeds south at least to Glacier Park, and probably occurs farther south along the mountains. There are few Alberta breeding records (*Canadian Field-Naturalist* 96:46–51).

Habitats and Ecology: Dense bottomland woods are this owl's favorite haunts, and it occurs in both coniferous and deciduous woods, possibly preferring the former. It often nests in tree cavities, but at times also uses old hawk or crow nests.

Seasonality: Resident where it occurs. Very few actual breeding records exist, but in Colorado eggs have been reported in March, and generally in the northern states it seems to breed in March and April.

Comments: This beautiful owl is widespread in eastern and northern North America, and is replaced by the very similar spotted owl in the west and southwest. Both are strongly nocturnal, and not seen during the day unless they are disturbed from their forest roosting sites.

Suggested Reading: Dunstan & Sample, 1972; Bent, 1938.

LATILONG STATUS

s	s	s	
	s	S	M
S	M	s	M

M			?
	V		
	V		

Great Gray Owl (*Strix nebulosa*)

Identification: An enormous grayish owl, with a very large, "earless" head. Rarely seen, it perches near the trunk of tall trees during the day, remaining motionless and silent. However, it sometimes appears at the edges of clearings, perched on a tall branch that allows it to survey the clearing for mice or other small rodents. Its call is a series of deep hoots. Its white "mustache" and black "bow tie" markings are also distinctive.

LATILONG STATUS

M	S	S
	M	M
R	M	M

R	s		?
M	R		
	R	M	

Status: A rare resident in the northern montane forest areas, south at least to the Wind River Range of Wyoming. Known to breed in several of the montane parks, but uncommon to rare in all.

Habitats and Ecology: In Alberta these birds usually nest in poplar woodlands, often near muskeg areas. Nests are usually in old hawk nests of various large species, from 10 to 80 feet above ground, in conifers or hardwood trees.

Seasonality: A permanent resident in some areas, but often moving in winter to regular winter feeding territories where mice and voles are abundant. There are few definite egg records for the region, but Alberta egg records are from March 22 to May 15, with a majority between April 9 and May 1.

Comments: This is one of the most beautiful and elusive of all North American owls; a single sighting provides a memory for a lifetime. The birds are superb at finding mice or voles in snowy areas, and crash down on them from above, with their talons outstretched, through the snowy cover.

Suggested Reading: Nero, 1980; Hoglund & Langren, 1968; Oeming, 1955; Brenton & Pittaway, 1971.

Great Gray Owl

Long-eared Owl (*Asio otus*)

Identification: This woodland owl is medium-sized, with long ear-tufts. It somewhat resembles a great horned owl, but is slimmer and smaller, and its calls consist of a variety of hooting notes. Perches during the day in shady tree roost sites, close to the trunk, where it is often overlooked.

Status: Resident over much of the region, except in Alberta, where relatively rare in montane areas, but known to have bred at Banff N.P. and a regular breeder in aspen parklands and prairie coulees.

LATILONG STATUS

s	M	s	
	s		M
R	S	s	M

R	?	M	M
s	R	M	M
	R	M	M

M	R	r	M
r	R	M	R
R	R	r	R

Habitats and Ecology: A widespread species, often associated with coniferous or deciduous forests, but also found in woodlots, orchards, large wooded parks, and even sagebrush or pinyon–juniper woodlands during the breeding season.

Seasonality: Generally resident in the montane parks, but apparently migratory in the northernmost parts of its range in Alberta. Egg records in Colorado are from April 15 to May 16, and in Wyoming and Montana egg records are from April 15 to June 17. Nestlings have been seen in Wyoming in June, and fledged young in Montana in early July.

Comments: Like the short-eared owl, this is a highly nocturnal species that is usually not seen by the casual observer unless he happens to flush one from its roosting site. Evidently their food consists almost entirely of mice and similar small mammals; thus, like most owls they are highly beneficial for agricultural interests.

Suggested Reading: Randle & Austing, 1952; Marti, 1974; Armstrong, 1958; Glue, 1977.

Short-eared Owl (*Asio flammeus*)

Identification: This open-country owl is often seen coursing over grasslands, where its moth-like flight and blackish wrist-marks are easily observed. Perched on a fence post, it appears "earless" (short ear-tufts are actually present), and it has a whitish, streaked belly. Its calls are rarely uttered, but wing-clapping noises are made during aerial courtship.

LATILONG STATUS

s	r	s	r
	R	M	S
V	S	r	s

r	s	M	R
s	R	M	M
S	R	M	R

M		R	M
r	R	M	R
S		S	R

Habitats and Ecology: This is a prairie-adapted species, usually breeding in areas of grassland, marshes, arctic (but not alpine) tundra, and low brushland. Nests are usually on the ground, but sometimes in burrows. More diurnal than most owls, and often seen hunting in mid-day.

Seasonality: Migratory in northern parts of the region, although sometimes overwintering even in Alberta. Migration peaks in Wyoming are in late March and again in October. In Montana egg records extend from April 3 to June 13, and in Wyoming nestlings have been observed in mid-June. Alberta and Manitoba egg records are from May 5 to June 20.

Comments: The spring courtship displays of this species are spectacular; the birds perform looping flights above their territories, with diving, wing-clapping, and even somersaulting maneuvers being performed, accompanied by a quavering, chattering call. Nearly all of this species' prey consists of rodents, making it a valuable addition to the prairie fauna and worthy of protection by farmers, who often kill owls.

Suggested Reading: Murray, 1976; Banfield, 1947; Clark, 1975.

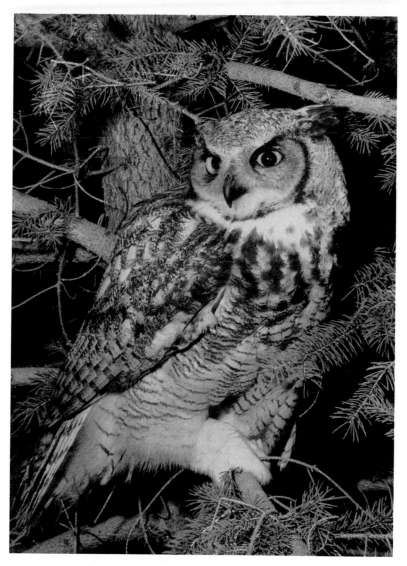

Plate 16. Great horned owl, adult. Photo by Alan G. Nelson.

Plate 17. Spotted sandpiper, adult in spring. Photo by author.

Plate 18. Northern saw-whet owl, adult. Photo by author.

Plate 19. Great gray owl, adult. Photo by Olaus Murie.

Plate 20. Calliope hummingbird, adult male. Photo by Kenneth Fink.

Plate 21. Rufous hummingbird, adult male. Photo by Alan G. Nelson.

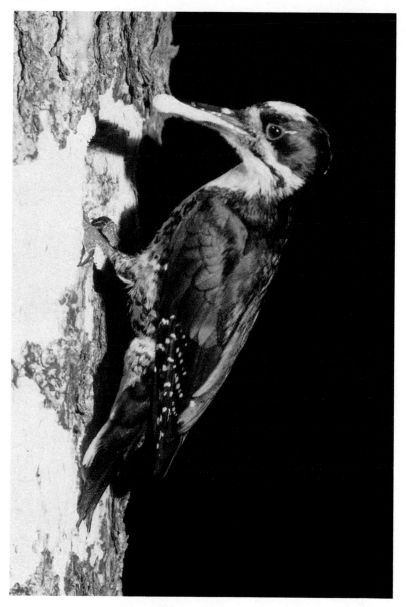

Plate 22. Black–backed woodpecker, adult male. Photo by Alan G. Nelson.

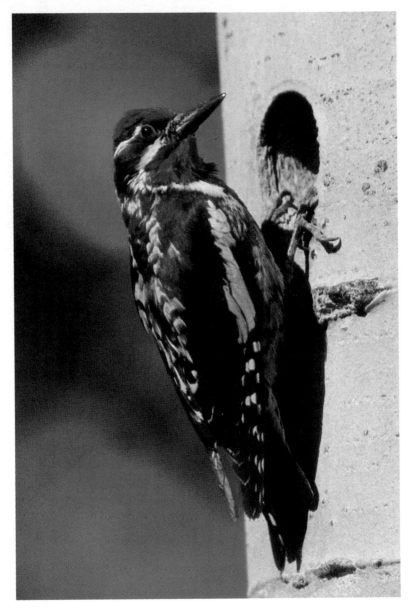

Plate 23. Yellow–bellied sapsucker, adult male. Photo by Alan G. Nelson.

Plate 24. Steller's jay, adult. Photo by author.

Plate 25. Clark's nutcracker, adult. Photo by author.

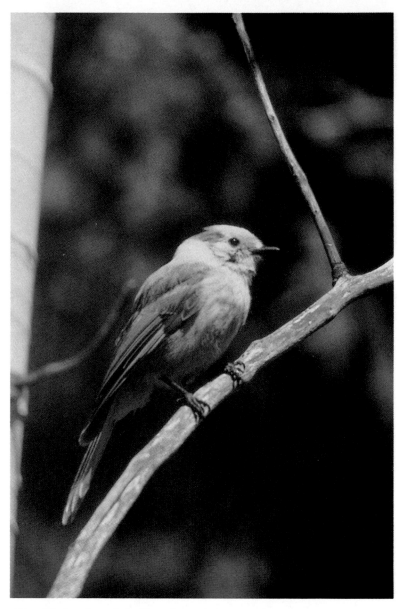

Plate 26. Gray jay, adult. Photo by author.

Boreal Owl *(Aegolius funereus)*

Identification: This small owl of the coniferous forests is "earless" and large-headed, with yellow eyes outlined by a grayish facial disk and a blackish crown spotted with snowflake-like dots. The chest is brownish, like a saw-whet's, but lacks distinct striping. Its usual call is a series of rapid hoots; bell-like tinkling calls have also been attributed to it.

Status: A rare to uncommon resident in Alberta's montane parks; local farther south and a rare breeder in Glacier N.P. Also nests locally in eastern Idaho (*Condor* 85:501) in the River of No Return wilderness area, in north-central Colorado (Cameron Pass area), and perhaps in the Jackson Hole and Togwotee Pass areas of Wyoming. Apparently a vagrant at Rocky Mountain N.P., with one park specimen record. However, it has been reported from Estes Park village (injured specimen), and has been heard calling at Bear Lake.

Seasonality: Virtually nothing is known of the status of this species regionally; a nest in east-central Idaho was found containing nestlings on June 2. The birds have also been found nesting in wood duck nesting boxes in Canada, but woodpecker holes are the more common site. Egg records from various localities in southern Canada are from April 11 to June 9.

Comments: One of the most attractive of all North American owls, and a very beneficial species, feeding on small mammals such as rodents up to the size of a squirrel, and rarely taking birds. In Europe this species is called the Tengmalm's owl.

Suggested Reading: Mikkola, 1983; Lundberg, 1979; Catling, 1972.

LATILONG STATUS

	W	S	
		W	
W			
V			
	V		
	r	V	
V			
		V	R

Northern Saw-whet Owl (*Aegolius acadicus*)

Identification: Similar in appearance to the preceding species, but smaller, and with a brownish rather than a grayish-black face and head, and without the "snowflake" crown pattern, which instead is streaked. The breast is also more distinctly streaked with rusty brown. The call is a long series of whoots or whistles, two or three per second, sounding something like the dripping of water or the tinkling of a bell.

LATILONG STATUS

S	R	S	
	S	s	
R	M	M	M

R	r		M
s	R	M	M
	r	R	

R		M	r
	M	M	R
	R	s	S

Status: Widespread permanent resident through the montane forests of the region, including the montane parks.

Habitats and Ecology: Occurs widely, from riparian woodlands through aspen groves to the coniferous forest zones, but not reaching timberline. The foothills and ponderosa pine zones are probably their favored habitats, where they nest in old woodpecker holes.

Seasonality: Probably a resident throughout the region. In Alberta eggs have been reported from April 18 to June 8, and in Colorado as early as April 15. The nesting season is quite prolonged; nestlings in Colorado have been seen as late as early July.

Comments: This tiny and attractive owl is a close relative of the boreal owl, but is more widespread and more southerly in its breeding distribution, the two species having more or less complementary distributions. Both are relatively tame when approached on roosting sites, and can be approached closely.

Suggested Reading: Mumford & Zusi, 1958; Forbes & Warner, 1974.

Saw-whet Owls

Common Nighthawk (*Chordeiles minor*)

Identification: This is a long-winged bird the size of a small hawk, but with pointed wings that have white patches near the tip resembling windows. The flight is light and butterfly-like, and often is accompanied by a swooping dive and a loud booming sound. Most active at dawn and dusk; during the day sometimes seen perched on fenceposts or on tree branches, almost the color of dead wood, but with a white throat and large dark eyes.

LATILONG STATUS

S	S	S	s
s	S	s	s
S	S	S	S

S	S	s	S
s	S	S	S
s	S	S	S

S	s	S	S
S	S	S	S
S	S	S	S

Status: A summer resident throughout the region, mainly below the zone of coniferous forests, but breeding in most of the montane parks, especially the more southern ones.

Habitats and Ecology: This species forages entirely in the air, on flying insects, and is especially common over grassland and urban areas, sometimes extending to sagebrush and desert scrub. Nesting occurs on the ground, usually in grasslands, or at the edges of woods, and sometimes on the asphalt rooftops of buildings.

Seasonality: Wyoming records are from May 23 to October 15, with migration peaks in early June and September. Montana records are from early June to late September, and in Colorado the birds have been reported from April 30 to November 5. Wyoming egg records are from June 15 to July 12, and Colorado records are from June 5 to July 17. Active nests in Montana have been seen from mid-June to August 18.

Comments: Nighthawks compete with swallows and swifts for aerial insects, especially the latter, but probably can find their prey under lower light conditions than either of these. Their loud aerial displays, penetrating calls, and distinctive white wing markings are probably also reflections of their daily activity patterns usually occurring under reduced light conditions.

Suggested Reading: Selander, 1954; Caccamese, 1974; Armstrong, 1965.

Common Poor-will (*Phalaenoptilus nuttallii*)

Identification: Similar to the common nighthawk in general shape and behavior, but lacking the white wing patches, and instead having white markings on the outer tail feathers. The species' call is a *poor-will* or *poor-will-uk* note that is often repeated during early summer evenings.

Status: A local summer resident in the region, mainly on drier plains toward the south. The only montane park supporting the species as a breeder is Rocky Mountain N.P., where the birds occur around rocky outcrops at lower elevations.

Habitats and Ecology: Generally this species is associated with rocky habitats having a cover of arid-adapted shrubs or low trees, such as pinyon–juniper, saltbush, greasewood, sagebrush, and dry grasslands. The birds nest on the ground, often under scrub oaks, whose leaves provide concealment for both adults and young.

Seasonality: Wyoming records are from May 3 to September 18, Montana records are from May 9 to September 15, and Colorado records extend from April 27 to October 13. Egg records in Colorado are from May 20 to June 14, and dependent young have been seen from June 6 to July 9. Similar dates are suggested for Wyoming and Montana.

Comments: This ghost-like species is seen far less often than are common nighthawks, and is more likely to be heard than seen. Sometimes they flush from gravel roads at night and one sees them momentarily in the beam of the headlights, but probably most go unseen by humans.

Suggested Reading: Bent, 1940.

LATILONG STATUS

		?	
	s	?	
	M	s	

M	M		S
		M	M
	s	s	s

M	M	S	s
s	S	S	S
S	s		s

Black Swift (*Cypseloides niger*)

Identification: This is the largest of the four swifts of the region, and the most uniformly blackish, both above and below. It is larger than the chimney swift, and has a slightly forked tail, and is darker below than either of the other two swift species.

LATILONG STATUS

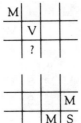

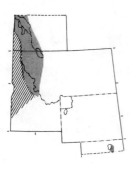

Status: A local summer resident in montane areas from Alberta south to central Montana, and again in northern Colorado. Rocky Mountain N.P. and Banff N.P. both support good populations.

Habitats and Ecology: Associated with mountains having steep, almost inaccessible cliffsides for nesting, often in narrow canyons, and almost always close to waterfalls. In Rocky Mountain National Park nests have been found at 10,500 feet at Loch Vale. Nesting in caves has been reported in other areas.

Seasonality: In Colorado the records extend from June 7 to September 12, and in Montana from mid-May to late August. Eggs in Colorado have been found from July 7 to August 17, and nestlings as late as early September. In Montana nestlings have been observed from July 25 to August 7. Nestlings have been seen near Banff (Johnston's Canyon) in early September.

Comments: This is one of the least known birds of the region, owing to its nesting in such extremely inaccessible locations. Bailey and Niedrach (*Birds of Colorado*) provided an extensive summary of nest studies in that state.

Suggested Reading: Knorr, 1961; Hunter & Baldwin, 1962.

Chimney Swift (*Chaetura pelagica*)

Identification: A dark gray swift, with a cigar-shaped body and long, tapering wings. It closely resembles the Vaux's swift, but is slightly darker on the undersides and has a louder, more chattering call.

Status: Rare in the area, and limited regionally to eastern Montana and northeastern Colorado. Absent from the montane parks except for Rocky Mountain N.P., where an accidental summer vagrant.

LATILONG STATUS

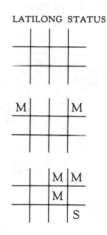

Habitats and Ecology: A familiar city bird over most of eastern North America, mainly in the vicinity of towns and cities where chimneys offer roosting and nesting sites. Caves and hollow trees were used before chimneys became available, and may occasionally still be used in some localities.

Seasonality: In Montana records are from May 10 to July 27, and Colorado records extend from April 28 to October 14. In that state eggs or nestlings have been noted in mid-July, and nesting in the region probably occurs from mid-May to late June.

Comments: This is the only species of swift in North America to have become relatively dependent on humans for nesting sites, and to have apparently benefited from that association. Perhaps as a result, it has slowly extended its range westwardly, and now has almost reached the range of the Vaux's swift in the Rocky Mountains.

Suggested Reading: Bent, 1940; Fisher, 1958.

Vaux's Swift (*Chaetura vauxi*)

Identification: Very similar to a chimney swift, but slightly paler below, and with a somewhat higher-pitched call. Unlikely to be seen entering or leaving chimneys.

LATILONG STATUS

s	s	S	
s	S	s	s
S	S	S	M

		?

Status: A summer resident in the northwestern parts of the region, including Glacier N.P., where it breeds commonly. Unreported for Wyoming or Colorado.

Habitats and Ecology: Generally similar to the chimney swift, but it seldom nests in chimneys, and instead uses hollow trees. Often found in woodlands near rivers and lakes. The nests are often placed in western hemlocks with dead or broken-off tops, or sometimes on cliffs.

Seasonality: Montana records extend from April 30 to September 1. There are no available egg records for the region, but in Montana nestlings have been observed from July 16 to August 10.

Comments: In Glacier National Park these birds are often seen near Avalanche Campground, as well as in the McDonald Valley, along the North Fork of the Flathead River, and near the Flathead Ranger Station.

Suggested Reading: Baldwin & Hunter, 1963; Baldwin & Zaczkowski, 1963.

White-throated Swift (*Aeronautes saxatilis*)

Identification: This is the only swift of the region with mostly white underparts, contrasting sharply with black flanks. The tail is long and fairly forked, and the birds are usually found near cliffs and canyons.

Status: A widespread summer resident in mountainous areas of the region, more common southwardly, and absent from the Canadian montane parks.

Habitats and Ecology: Associated with steep cliffs, deep canyons, and generally mountainous terrain, sometimes as high as 13,000 feet. Nesting occurs in crevices of canyon walls, in completely inaccessible locations.

Seasonality: Wyoming records are from April 4 to September 9, with peaks in late April and early September. Montana records are from April 19 to August 12, and Colorado records are from March 23 to October 12. In Colorado eggs have been reported in mid- to late June, and in Wyoming and Montana nesting extends from mid-June (eggs) to mid-July (nestling young).

Comments: The nests of this species of swift are composed largely of feathers, which are mixed with grasses and glued together in usual swift fashion, and attached to the sides of rock walls. Copulation evidently occurs in the air, as has also been reported for other swifts. On their wintering areas the birds are highly social, and sometimes hundreds will roost in rock-crevice sites. In one case the birds were observed to enter the site by flying through an entry only two or three inches wide and less than three feet long.

Suggested Reading: Bent, 1940; Bartholomew et al., 1957.

LATILONG STATUS

s	M	s	
s	s	s	
S	S	S	s

s	S	M	S
	S	S	S
	M	S	S

s	M	S	M
M	S	M	M
S	S	s	S

Magnificent Hummingbird (*Eugenes fulgens*)

Identification: The largest hummingbird of the region, and the only one in which males have a blackish breast and a green gorget. Females are much duller, but also are rather dark below, and like the male have a small oval white spot behind the eye.

LATILONG STATUS

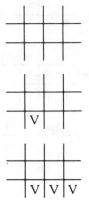

ACCIDENTAL VAGRANT

Status: An accidental summer vagrant in the southern part of the region, nesting to the south in Arizona and casually in the mountains of Colorado. Reported in Rocky Mountain N.P. on several occasions, and one nesting occurred within 15 miles of the southern boundary of the park in 1965.

Habitats and Ecology: This hummingbird often occurs along lower mountain streams, especially in the ponderosa pine zone, but sometimes reaching the fir belt. Open areas of forest are preferred, where there are many flowering herbs such as penstemons and the like.

Seasonality: There are relatively few regional records, but in Colorado the birds have been seen from May 21 to August, and eggs have been reported in late July. In Arizona there are egg records from May 6 to July 28, with a maximum between mid-June and mid-July.

Comments: This species until recently was known as Rivoli's hummingbird, and it has an extensive breeding range reaching south to Panama. Apparently the Colorado occurrences represent a recent range extension, and the birds have been seen as far north as Jackson Hole, Wyoming.

Suggested Reading: Johnsgard, 1983a.

Ruby-throated Hummingbird (*Archilochus colubris*)

Identification: This eastern species of hummingbird has (in adult males) a ruby-red throat gorget, a slightly forked tail, and rather dusky flanks. In flight it lacks the loud buzzing noise of the similar broad-tailed, and the black to violet throat of the black-chinned. Females and immature males are too difficult to identify with certainty in the field except perhaps by experts.

Status: A rare summer resident in eastern Montana; elsewhere in the region a rare migrant or vagrant; the only montane park reporting the species is Watertown Lakes N.P., where it is rare.

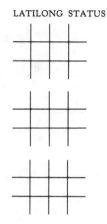

Habitats and Ecology: This is the familiar hummingbird of the eastern half of North America, where it is the only breeding species. It occupies a variety of woodland habitats, but is mostly found in open hardwood forests or forest edges, and in similar habitats such as orchards or city parks where a variety of nectar-bearing flowers are to be found.

Seasonality: There are few records for the area, but the birds are usually found in the northern plains from mid- or late May onward, with active nests in late June and early July. In Alberta they return about mid-May, and remain until about the end of September or early November.

Comments: This is the most widely distributed of North American hummingbirds, and one with an extremely long migration pattern, with migrants sometimes reaching Panama. At least some birds fly across the Gulf of Mexico during migration, an incredible feat for a bird weighing only slightly more than three grams, and one of the smallest warm-blooded animals in the world.

Suggested Reading: Johnsgard, 1983a.

Black-chinned Hummingbird (*Archilochus alexandri*)

Identification: Adult males of this species are unique in having a black chin, with a purplish rear border, and a whitish band between the gorget and the rather grayish flanks. Females and immature males cannot be separated in the field from those of ruby-throats, which in this region are rare or accidental except in the northeastern areas.

s	S	V
s	s	s
S	S	S

M		
	M	
	M	M

S	s	s	V

Status: A summer resident in the western and southwestern parts of the region, mainly at lower elevations. A rare migrant or vagrant in the montane parks, but common and probably breeding at Dinosaur N.M.

Habitats and Ecology: Typically associated with riparian habitats in dry canyons, but also occurring in oak–juniper woodlands, edges of aspen groves, and other habitats that usually are near water and offer open areas with many flowering plants.

Seasonality: Montana records extend from May 27 to late July, and in Colorado there are records from April 21 to September 22. Eggs in Colorado have been found as early as May 16, and nestlings to July 11; in Montana nestlings have been seen from June 27 to July 19.

Comments: This is a close relative of the ruby-throated hummingbird and they have generally complementary distributions, although the black-chin is associated with relatively drier habitats. They also have a much shorter migratory pattern, wintering almost entirely in Mexico.

Suggested Reading: Johnsgard, 1983a.

Calliope Hummingbird (*Stellula calliope*)

Identification: The smallest hummingbird of the region, and adult males are unique in having a striped gorget of iridescent scarlet and white. Females are best recognized by their tiny size, slightly streaked throat, and pale cinnamon underparts.

Status: A common summer resident over most of the area west of the plains; probably the commonest breeding species in most of the montane parks.

Habitats and Ecology: Open meadow areas near coniferous forests, such as low willow or sage areas rich in plants such as Indian paintbrush or gilia, are favored areas for this species in the Jackson Hole area. Openings in woodlands, sometimes as high as timberline, are also frequented, and in late summer alpine meadows are commonly used by migrating birds.

Seasonality: Wyoming records are from May 20 to August 12, and those from Colorado are from May 9 to August 9. Egg records in Montana are from June 18 to 20. In the Jackson Hole area nesting occurs in mid- to late June, with hatching near the end of July.

Comments: The nests of this species, like most hummingbirds, are extremely difficult to locate, but often are placed in lodgepole pines, in a clump of old cones, where their size, shape, and color blend in perfectly with the cones. They also are usually placed under a large branch for protection from the elements, but with open sky to the east, where the rays of the morning sun will help warm the incubating bird after a cold night.

Suggested Reading: Johnsgard, 1983a; Calder, 1971.

LATILONG STATUS

S	S	S	s
s	S	s	s
S	S	S	s

s	S	s	s
s	S	S	
s	S	S	

		M	
S			M
M		M	M

Broad-tailed Hummingbird (*Selasphorus platycercus*)

Identification: Adult males of this species can usually be recognized by the loud, buzzy, and insect-like trill made by their wings while in flight. They resemble large ruby-throats, which are very rare in the region. Females and immatures are more difficult to identify, but have speckled throats, pale cinnamon flanks, and are fairly large for hummingbirds.

LATILONG STATUS

	M	S	
?	?	?	
V	?	M	

?	M		
s	S	S	
	S	s	M

M	s	s	S
M	S	S	S
S	s	S	S

Status: A summer resident in the southern parts of the region, mainly west of the plains and south of Montana. The common breeding hummingbird in Rocky Mountain N.P., but rare or absent from the more northerly ones.

Habitats and Ecology: Typically associated with ponderosa pine forests and aspen groves, but also extending into mountain meadows and pinyon–juniper woodlands in this region. In Colorado breeding birds are abundant in moist canyons with aspens, pines, or Douglas fir at about 6500 to 7500 feet elevation. During the summer the birds gradually move upwards, finally reaching alpine meadows in late summer.

Seasonality: In Wyoming records extend from May 9 to September 16, and in Montana from May 31 to mid-August. The span of Colorado records is from April 27 to October 24, with eggs reported from June 7 to July 20. Wyoming egg records are from June 15 to July 13.

Comments: The distinctive noises made by males of this species are produced by the outermost primary feathers, which are pointed and vibrate rapidly during flight. This noise probably adds impact to the impressive visual displays of territorial males, which involve vertical dives from 30 or 40 feet, often directly in front of a perched female.

Suggested Reading: Johnsgard, 1983a.

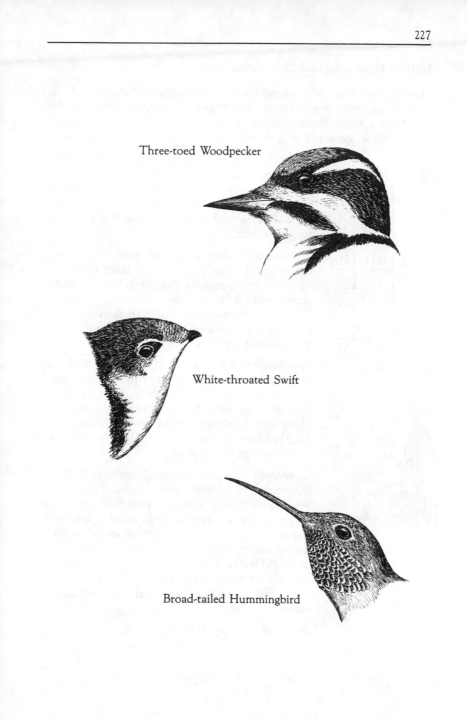

Three-toed Woodpecker

White-throated Swift

Broad-tailed Hummingbird

Rufous Hummingbird (*Selasphorus rufus*)

Identification: The only rusty-colored hummingbird of the area; both sexes are distinctly orange-red on the flanks and upperparts. Adult males also have a bright scarlet gorget, while females and immature males usually show some red or greenish spotting on the throat region.

LATILONG STATUS

S	S	S	M
s	S	S	S
S	S	S	s

M	M	M	M
M	S	S	M
	S	M	

M	M	M	M
	M	M	M
M	V	M	M

Status: A summer resident in the northern parts of the area, becoming less common southwardly and generally rare in Wyoming. The common breeding hummingbird of the Alberta montane parks, but not known to breed south of the Jackson Hole area.

Habitats and Ecology: In general, coniferous forests are used for breeding, but the birds occupy a variety of forest-edge habitats including mountain meadows and burned over forest areas where flowers are abundant. Brushy areas in the foothills are also used on migration, as are urban gardens and alpine tundra.

Seasonality: In Alberta and Montana the birds are present from late May or early June to mid-September. Migrants in Colorado and Wyoming are present from mid-May to about June 10, and again from late July to mid- or late September. Eggs in Montana have been observed in mid-July, and nestlings in early July.

Comments: Rufous hummingbirds are, like all hummingbirds, highly territorial, and males defending territories often advertise them with display flights that describe large aerial ovals, during which a mechanical buzzing sound is produced, ending in a distinctive rattling noise. The outer tail feathers of this species, rather than the primary feathers of the broad-tailed hummingbird, are distinctly narrowed and evidently vibrate in flight.

Suggested Reading: Johnsgard, 1983a.

Belted Kingfisher (*Ceryle alcyon*)

Identification: This large and conspicuous bird is easily identified by its bluish, crested head, a bluish breastband, and white underparts. It is always found near water, and in flight usually utters a dry rattling note. Often observed hovering above water, sometimes followed by a plunge into it.

Status: A resident in southern parts of the region, but more migratory farther north. Common in all the montane parks and probably breeding in all of them.

LATILONG STATUS

R	R	S	s
s	R	S	S
R	R	R	r

R	R	S	R
r	R	R	R
s	R	R	R

R	R	r	R
R	M	r	M
R	R	R	R

Habitats and Ecology: Found near water rich in small fish populations, usually where road cuts, eroded banks, gravel pits, or other exposed earthen surfaces provide opportunities for nesting, and usually also where nearby trees provide convenient perching sites between flights.

Seasonality: Resident in Colorado and probably much of Wyoming; in Montana migration is evident in late March and again in September or October. In Alberta the birds may arrive as late as May, or as soon as fishing areas become ice-free. They usually can be seen that far north as late as the end of October. Egg records in Colorado are from May 10 to June 15, and in Montana nesting has been reported from April 17 to June 4. In Jasper N.P. nesting has been observed as late as June 14, but by July there are usually dependent young out of the nest in Alberta.

Comments: Like terns, kingfishers hunt their prey by sight, hovering directly above the water and then plunging down and grasping a fish in their long and pointed beaks. Such fishing abilities must be learned, and adult birds teach their fledged young to catch fish by first capturing a fish, then beating it nearly senseless, and finally dropping it back into the water so that their young can capture it for themselves.

Suggested Reading: Salyer & Lagler, 1946; Kilham, 1974.

Lewis's Woodpecker (*Melanerpes lewis*)

Identification: This is the only woodpecker that is mostly grayish black, with a dark red face, and reddish gray underparts. Young birds are nearly uniformly blackish, resembling a small crow.

LATILONG STATUS

S	S	S	
s	S	s	S
S	S	S	S

S	s	S	S
s	S	M	S
s	S	M	

		M	S
M	M	M	M
M	S	s	R

Status: A local summer resident in forested montane areas of the region, especially at lower altitudes; probably breeding rarely in most of the montane parks.

Habitats and Ecology: This unusual woodpecker is especially associated with pine forests that are rather open, with burned over areas having abundant dead snags or stumps. Streamside cottonwood groves are also used, as are pinyon–juniper and oak–mountain mahogany woodlands. The birds are mainly adapted to catching free-living insects rather than excavating for insects in wood, and are often observed fly-catching in a rather surprising manner for persons not used to seeing woodpeckers feeding this way.

Seasonality: A distinctly migratory species, with the birds arriving in Montana in early May and leaving in mid-September or October, rarely remaining longer. In Wyoming the migration peaks are in May and September. Some birds overwinter as far north as Colorado. Colorado nesting records are from May 15 to June 20. Farther north in the region the egg records are for June, and throughout the region nestlings have been observed from late June to the end of July.

Comments: This species is a close relative of the red-headed woodpecker of the more eastern parts of North America, and does not do much drilling for prey. Perhaps because it is poorly adapted to wood-drilling, it usually nests in dead trees, or the dead portions of live trees, where excavating is easier. Or, old nest sites of other woodpeckers may be used.

Suggested Reading: Bock, 1970; Short, 1983.

Red-headed Woodpecker (*Melanerpes erythrocephala*)

Identification: Adults of this species are easily identified by their entirely red heads and black-and-white back and wing pattern; the white wing patch is especially evident in flight. Immatures are mostly brown and white, but have a similar white wing-patch. Like most woodpeckers, it has a raucous call.

Status: A local summer resident at lower elevations east of the mountains; rare or accidental in the montane parks except for Rocky Mountain N.P., where it is sometimes seen in summer but still not known to breed.

LATILONG STATUS

	M	M	
		M	M
V			

M	M	S	S
M	M	M	S
	M	M	S

		s	S
		M	M
s	V	V	S

Habitats and Ecology: Associated with open deciduous forests, woodlots, and riparian areas, sometimes extending into the ponderosa pine zone. Aspens and riparian hardwood forests are the species' major habitats in this region. Like the Lewis' woodpecker, this species tends to nest in dead trees or the dead portions of living trees, and does less excavating for insects in wood than do most woodpeckers.

Seasonality: In Wyoming the records extend from May 15 to September 14, with the birds usually arriving about the first of June and leaving by the end of August. Colorado records are from April 22 to October 23, with one December record.

Comments: These birds are effective insect-catchers, sometimes being attracted to insects on or over highways, and thus can often be observed on telephone posts or fenceposts along roadsides. Because of their relative independence from tree-associated insects, they are often found far from forests in plains country.

Suggested Reading: Kilham, 1977a; Reller, 1972; Bock et al., 1971.

232

Acorn Woodpecker (*Melanerpes formicivorus*)

Identification: Somewhat similar to the red-headed woodpecker, but mostly black, except for a white rump, a small white wing patch at the base of the primaries, and a white throat, with red limited to the crown.

LATILONG STATUS

Status: An accidental vagrant in the region. Photographed in June of 1975 in the Jackson latilong of Wyoming. The nearest breeding areas are in Arizona and New Mexico.

Habitats and Ecology: Acorn woodpeckers are closely associated with pine–oak woodlands in the southwestern states.

Seasonality: No regional information. In Arizona the egg records are from May 10 to June 10, and the birds are permanent residents there.

Comments: This woodpecker is unique in that it gathers and stores acorns, which it typically caches in specially drilled holes in telephone posts or other convenient locations, and later consumes. The most common site used is a tree, often an oak, but many other kinds of trees are also used. One large oak was estimated to have some 20,000 acorns imbedded in it, while a ponderosa pine was judged to have some 50,000 acorns stored in its trunk. Acorns and similar nuts provide more than half the food of this woodpecker, especially in winter, when few insects are available.

Suggested Reading: Gutierrez & Koenig, 1978; MacRoberts & MacRoberts, 1976; Swearingen, 1977.

Yellow-bellied Sapsucker (*Sphyrapicus varius*)

Identification: In this region, males have a red crown, a red nape patch, and a red throat, with these areas separated by black and white stripes. Females have red only on the crown. Both sexes have large white wing patches, and exhibit white rumps in flight. Their calls are weak by comparison with most woodpeckers. Their presence in an area is often evident by the parallel rows of holes that they drill in aspens.

Status: A breeding summer resident throughout the wooded portions of the region, common in montane areas and probably breeding in all the montane parks.

Habitats and Ecology: Coniferous forests, deciduous forests, and mixed woodlands are all used by these birds, but aspens are a favorite habitat in this region. The birds excavate holes in these trees to drink the sap, and also nest in aspens, either in dead trees or living ones that have dead and rotting interiors.

Seasonality: Wyoming records are from April 12 to December 5, with peaks in May and October. In Montana the birds usually arrive in mid-April and leave in mid-October. In Alberta they are usually seen from early May to mid-August. Colorado records are from March 30 onward, with the birds sometimes wintering. Nest-drilling in Rocky Mountain N.P. has been seen from mid-May to early June, and Colorado egg records are from June 8 to 23. In Montana and Wyoming nestlings have been noted from late June to mid-July.

Comments: After this book went to press the American Ornithologists' Union taxonomically separated the Rocky Mountain population of the yellow-bellied sapsucker as a distinct species, to be called the red-naped sapsucker (*Sphyrapicus nuchalis*). In this montane population both sexes exhibit a red nape patch, and females additionally have a partially red throat.

Suggested Reading: Howell, 1952; Devillers, 1970; Kilham, 1971b.

LATILONG STATUS

S	S	S	s
s	S	s	s
S	S	S	S
S	S	S	s
S	S	s	s
s	S	S	S
M	S	S	S
M	S	S	S
M	S	S	S

Williamson's Sapsucker (*Sphyrapicus thyroides*)

Identification: This sapsucker resembles the preceding species, but males are much darker, with more black and no red on the crown or nape, and with an extensive black breastband. Females also have a blackish breastband, and are barred with black on the flanks and back.

LATILONG STATUS

S	S	S	
	s	s	
S	s	S	

S	S	S	
s	S	M	
	S	s	s

			M
	M	S	s
S	S	S	S

Status: A local and usually uncommon summer resident over much of the region; variably common in the U.S. montane parks, generally rare or absent in the Alberta parks.

Habitats and Ecology: Breeding in this region usually occurs in the aspen or coniferous zones, especially in burned areas of ponderosa pine forests, mainly between about 7000 and 8500 feet in Colorado, and extending somewhat higher than yellow-bellied sapsuckers in the same region. However, their ecological patterns are very similar, and they often nest in close proximity to one another.

Seasonality: Wyoming records are from April 29 to October 18, and in Montana the birds usually are present from mid-April to mid-September. Colorado records are from February 26 to September 24, and egg records in that state are from May 24 to June 16. In Wyoming and Montana eggs have been found in late May and early June, and nestlings seen as late as early July.

Comments: These birds are more prone to nest in pines, and to excavate for insects in pines, than are yellow-bellied sapsuckers. In both species it is rather easy to find nests after the young have hatched, because of the loud buzzing sounds made by the nestlings as they beg to be fed. Both species move up the mountains as summer progresses, sometimes being seen as high as 10,000 feet in late summer.

Suggested Reading: Howell, 1952; Michael, 1925; Crockett & Hadow, 1975; Crockett & Hansley, 1978.

Downy Woodpecker (*Picoides pubescens*)

Identification: This species closely resembles the hairy woodpecker, but is a good deal smaller, and has a shorter and weaker beak, about half as long as the head. Both sexes are mostly black and white, with white on the back and rump, and white spotting on the wings. Males have a small red nape patch.

Status: A resident throughout the region in wooded habitats; relatively common and probably breeding in all of the montane parks.

LATILONG STATUS

r	R	R	r
r	R	r	R
R	R	R	r

R	R	r	R
r	R	R	R
	R	R	R

R	r	r	M
M	r	M	R
s	R	R	R

Habitats and Ecology: A wide variety of wooded habitats are used by downy woodpeckers, including farmlots, orchards, city parks, and natural habitats ranging from riparian forests to pinyon–juniper woodlands, oak–mountain mahogany scrub, and aspen or coniferous forests.

Seasonality: A permanent resident throughout the region. Egg records in Colorado are from May 24 to June 30, and eggs have been reported as early as April 15 in Wyoming, with nestlings seen as late as July 20.

Comments: This is one of the most widespread and common of the North American woodpeckers, and has a range that essentially coincides with that of its larger relative, the hairy woodpecker. Competition between them is evidently reduced by size differences of the bill, and associated differences in foraging locations.

Suggested Reading: Kilham, 1962a, 1974; Lawrence, 1967; Jackson, 1970.

Hairy Woodpecker (*Picoides villosus*)

Identification: Larger than the downy woodpecker (about 7–8 inches long), and with a heavier and longer beak that is somewhat more than half as long as the head. Both species utter a harsh rattling call, but that of the hairy tends to remain constant in pitch, rather than descending toward the end.

LATILONG STATUS

R	R	R	r
r	r	r	R
R	R	R	R

R	R	R	r
r	R	R	R
	R	R	R

R	r	r	R
M	R	R	R
M	R	R	R

Status: A resident in wooded areas throughout the region, occurring in all the montane parks and probably breeding in all.

Habitats and Ecology: Optimum breeding habitat consists of fairly extensive areas of woodlands of conifers or hardwoods, but nesting occurs in riparian forests, in aspen groves, and in various coniferous forests nearly to timberline. Generally aspens and other hardwoods are preferred over conifers for breeding.

Seasonality: Resident throughout the region. Colorado egg records are from May 8 to June 13, and in Wyoming and Montana there are egg records from April 1 to May 22, with nestlings observed as late as July 18.

Comments: Hairy woodpeckers are somewhat less tame than downy woodpeckers, and thus somewhat less easily observed at length. They tend to forage in larger trees, and on the larger branches of small trees, than do downy woodpeckers, and rarely can be observed feeding on dead weeds or small shrubs as is sometimes typical of downy woodpeckers. Both species are attracted to suet in winter, when insect life is hard to find.

Suggested Reading: Lawrence, 1967; Kilham, 1966; Jackman, 1974; Staebler, 1949.

White-headed Woodpecker (*Picoides albolarvatus*)

Identification: This is the only North American woodpecker with a nearly entirely white head; it is otherwise mostly black except for white wing patches on the primary feathers.

Status: An accidental vagrant over most of the region; reported from at least three Montana latilongs, and there have been several unsubstantiated sightings in Wyoming. There is a sight record from Jackson Hole for October of 1924, but no other records for the montane parks. The nearest breeding areas are in northwestern Idaho and southeastern British Columbia.

LATILONG STATUS

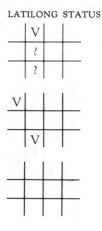

Habitats and Ecology: Associated with pine and fir forests of the Pacific northwest, and extending southward into interior California. It feeds on a combination of pine seeds and insects, sometimes hanging on pine cones while extracting the seeds, and occasionally seen upside-down, in nuthatch-like positions.

Seasonality: Regional records for this species are too incomplete to provide an idea of its occurrence. In Idaho it occurs as a permanent resident only in extreme western areas, and no good seasonality information on nesting is available.

Comments: In Idaho this species seems to be associated with open ponderosa pine forests. Over half of its foods consist of pine seeds, but it also eats ants and flying insects. It pries off bark rather than drilling for insects.

Suggested Reading: Jackman, 1974; Ligon, 1973.

Three-toed Woodpecker (*Picoides tridactylus*)

Identification: This and the following species are the only two North American woodpeckers with yellow crowns; additionally, this species has a distinctive ladder-like appearance of black and white barring on the back and flanks. Females lack yellow on the crown, but do show the barred back and flank pattern. Usually found in areas of recently burned forest.

LATILONG STATUS

R	R	R	r
	R	r	
R	M	r	R

R	r	?	
	R	M	
	R	R	M

M		M	M
M	M	M	R
		R	R

Status: A local and variably common resident in montane areas of the region; present and reported breeding in all the montane parks.

Habitats and Ecology: Like the following species, this is a fire-adapted form, typically moving into a burned forest area immediately after the fire, breeding, and dispersing four or five years later. The nest cavities that they excavate are then used by bluebirds, nuthatches, and other cavity nesters until the snags eventually topple.

Seasonality: A permanent resident throughout the area. Active nests in Colorado have been observed from June 2 to July 23, and in Montana nest-building has been observed in June, with young out of the nest by early August.

Comments: Like the following species, this woodpecker exploits newly burned forests in order to feed on bark-boring beetles that attack the recently killed trees. About three-fourths of the food of the woodpeckers comes from this source, which the birds often expose by tearing away bark rather than drilling holes. The holes they excavate for nests set into motion a nesting cycle for other species of birds that may last for several decades.

Suggested Reading: Bock & Bock, 1974; Taylor & Barmore, 1980.

Black-backed Woodpecker (*Picoides arcticus*)

Identification: Similar to the preceding species, but with an entirely blackish back; the flanks are similarly barred. Only males have the yellow crown patch; as in the other species females have a black and white head pattern.

Status: A resident in wooded areas throughout the mountains south to about Jackson Hole, Wyoming, and breeding in most or all of the montane parks from Grand Teton N.P. northward, but rare in the Alberta parks. Of uncertain occurrence in eastern Wyoming, but breeds in the adjacent Black Hills of South Dakota.

Habitats and Ecology: This species appears essentially identical to the preceding one as to its general adaptations to feeding and breeding in burned coniferous forest areas. As such, it is highly local and eruptive. Apparently the birds feed exclusively on conifers, but sometimes nest in nearby aspens.

Seasonality: A permanent resident in the region. There are few nest records, but in Colorado active nests have been seen from late May to early July, and young out of the nest have been observed in mid-July in Montana.

Comments: The blackish plumage of this and the preceding species makes them hard to see in a fire-blackened forest, and additionally both are very quiet for woodpeckers. However, accumulations of bark, flaked or torn from dead trees, accumulate at their bases, and often provide evidence of the birds in a particular locality.

Suggested Reading: Mayfield, 1958; Bock & Bock, 1974; Taylor & Barmore, 1980.

LATILONG STATUS

R	R	R	
r	r	r	R
R	r	r	r

r	R		?
r	R		
	R		

| | | M | |

Northern Flicker (*Colaptes auritus*)

Identification: This woodpecker is quite easily identified by the salmon-to golden-yellow color of its underwing surface when it flies; additionally, it has a white rump patch and a black breastband, with the rest of the underparts spotted heavily with black. A loud *flicker* call is commonly uttered.

LATILONG STATUS

R	R	R	s
s	R	R	R
R	R	R	r

R	R	S	R
S	R	R	R
R	R	R	R

R	R	r	R
R	R	R	R
R	R	S	R

Status: A common resident or migrant throughout the region, more abundant in wooded areas, but also in open country where trees are scattered. Present in all the montane parks, and probably breeding in all of them.

Habitats and Ecology: Flickers are unusual among woodpeckers in that much of their food consists of insects such as ants that are obtained by probing in the ground rather than by excavating trees. However, they do excavate holes in trees for nesting, usually those that are already dead or have decaying interiors. Open woodlands, such as orchards, parks, and similar areas offering foraging opportunities on grassy areas nearby are preferred over dense forests.

Seasonality: Although locally resident, there are some migrations, and in Wyoming there are migration peaks in April and October. In Montana the birds usually arrive in April and leave in September, as is also true in Alberta. Colorado egg records are from May 3 to July 3, and in Wyoming eggs have been reported from May 1 to June 4. Eggs have been noted as early as May 12 in Jasper N.P.

Comments: The flickers of this region are in a zone that includes both "red-shafted" and "yellow-shafted" types, as well as all possible hybrid combinations between them. Besides the shaft color differences, males of the more westerly form (the red-shafted) have reddish "mustaches," while males of the eastern form have black feathers in this area.

Suggested Reading: Kilham, 1959; Lawrence, 1967; Short, 1965, 1983.

Pileated Woodpecker (*Dryocopus pileatus*)

Identification: The largest regional woodpecker, this crow-sized bird has a mostly black plumage with white underwing coverts and a shaggy crest that is red in both sexes but less extensive in females. Its flight is heavy and undulating; like other woodpeckers it alternates flapping and gliding regularly.

Status: A local resident in the northern parts of the region, south to southwestern Montana and adjacent Idaho. Common in the montane parks of Alberta and also Glacier N.P., but only vagrants occur in the Yellowstone area.

LATILONG STATUS

r	R	R	
r	R	r	s
R	R	R	s

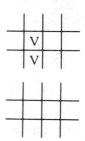

Habitats and Ecology: This magnificent bird is associated with mature forests of various types. Preferred habitats are usually near water and include mature trees among which there are tall trees having dead stubs, where nesting occurs. Nests are usually in trees that are 15 to 20 inches in diameter at the place of excavation, and are from 15 to 70 feet above ground.

Seasonality: A permanent resident where it occurs. Territorial drumming begins in April in Alberta, and egg dates there extend from May 10 to June 22, with half of the records for the latter half of May. In Minnesota there are egg records for early to late May, which probably corresponds to the nesting time in Montana and southern Alberta.

Comments: These large birds are quite sedentary, with pairs maintaining the same territory year after year. Territorial drumming is performed by both sexes, but mainly by the male. Incubation is by both sexes, and a female has been observed removing her eggs from a damaged nesting tree and carrying them off to some other location.

Suggested Reading: Bull & Meslow, 1977; Bull, 1975; Hoyt, 1957.

Olive-sided Flycatcher (*Contopus borealis*)

Identification: This flycatcher is usually seen perched on an outer branch of a conifer, where its white underparts contrast strongly with dark gray flanks and upperparts, producing a distinctive two-toned and "vested" appearance. It has a three-parted whistled call, sounding like "Look, three bears!"

LATILONG STATUS

s	s	S	
s	s	S	s
S	s	s	s

s	s	s	s
s	S	s	
	S	S	S

M	M	s	s
M	S	s	S
S		S	S

Status: Widespread in the region west of the plains, primarily in coniferous forests, but also in riparian forests. It occurs in all the montane parks, and probably breeds in all.

Habitats and Ecology: Associated with coniferous montane forests, burned over forests, and muskeg areas in the region. Typically tall conifers and open, often boggy or meadow-like areas are present in their territories. Muskeg areas in northern Alberta are favored breeding sites.

Seasonality: Wyoming records extend from May 16 to September 18, with migration peaks in late May and early September. Similarly, in Alberta they usually arrive the third week of May, and leave again in late August. Colorado records are from May 7 to September 20, and egg dates from there are from June 16 to July 20, with nestlings seen to August 1. Fledged young have been seen in Jasper N.P. as early as July 1.

Comments: This flycatcher rarely occurs as low in trees as kingbirds or some of the smaller flycatchers, and instead is usually seen outlined against the sky as it waits for insect prey to come into view. Nests are usually also fairly well elevated, on limbs from about 15 to 50 feet above ground, generally well hidden in a cluster of needles on a horizontal conifer branch.

Suggested Reading: Tvrdik, 1971; Bent, 1932.

Western Wood-pewee (*Contopus sordidulus*)

Identification: This inconspicuous brownish flycatcher is much more likely to be heard than seen; its descending *peer* and two-syllable *pee-a* notes are likely to betray its presence. If seen, the generally dark grayish upperparts, without strong wing-barring or definite pale eye-ring, help to identify it.

Status: Widespread throughout nearly the entire region in summer; breeding occurs in most and probably all of the montane parks.

Habitats and Ecology: Breeds in most coniferous forest types, and also to a varying extent in aspens, riparian forests, and various open deciduous or mixed woodland habitats. Open forests are favored, especially those dominated by conifers. Nests are well hidden, on horizontal branches of trees, or sometimes on a fork, and are usually well covered with spider webs, to which lichens may be attached for camouflage. However, this behavior is seemingly not so common in this species as in the eastern wood-pewee.

Seasonality: Wyoming records are from May 18 to September 30, and Montana records extend from mid-May to mid-September. In Colorado the extreme dates are April 11 and October 21. There are relatively few egg dates, but in Wyoming they extend from June 20 to July 15, and nestlings have been seen as late as August 26. Colorado egg dates are from May 6 to July 27, and nests with hatched young have been seen in Colorado during May.

Comments: This is the western member of a species-pair that includes two extremely similar birds. So far, the eastern species has not been found to nest within the region covered by this book, but does extend to the western part of the Dakotas.

Suggested Reading: Beaver & Baldwin, 1975; Eckardt, 1976.

LATILONG STATUS

S	S	S	r
s	S	s	S
S	S	S	S

S	S	s	s
s	S	S	S
s	S	S	S

M	s	S	S
	S	S	S
S	S	S	S

Yellow-bellied Flycatcher (*Empidonax flaviventris*)

Identification: One of the small flycatchers of the region; this one is notable for its yellowish underparts, its wide yellow eye-ring, and white to yellowish wing-bars. Its call is a *che-bunk* similar to that of the least flycatcher.

LATILONG STATUS

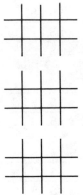

Status: Reported as a migrant or vagrant in May at Banff and Jasper parks; not recorded elsewhere in the region.

Habitats and Ecology: Associated with spruce–fir forests and bogs; in Alberta generally found near receding muskegs and at the edges of forests where thick shrubbery shades the forest floor. Alder-edged swamps are another nesting habitat there.

Seasonality: There is little information, but in Alberta the birds do not become very evident until the end of June or early July, when singing becomes apparent. Probably their migration dates are much like those of the other small *Empidonax* species of the region.

Comments: This species breeds across much of northern Canada, but is extremely elusive and rarely seen either on the breeding grounds or during migration.

Suggested Reading: Walkinshaw & Henry, 1957.

RARE MIGRANT

Alder Flycatcher (*Empidonax alnorum*)

Identification: A small flycatcher with a fairly inconspicuous eye-ring, definite wing-bars, and somewhat yellowish underparts. Its song, a buzzy, falling *fee-beeo*, provides the best field identification guide. Not separable from the willow flycatcher visually.

Status: An uncommon summer resident in the northwestern portions of the region, breeding in the montane parks of Alberta south to at least Kootenay and Banff. Not reported elsewhere in the region.

LATILONG STATUS

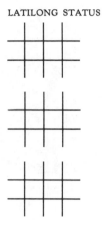

Habitats and Ecology: This species and the willow flycatcher are so similar in most regards that the literature on them is greatly confused. However, this species seems to breed over most of the same general area as the willow flycatcher, in birch or willow thickets, along the edges of muskegs, forest margins, streamside shrubbery, and wooded lakeshores.

Seasonality: There is little information, but in Banff and Jasper parks the species have been reported from mid-May to the latter part of August.

Comments: Much more work is needed to determine the relative abundance and ranges of this species and the willow flycatcher where they both occur.

Suggested Reading: Aldrich, 1953; Robbins, 1974.

Willow Flycatcher (*Empidonax trailli*)

Identification: A small flycatcher with a faint eye-ring, white wing-bars, and pale olive-gray to yellowish underparts. Best recognized by its song, a sneeze-like *fitz-bew*.

LATILONG STATUS

s	S	S	s
s	S	s	M
S	S	S	s

s	s		s
s	S	s	
s	S	s	S

s	s	M	s
M	S		M
s	s	S	S

Status: A summer resident over nearly all of the region except for some of the low plains areas in the east. Present and probably breeds in all the montane parks.

Habitats and Ecology: Especially associated with riparian or wetland habitats in this region, including willow thickets, low gallery forests along streams, prairie coulees, and, farther north, in woodland edge habitats such as muskegs and boggy openings.

Seasonality: Colorado records extend from May 18 to September 19, and in Montana the birds usually arrive in mid-May and leave by the first week of September. Banff and Jasper records are from mid-May to mid-August. Egg records in Colorado are from late June to July 8, and in Montana active nests have been reported throughout July.

Comments: This is a typical edge-adapted species, which usually nests no more than about six feet above ground in the crotches of shrubs or small trees, usually at the edge of a thicket or shrub so that easy entrance and exit are possible. The males typically sing from the highest point in the territory, sometimes up to 30 times per minute.

Suggested Reading: Walkinshaw, 1966; Holcomb, 1972; King, 1955; Ashmole, 1968.

Least Flycatcher (*Empidonax minimus*)

Identification: A small, grayish flycatcher with buffy wing-bars, a distinct eye-ring, and a rather more grayish cast on the upperparts than typical of the similar willow flycatcher. Its song is a sharp *che-bek* with the second syllable emphasized, and frequently repeated.

Status: A common summer resident in the northern and northeastern parts of the region, south through Montana and central Wyoming. Most common in the montane parks of Alberta, where it regularly breeds. Possibly a rare breeder in Glacier N.P.

Habitats and Ecology: Associated with open and edge-dominated habitats such as floodplain forests in prairie areas, scattered prairie grovelands, shelterbelts, woody lake margins, and urban parks or gardens.

Seasonality: Reported in Wyoming and Montana from early May to late August, and in Colorado from May 3 to September 23. Noted in Banff and Jasper parks from early May to early September.

Comments: This is a very common and widespread species of *Empidonax* which breeds in company with four other species of the same genus in Banff and Jasper parks. Presumably minor habitat differences help to reduce competition among all of these very similar forms.

Suggested Reading: Davis, 1959; Nice & Collias, 1961; Breckenridge, 1956; McQueen, 1950.

LATILONG STATUS

M	s	s	
	s	s	S
V	M	s	

S	s	S	s
	?	s	
	M	s	S

		s	M
	s		
			M

Hammond's Flycatcher (*Empidonax hammondi*)

Identification: This tiny flycatcher is typically found in tall coniferous forests having a closed canopy. Its eye-ring and wing-bars are well developed, and it is generally ashy gray to dusky in coloration. Its song is an emphatic, variable sequence of burry and low-pitched notes, usually in three parts.

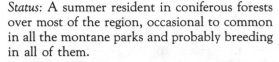

LATILONG STATUS

S	S	S	
s	S	s	?
S	S	S	s

S	S		s
s	S		
s	s	S	S

M	s	s	M
	s	M	M
s	S	S	S

Status: A summer resident in coniferous forests over most of the region, occasional to common in all the montane parks and probably breeding in all of them.

Habitats and Ecology: Associated with tall, mature montane coniferous forests, probably from about 7000 to 10,000 feet in this region. Often found along willow and alder-lined mountain streams, with males sometimes singing at elevations above timberline.

Seasonality: Records for Wyoming are from May 13 to October 9, and for Colorado from May 13 to September 19. In Montana the birds are usually present from early May to late August, and in Banff and Jasper parks they have been seen from mid-May to mid-August. There are Colorado nest records for the latter half of June, and eggs reported from Montana from June 26 to July 2.

Comments: This is another of the small and rather elusive flycatchers of the western states, which often demand great patience and expertise in observation and identification.

Suggested Reading: Davis, 1954; Johnson, 1963; Manuwal, 1970; Beaver & Baldwin, 1975.

Dusky Flycatcher (*Empidonax oberholseri*)

Identification: A small grayish flycatcher that has a faint eye-ring, very faint wing-bars, dull olive-gray underparts, and a song that is a combination of low-pitched, burry notes and clear ascending notes, often of three or four phrase elements and ending on a clear, high note.

Status: A summer resident in most areas west of the plains; present but variably common in all the montane parks, and breeding in most.

LATILONG STATUS

S	S	S	s
s	S	S	s
S	S	S	S

S	S	s	s
S	S		
	S	S	S

s	s	S	M
S	S	S	S

Habitats and Ecology: Associated with open woodland and shrubby habitats, ranging from riparian edges through oak–mountain mahogany woodlands, to aspens and open ponderosa pine woods. In Montana, brushy, logged-over slopes seem to be favored habitats.

Seasonality: Wyoming records are from May 6 to September 15, with migration peaks in May and early September. In Montana the birds arrive in early May, and depart in late August. Banff and Jasper park records are from mid-May to late August. Colorado egg records are from May 30 to July 27, and in Montana egg records extend from July 15 to late July.

Comments: Dusky flycatchers tend to occupy rather drier habitats than many of the other *Empidonax* species of the region, and they frequently nest in low trees or shrubs, from about three to eight feet above ground.

Suggested Reading: Johnson, 1966; Sedgewick, 1975; Bent, 1942.

Gray Flycatcher (*Empidonax wrightii*)

Identification: This is one of the nondescript *Empidonax* flycatchers, which is distinctly grayish above (slightly more olive-colored in fall), with an inconspicuous eye-ring and rather faint wing-bars. Its upper bill is blackish, while the lower one is mostly yellow, but becomes darker toward the tip. Its call is a double *chip-wip* or *chi-bit*, usually followed by a higher and more liquid note. The song is uttered emphatically, often with intervals of only one or two seconds between phrases.

LATILONG STATUS

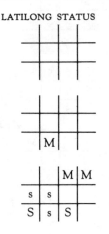

Status: A local summer resident in the southwestern portions of the region, especially where pinyon pines and junipers form an open woodland habitat type, often with interspersed sagebrush. Less frequently they occur in sagebrush scrub, or in dense brush near streams in semi-arid areas.

Habitats and Ecology: In western Colorado these birds are typically found in pinyon-juniper woodlands, where they are a characteristic breeding species, nesting either in forks of junipers or sometimes in sagebrush. The nests are typically constructed of the weathered outside strands of juniper bark, which makes them very difficult to find in the junipers.

Seasonality: Colorado records extend from April 27 to September 5, with egg records for late May and early June.

Suggested Reading: Bent, 1942; Johnson, 1963.

Western Flycatcher (*Empidonax difficilis*)

Identification: A small, yellow-tinted flycatcher with faint eye-rings and wing-bars, and with a distinctly yellow throat and underparts. The song is high-pitched, and consists of a series of thin, squeaky whistles and more snappy notes.

Status: A summer resident over much of the region, mainly near streams; probably present in all the montane parks, but a common breeder only in Rocky Mountain N.P., where it is the most common small flycatcher.

LATILONG STATUS

S	M	s	
	s		?
S	S	s	

s	s		s
M	S	s	
	s	S	S

	S	s	S
S	S	M	M
s	S	S	S

Habitats and Ecology: A widespread and adaptable small flycatcher, ranging from riparian woodlands through aspens into the coniferous forest zones, all the way to the upper spruce–fir zones. Extends out into sagebrush areas during the non-breeding season.

Seasonality: Colorado records are from April 7 to September 9, while Wyoming records extend from May 20 to September 13. In Montana the birds are usually present from late May to August. Colorado egg records are from June 3 to July 23, and a few records for Wyoming and Montana are from July 17 to July 6.

Comments: In Rocky Mountain N.P. this flycatcher nests in a wide variety of sites, ranging from rocky ledges to open garages and underneath porch roofs. In the Black Hills region the birds seek out canyons that offer a combination of shady sites and where streams or other moist habitats are nearby.

Suggested Reading: Beaver & Baldwin, 1975; Verbeck, 1975; Davis et al., 1963.

Eastern Phoebe (*Sayornis phoebe*)

Identification: This flycatcher is distinctly grayish, with a nearly black head, a blackish tail that is often pumped up and down while the bird is perching, and has whitish underparts that are paler than those of any other flycatchers of the area. The usual song is a spoken *phe-be*, unlike the whistled *phe-bee* calls of chickadees.

LATILONG STATUS

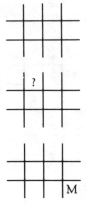

Status: A summer resident in eastern Alberta, and a migrant through the other eastern portions of the region east of the mountains. Rare in Banff N.P., and unreported for the other montane parks.

Habitats and Ecology: Associated with woodland edges, wooded ravines near water, woodlots, and lakes or streams in partially wooded areas. Often breeds close to humans, nesting on building ledges, or on the understructure of bridges.

Seasonality: Colorado records extend from March 11 to October 1, with egg records from May 10 to May 30. In Alberta these birds usually arrive shortly after the middle of April, and remain until August. By May they have established territories, and although specific egg records are unavailable the active period of breeding is probably from early May to late June.

Comments: This eastern species of flycatcher is one of the earliest of its group to migrate north, often arriving long before insects are abundant, and the males establish territories well before the potential egg-laying period. The laying period is fairly long, and two broods are usually raised per pair in this species.

Suggested Reading: Faanes, 1980; Smith, 1969; Bent, 1942.

Say's Phoebe (*Sayornis saya*)

Identification: This species somewhat resembles the eastern species, but is strongly tinted with rusty brown on the lower breast and belly. Like the eastern species, it often pumps its tail up and down. Its song is a frequently repeated *chu-weer*.

Status: A summer resident over most of the region, but infrequent in the mountains and generally rare or absent from the montane parks.

Habitats and Ecology: Generally associated with grasslands, sagebrush, and agricultural areas in the region, especially prairie coulees and steep, eroded river banks. They sometimes reach foothill areas, but do not breed in the wooded mountain zones.

Seasonality: Wyoming records are from April 7 to October 3, with migration peaks in April and September. In Alberta they are usually seen from the last week of April to early or mid-September. Colorado egg records are from March 23 to June 17, while Montana nesting records are from May 11 to July 2.

Comments: These birds are typically found in rather dry regions, where their brownish earth colors seem especially appropriate. They often use sunny canyons, open areas near buildings, and lower montane meadows, sometimes well away from any water. Nesting is often done on rock ledges, or even in caves or old mine shafts.

Suggested Reading: Schukman, 1974; Ohlendorf, 1976.

LATILONG STATUS

M	M	s	S
M	S	s	S
V	s	S	S

S	M	S	S
M	S		S
	M	s	S

M	M	S	s
S	S	S	S
S	S	s	S

Vermilion Flycatcher (*Pyrocephalus rubinus*)

Identification: Males of this species are unmistakably brilliant crimson on the crown and most of the underparts, while the rear part of the head and back are blackish. Females are more brownish, but are also heavily washed with reddish on the underparts.

LATILONG STATUS

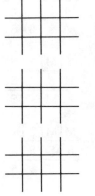

Status: An accidental vagrant in the area, with a record for Watertown Lakes N.P. one of the few in the region. A rare migrant in Colorado, with a reported case of breeding in the Sterling latilong of northeastern Colorado.

Habitats and Ecology: In Texas, where this species breeds regularly, it is associated with open grasslands having scattered junipers and oaks, as well as water areas lined by cottonwoods, willows, oaks, and other trees. It is usually associated with water, but is not associated with deep canyons.

Seasonality: There are not many regional records, but in Colorado the birds have been reported between April and December.

Comments: This is a desert-adapted species, extending from the American southwest well into Mexico in the Sonoran Desert region.

Suggested Reading: Smith, 1970; Taylor & Hanson, 1970.

ACCIDENTAL VAGRANT

Ash-throated Flycatcher (*Myiarchus cinerascens*)

Identification: This flycatcher is of the same general size and shape as are kingbirds, but are rusty brown on the wings and tail, and have whitish throats and underparts, becoming yellowish toward the belly. The usual song is a series of *pe-reer* notes, and its usual calls are *ha-whip* and *pe-reer* notes.

Status: A summer resident in the southwestern portions of the region, rare or absent from the montane parks. However, common in summer at Dinosaur N.M., and an almost certain breeder there. A local breeder in southwestern Wyoming (*American Birds* 35:964).

LATILONG STATUS

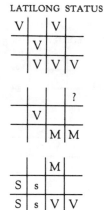

Habitats and Ecology: A desert species associated with mesquite and cactus deserts and, in this region, open pinyon–juniper woodlands, grasslands with scattered trees, and gulches or riparian edges in dry country.

Seasonality: Colorado records are from March 25 to September 17, and eggs have been reported from there in May.

Comments: This is a part of the southwestern desert avifauna that has moved north through the basin-and-range areas of Utah to the Snake River valley of Idaho, where it reaches its northernmost limits.

Suggested Reading: Lanyon, 1961; Bent, 1942.

Cassin's Kingbird (*Tyrannus vociferus*)

Identification: This rather large flycatcher closely resembles the more widespread western kingbird, but lacks white outer tail feathers, which instead are narrowly tipped with white. It is found in dry, semidesert regions that are more brushy or wooded than those usually used by western kingbirds, and its call is a rough, nasal *chic-queer*, with the second note accented.

LATILONG STATUS

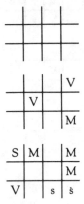

Status: A local summer resident in the high plains east of the mountains, from central Montana southward. Once reported from Rocky Mountain N.P., but apparently now extirpated, with no recent records.

Habitats and Ecology: Associated with open country, usually with scattered trees, or with open woodlands, and extending out into grasslands and agricultural lands where there are locally available trees for nesting. However, bushes and posts may at times also be used as nesting sites.

Seasonality: Reported in Colorado from April 17 to September 23, in Wyoming from May 6 to September 18, and in Montana from May 22 to September 3. Active nests in Colorado have been noted between May 15 and July 1.

Comments: In some areas both western kingbirds and Cassin's kingbirds occur locally, and in one such area of Arizona it was found that the Cassin's kingbird was most abundant locally where pine or oak woodlands were transitional with deserts, while western kingbirds were in more desertlike habitats.

Suggested Reading: Ohlendorf, 1974; Hespenheide, 1964.

Western Kingbird (*Tyrannus verticalis*)

Identification: This common flycatcher has an ashy gray head and back, a black tail that is edged on each side with white, and a yellowish belly. Its song is a complex and loud mixture of notes, often uttered in flight.

Status: A summer resident in most of the region, but infrequent in montane areas, and rare or absent from the northern montane parks.

Habitats and Ecology: This species is always associated with edge habitats near open country, such as shelterbelts, hedgerows, margins of forests, tree-lined residential districts, riparian forests, and the like.

Seasonality: Reported in Colorado from April 6 to September 22, and in Wyoming from May 5 to September 13, with peak migrations in May and August. In Montana and Alberta the birds usually arrive about mid-May, and are mostly gone by the end of August. Colorado egg records are from May 29 to July 2, and in Wyoming and Montana the egg records are for mid- to late June, with nestlings observed as late as late August.

Comments: A highly conspicuous species, owing to its high level of territorial behavior, and its generally fearless behavior, often threatening or chasing much larger birds away from its nesting area.

Suggested Reading: Smith, 1966; Ohlendorf, 1974; Whedon, 1938; Hespenheide, 1964.

LATILONG STATUS

S	S	s	s
S	S	?	s
S	S	S	s

S	s	S	S
s	S	s	S
s	M	M	S

s	s	S	s
M	M	M	M
S	S	S	S

Eastern Kingbird (*Tyrannus tyrannus*)

Identification: Similar in size and shape to the western kingbird, but with more black and white underparts and a white-tipped black tail. It is as noisy as the western kingbird, and its song is a long series of chirps and twitters.

LATILONG STATUS

S	S	S	s
s	S	S	S
S	S	S	S

S	S	S	S
S	S	s	S
s	M	S	S

s	S	S	s
M	s	M	M
s	s	s	S

Status: A summer resident nearly throughout the region, except in the drier areas of the southwest. Present in all the montane parks, and probably breeding in all of them.

Habitats and Ecology: Associated with open areas with scattered trees or tall shrubs, such as forest edges, fencerows, riparian areas, agricultural lands, farmsteads, etc.

Seasonality: Colorado records are from April 24 to September 19, and Wyoming records extend from May 5 to September 18. In Montana and southern Alberta the birds are usually present from early May to mid-September. Colorado egg records are from June 7 to July 24, and similar dates are typical for farther north in the region.

Comments: Like the western kingbird, this is one of the noisiest small birds of the region, and from the time of its arrival until nesting is well underway its screaming calls and chases of other birds are familiar sights. Sometimes birds such as orioles seem to nest preferentially in the same trees as kingbirds, perhaps gaining protection from possible predators such as jays or crows because of the watchful behavior of the kingbirds.

Suggested Reading: Smith, 1966; Morehouse & Brewer, 1968.

Scissor-tailed Flycatcher (*Muscivora forficata*)

Identification: This highly distinctive flycatcher is unique in having a long, forked tail that is longer than the body, as well as a pale grayish body and head, with pinkish to red underwings. Females have a tail that is only slightly forked, but also exhibit pale pink on the underwings and "armpit" area.

Status: An accidental vagrant in the region. Reported for at least three Montana latilongs, once from Wyoming, and from numerous locations in Colorado, where it breeds in the extreme southeastern corner of the state.

LATILONG STATUS

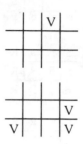

Habitats and Ecology: This species is found in open to semiopen habitats having a scattering of trees or other elevated nesting sites, such as buildings, and in riparian areas. Where trees are lacking, windmills, utility poles, or other structures are used as substitutes for nest sites.

Seasonality: Colorado records extend from April 24 to November 9, and active nests have been seen from May 31 to June 23.

Comments: Like other flycatchers, this species is highly territorial, but nonetheless males breeding in a particular area often congregate for roosting together, and return in the morning to their respective nests to help feed their offspring. After fledging, the young birds join these roosting groups, which eventually may number more than 200 birds.

ACCIDENTAL VAGRANT

Suggested Reading: Fitch, 1950; Bent, 1942.

Horned Lark (*Eremophila alpestris*)

Identification: This prairie and grassland species has feathered "horns" that are visible only at close range; the white-edged black tail and somewhat pinkish brown upperpart coloration are more useful as fieldmarks, as is the black breastband. In spring, the male often sings an extended flight song above his territory that is a high-pitched assortment of tinkling notes.

LATILONG STATUS

M	M	S	r
	r	r	S
R	R	R	r

S	R	s	R
S	R	R	R
r	r	R	R

r	R	R	R
R	R	R	R
R	R	R	R

Status: Present throughout the entire region, and resident in most areas, although migratory movements do occur. Present in nearly all the montane parks, and breeding in several.

Habitats and Ecology: Open-country habitats, ranging from shortgrass plains through agricultural lands such as pastures, desert scrub, mountain meadows, and alpine tundra, are the basic requirements for this species, which has an enormous ecological and geographic range in North America.

Seasonality: Essentially resident in the southern parts of the region, and among the earliest of spring migrants in Alberta, often arriving in small numbers by mid-February, but with large flocks appearing in April. The major southward movement is in September, with small groups usually wintering in the southern parts of the province. Nest records in Colorado are from mid-April to early July, and in Wyoming egg records are from April 15 to July 23. A large sample of nest records from Montana extend from April 10 to July 19.

Comments: This is the only native North American member of the true lark family, which is well represented in the Old World, and includes such famous singers as the skylark.

Suggested Reading: Behle, 1942; Pickwell, 1931; Beason & Franks, 1974; Verbeek, 1967.

Horned Lark

American Crow

Sage Thrasher

Purple Martin (*Progne subis*)

Identification: The largest of the swallow family in North America, and one of the most familiar, owing to its tendency to nest in man-made birdhouses. Adults are mostly bluish black above, and bluish to grayish below, depending on sex and age. The tail is somewhat forked, and the frequently uttered calls include extended twittering notes. Nests colonially, and usually seen near cities or human habitations.

LATILONG STATUS

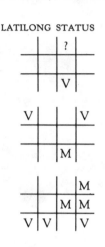

Status: Limited as a summer resident to extreme northeastern Montana and the central portions of Alberta in this region, but migrating through eastern Montana and Wyoming. A rare migrant in eastern Colorado, but breeds locally in western Colorado to the south of this book's coverage. Absent from most of the montane parks, but a vagrant at Rocky Mountain N.P.

Habitats and Ecology: Restricted in our region almost entirely to areas where "martin houses" have been erected. Like the other swallows, it obtains all of its food by aerial foraging for insects.

Seasonality: Colorado records extend from May 20 to August. In Alberta they arrive about the first day of May, and few are seen after the end of August. There are few regional egg records, but in North Dakota eggs have been reported from April 29 to August 24, an extended breeding period apparently related to persistent renesting efforts.

Comments: This species has probably benefitted greatly from human interest in it, and especially at the western edge of its range is essentially limited to those locations where nesting boxes are provided for it. However, the birds sometimes also use old woodpecker holes or crevices in old buildings for nesting.

Suggested Reading: Allen & Nice, 1952; Johnston & Hardy, 1962.

Tree Swallow (*Iridoprocne bicolor*)

Identification: This attractive swallow is a two-toned iridescent bluish black above and immaculate white below, and has a somewhat forked tail. It closely resembles the violet-green swallow, but lacks that species' large white flank patches.

Status: A local summer resident throughout the region, but lacking from treeless areas and the high montane communities. Present and breeding in all the montane parks.

Habitats and Ecology: Breeding in the region extends from riparian woodlands through the aspen zone, and into the lower levels of the coniferous forest zone. Outside the breeding season often seen over lakes or rivers, as well as over other open habitats. Nesting is especially prevalent in the aspen areas, where old woodpecker holes are available, but also occurs at times in birdhouses.

Seasonality: Colorado records are from April 6 to October 9, and in Wyoming range from April 11 to October 1. In Alberta the birds appear the latter half of April, and are usually gone by the end of August, with a few sometimes extending into September.

Comments: This is one of the earliest swallows to arrive in northern areas, often arriving at least a month before nesting gets underway. This may reflect the limited number of suitable nesting sites available to these hole-nesters, and the advantages of taking early possession of the available sites.

Suggested Reading: Chapman, 1955; Stocek, 1970; Kuerzi, 1941; Paynter, 1954.

LATILONG STATUS

S	S	S	s
s	S	S	S
S	S	S	S

S	S	S	S
s	S	s	M
s	S	S	s

S	S	S	s
S	S	S	S
S	S	S	S

Violet-green Swallow (*Tachycineta thalassina*)

Identification: Very similar to the preceding species, but the white sides are continued up behind the wings to form large flank patches that nearly meet on the lower back. There is also somewhat less black on the head, with white extending slightly above the eye.

LATILONG STATUS

S	S	S	
S	S	S	M
S	S	s	s

S	S	S	S
s	S	S	M
s	S	S	S

s	s	S	S
S	S	s	S
S	S	S	S

Status: A summer resident in mountainous areas, occurring locally throughout the entire region; present and probably breeding in all of the montane parks.

Habitats and Ecology: Generally associated with open coniferous forests, such as ponderosa pines, but also breeds in aspen groves, in riparian woods, and sometimes also in urbanized areas. Nesting sites are rather variable, and include old woodpecker holes, natural tree or cliff cavities, and occasionally in birdhouses.

Seasonality: Colorado records are from April 1 to October 9, and Wyoming records are from May 7 to September 8. In Montana and southern Alberta the birds usually arrive in April and leave in late August or early September. Egg or nest records in Colorado are from early June to mid-July, and in Montana active nests have also been seen from early June to early July, with fledged young out of the nest by July 23.

Comments: Like the hole-nesting tree swallow, this is a relatively early migrant in the spring, and it also tends to leave fairly early in the fall, moving to wintering areas from Mexico southward.

Suggested Reading: Edson, 1943; Combellack, 1954.

Northern Rough-winged Swallow (*Stelgidopteryx ruficollis*)

Identification: This brownish swallow is easily confused with the similar bank swallow, but unlike that species it has a dingy gray throat that grades into a somewhat lighter belly. Its tail is slightly forked, like that of the bank swallow, but unlike that species it rarely if ever nests in large colonies. Instead it is likely to be seen as single individuals or pairs during the breeding season.

Status: A local summer resident throughout the region, mainly at lower elevations and in open habitats. Present in all the montane parks, and breeds in most, but rare in the more southerly parks except at the lowest elevations.

LATILONG STATUS

S	S	S	s
s	S	s	s
S	S	s	s

Habitats and Ecology: Associated with open areas, including agricultural lands, rivers and lakes, and grasslands near water, and breeding almost exclusively in cavities dug in earthen banks of clay, sand, or gravel.

S	S	s	S
s	S	S	S
s	S	S	S

Seasonality: Colorado records extend from April 4 to October 4, and Wyoming records are from May 1 to September 6. In Montana and southern Alberta the birds are usually present from late April or early May until the end of August. Egg records are limited for the region, but active nests in North Dakota have been seen from May 10 to July 15, and Montana egg records are from June 14 to July 6. Nest excavation has been observed in Jasper N.P. as early as May 31.

S	S	S	M
M	S	S	S
S	S	S	S

Comments: The "rough-winged" condition of this species refers to the unusually roughened leading edges of the outer primary feathers. The function of this condition is unproven, but perhaps it serves as a sound-damping device, making the birds more efficient aerial hunters, in a similar manner to the specialized feathers of owl wings.

Suggested Reading: Lunk, 1962; Bent, 1942.

Bank Swallow (*Riparia riparia*)

Identification: Very similar to the previous species, but in this one the throat is a clear white, which area is separated from the white belly by a brownish breastband. The birds nest in cavities dug in exposed banks, usually in rather large colonies.

LATILONG STATUS

S	S	S	s
s	S	S	s
S	S	S	s

S	S	S	S
S	S	s	s
s	S	S	S

S	S	S	M
s	s	s	S
S	S	S	S

Status: A summer resident in suitable habitats throughout the region, mainly at lower elevations. Breeds in several of the montane parks, but not common in any.

Habitats and Ecology: Breeding almost always occurs near water, such as in steep banks along rivers, roadcuts near lakes, gravel pits, and similar areas with steep slopes of clay, sand, or gravel. Outside the breeding season the birds are of broader distribution, sometimes foraging over agricultural lands.

Seasonality: Colorado records are from April 7 to October 20, and Wyoming records are from April 21 to October 4. In Montana and southern Alberta the birds usually arrive in late April or early May, and are often gone by the end of August.

Comments: Territoriality in these highly colonial birds is limited to the burrow itself, which is sometimes used in subsequent years by the same individuals. When burrows need to be dug or deepened, both members of the pair participate, scratching at the earth with their claws, often until the burrow is a full two feet in length.

Suggested Reading: Beyer, 1938; Gaunt, 1965; Peterson, 1955.

Cliff Swallow (*Petrochelidon pyrrhonota*)

Identification: This swallow is easily recognized by its golden-orange rump patch, its square rather than forked tail, and its pale yellowish forehead patch. It is a highly social species, and is usually seen in large groups on the breeding grounds.

Status: A summer resident throughout the region, including all of the montane parks, where it is a common to abundant breeder.

Habitats and Ecology: A wide variety of nesting areas are used by this species, but in the region under consideration vertical cliff-sides and the sides or undersides of bridges are perhaps most commonly used. The nests are gourd-like structures of dried mud, made of small mud globules that are gathered by the birds and carried back in the bill.

Seasonality: Colorado records are from April 16 to October 10, and Wyoming records extend from May 7 to September 4. In Alberta and Montana the birds are usually present from early May to late August. Egg records in Wyoming are from June 11 to July 15, and in Montana from May 20 to mid-July.

Comments: Cliff swallows usually have to rebuild most or all of their mud nests each year, a job that may occupy both members of the pair for nearly two weeks. The nesting season is quite prolonged, and at least in some areas a proportion of the females produce two broods, often changing mates between broods.

Suggested Reading: Samuel, 1971; Emlen, 1952, 1954; Grant & Quay, 1977; Mayhew, 1958.

LATILONG STATUS

S	S	S	S
S	S	S	S
S	S	S	S

S	S	S	S
S	S	S	S
S	S	S	S

S	S	S	S
s	S	S	S
S	S	S	S

Barn Swallow (*Hirundo rusticola*)

Identification: This is the only North American swallow with a deeply forked tail, and also the only one with orange to rusty brown underparts. Above it is mostly dark bluish black, like several of the other swallows. Often found near barns or other farm buildings, which provide nest sites.

LATILONG STATUS

S	S	S	s
S	S	S	S
S	S	S	S

S	S	S	S
S	S	S	S
S	S	S	S

S	S	S	S
s	S	S	S
S	S	S	S

Status: A summer resident throughout the region, including the montane parks, where it is a common to occasional breeder in all.

Habitats and Ecology: Except for the purple martin, this species is the swallow that is most closely associated with humans in the Rocky Mountain region. Although it may still occasionally nest on cliff or cave walls, its normal nesting sites are the horizontal beams or upright walls of buildings and similar structures.

Seasonality: Colorado records are from April 7 to October 20, and Wyoming records extend from April 21 to October 4. In Montana and southern Alberta the birds usually arrive in late April and depart by late September. Wyoming egg records are from May 16 to July 15, and similar dates seem to apply elsewhere in the region.

Comments: Barn swallows are persistent nesters, and usually begin a second clutch within a few days of fledging the first family. There is typically a month's span between nesting cycles, and in northern areas second broods may be difficult to bring to fledging before the onset of cooler weather.

Suggested Reading: Snapp, 1976; Samuel, 1971.

Gray Jay (*Perisoreus canadensis*)

Identification: This common species is the size of a blue jay, but lacks a crest, is mostly grayish black and whitish, with a black nape, and white cheeks. Young birds are almost uniformly dark gray, except for a white "whisker" patch. This is the species that most often appears at picnic areas and campgrounds, boldly looking for food.

Status: A resident in montane forests throughout the region; present in all the montane parks and relatively common in all.

Habitats and Ecology: This species is associated with a wide variety of boreal and montane coniferous forest types, and occasionally extends into aspens and riparian woodlands outside the breeding season. Nesting almost always is in coniferous vegetation.

Seasonality: A permanent resident throughout the region in forested areas, with little seasonal movement. Eggs in Colorado have been reported from March 17 to May 2, and in Wyoming nestlings have been seen as early as April 11. In Jasper N.P. fledged young have been seen by mid-May, and nesting in Alberta sometimes begins as early as March.

Comments: This familiar bird is well known to every camper in the area; it is also commonly called the "whiskey jack" and the "Canada jay" as well as "camp robber." Like the Clark's nutcracker, the birds may cache any excess tidbits that they obtain, and return to the food supply at a later time.

Suggested Reading: Rutter, 1969; Dow, 1965; Goodwin, 1976.

LATILONG STATUS

R	r	R	
r	R	r	R
R	r	r	r

R	R	M	r
r	R	R	
	R	r	M

r		r	r
M	M	r	r
	R	R	R

270

Steller's Jay (*Cyanocitta stelleri*)

Identification: The same shape and size as the familiar blue jay, but with a more distinct crest and a nearly uniformly deep blue color, with white markings limited to the area above and in front of the eyes. The species' calls include various loud and raucous notes, including a repeated *shook* call.

LATILONG STATUS

r	r	R	
r	R	r	R
R	R	r	r

R	R	r	M
r	R	R	
s	R	M	M

M		r	M
M	r	r	R
R	R	R	R

Status: A resident in coniferous woodlands throughout the region. Present in all the montane parks, and breeding in most or all, but less common northwardly.

Habitats and Ecology: This species is centered in the ponderosa pine zone, but also extends down into the pinyon–juniper zone, and as high as the Douglas fir zone. During the non-breeding season the birds may wander well away from their coniferous forest habitats, sometimes coming into cities and feeding at bird feeders during winter.

Seasonality: A permanent resident throughout the region. Colorado egg records are from April 23 to June 3, and egg records from Wyoming and Montana are from May 15 to June 28. Recently fledged nestlings have been seen in Montana as late as August 17.

Comments: The Steller's jay replaces the blue jay west of the Great Plains, and occupies much the same niche as does that species in the deciduous forests of eastern North America. Only in Colorado and perhaps eastern Wyoming do these two species possibly come into any natural contact, and some hybridization has been reported between them.

Suggested Reading: Brown, 1964; Goodwin, 1976.

Blue Jay (*Cyanocitta cristata*)

Identification: This jay is crested similarly to the Steller's jay, but is grayish underneath, has white barring and spotting on the wings, and white tips on the tail feathers. It has a loud call, including a shrill *thief* call, and a mellow and whistled *too-weedle* note that is repeated several times.

Status: A local resident in eastern and northern portions of the region, mainly in deciduous habitats. A rare vagrant in the montane parks, but gradually extending its range westwardly and perhaps becoming more common in the Rocky Mountain area.

LATILONG STATUS

		M	M	
?		M		
V		M	M	M

M	M	M	M
		R	
		M	M

M		R	M
		M	r
		M	R

Habitats and Ecology: Widely distributed in deciduous woodland, city parks, suburbs, and almost anywhere there is an intersection of woods and open grassy areas. Riparian woods, with large willows or cottonwoods, are favored habitats on the western plains.

Seasonality: A permanent resident in southern parts of the region; more northern birds may move to cities or other protected sites during severe weather. Most records for Wyoming are between May 6 and October 26, suggesting some migration in that state. There are few regional egg records, but in North Dakota active nests have been reported from May 7 to June 2.

Comments: This adaptable and somewhat flamboyant bird adds a good deal of color and sound to any area, but it is also a notorious stealer of eggs and small nestling birds, and so is often a hazard to breeding songbirds of a particular area. It has now reached eastern British Columbia in its western range extension, and is starting to appear in Oregon and California during winter.

Suggested Reading: Hardy, 1961; Goodwin, 1976.

Scrub Jay (*Aphelocoma caerulescens*)

Identification: This crestless jay has bluish upperparts except for a brown back, a white throat bounded by a blue "necklace," and pale gray underparts. It is most similar to pinyon jay, but that species is almost uniformly dull blue above and below.

LATILONG STATUS

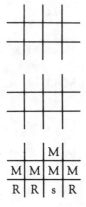

		M	
M	M	M	M
R	R	s	R

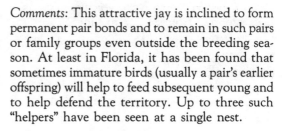

Status: A local migrant or summer resident in the southwestern parts of the region, south of the Snake River in Idaho and in northwestern Colorado. Rare or absent from the montane parks, but occasional at Dinosaur N.M., where breeding is possible.

Habitats and Ecology: Associated with low arid woodlands, including pinyon–juniper and oak–mountain mahogany, and less frequently extending into the ponderosa pine zone where oaks are also present. Often found along brushy ravines or wooded creek bottoms.

Seasonality: Probably a resident where it occurs, except at the extreme northern edge of its range, where migrant or vagrant non-breeders may be seen. In Colorado egg records extend from May 2 to June 16. Utah records are from April 6 to May 20.

Comments: This attractive jay is inclined to form permanent pair bonds and to remain in such pairs or family groups even outside the breeding season. At least in Florida, it has been found that sometimes immature birds (usually a pair's earlier offspring) will help to feed subsequent young and to help defend the territory. Up to three such "helpers" have been seen at a single nest.

Suggested Reading: Woolfenden, 1975; Hardy, 1961; Ritter, 1972; Pitelka, 1951.

Pinyon Jay (*Gymnorhinus cyanocephalus*)

Identification: Like the preceding species, a crestless jay, but in this species the plumage is almost entirely pale to dark blue, with slight whitish streaking on the throat but no definite "necklace." As with the scrub jay, it also has a relatively short tail but has an unusually long bill. It also flies in a direct line, without the undulations typical of the scrub jay. Its usual flight call is a loud, cat-like *mew*.

Status: A resident over much of the southern parts of the region, including southern Idaho, southern Montana, and most of Wyoming, especially at lower elevations. Rare in the montane parks of Wyoming and Colorado.

Habitats and Ecology: Generally associated with pine forests growing on dry substrates, especially the pinyon–juniper association, but extending during the non-breeding period into the oak–mountain mahogany, sagebrush, and desert scrub habitat types.

Seasonality: Resident in most of the area concerned, but somewhat migratory at the northern parts of the region. Colorado egg records are from March 23 to May 19, which range encompasses the records for elsewhere in the region.

Comments: These are relatively gregarious jays, remaining in flocks for much of the year, and probably establishing fairly permanent pair bonds within such flocks. Caching behavior, mainly of pine seeds, is important in this species, and part of courtship consists of the passing of pine seeds from one bird to another.

Suggested Reading: Balda & Bateman, 1971, 1973; Balda et al., 1972; Bateman & Balda, 1973.

LATILONG STATUS

M	M		
V	M	M	M
R	S	s	R
	M	r	r
	M	r	r
R	M	R	M
r	R		M
R	R		R

Clark's Nutcracker (*Nucifraga columbiana*)

Identification: Another "camp-robbing" species that frequents picnic areas and other food sources. It somewhat resembles the gray jay, but has an entirely pale gray head, and has white outer tail feathers and large white spots on the trailing edge of the inner wing feathers. One of the common calls is a loud, grating crow.

LATILONG STATUS

r	R	R	r
r	R	r	R
R	R	R	r

R	R	r	r
R	R	R	R
	R	r	r

r	R	R	r
M	M	r	r
R	R	R	R

Status: A resident in wooded areas over much of the region, common in all the montane parks and breeding in all of them.

Habitats and Ecology: Widespread in coniferous habitats, from the ponderosa pine zone to timberline. More common in the higher coniferous zones in summer, but descending during winter to the pinyon zone and sometimes out onto the plains areas.

Seasonality: A permanent resident throughout its range in the region. Colorado egg records are from March 15 to April 16, and in Wyoming eggs have been seen from the end of February onward, with nestlings noted as late as May 13. Eggs have been seen from March 19 to April 27 in Montana, and nest-building observed in Banff N.P. during the first half of April.

Comments: "Nutcracker" is perhaps not the very best name for this species, which favors the large seeds of the pinyon pine where they are available, but otherwise take a large variety of foods. The species is somewhat irruptive, and in some winters appear in large numbers in the desert and lowland areas of the western states. Like all jays, it is highly adaptable and efficient at finding new food sources.

Suggested Reading: Dixon, 1934; Mewaldt, 1956; Tomback, 1977.

Black-billed Magpie (*Pica pica*)

Identification: Easily identified by its long, flowing tail, its black-and-white body color, and its black wings with flashing white inner markings that are exposed in flight.

Status: A resident throughout the region, in most habitats. Present and a common breeder in all the montane parks.

Habitats and Ecology: Of widespread occurrence, but especially common in riparian areas with thickety vegetation, agricultural areas with scattered trees, sagebrush, aspen groves, and the lower levels of the coniferous forest zones. Small, thorny trees are especially favored nest sites, but junipers and similar trees are also used.

Seasonality: A permanent resident in the area. Egg records for Colorado are from March 26 to May 29, with a peak from April 24 to May 8. Montana records are from March 28 to May 26, and are centered between May 6 and 12.

LATILONG STATUS

r	R	R	r
R	R	R	R
r	R	R	R

R	R	R	R
R	R	R	R
R	R	R	R

R	R	R	R
R	R	R	R
R	R	R	R

Comments: One of the characteristic birds of the western states, the black-billed magpie is an opportunistic scavenger that feeds on road-killed rodents, the eggs and nestlings of birds, and sometimes even pecks at open sores of livestock and other ungulates. Their conspicuous nests are well protected by several layers of twigs that enclose an egg chamber consisting of a mud-cup lined with soft materials.

Suggested Reading: Linsdale, 1937; Erpino, 1968; Jones, 1960; Goodwin, 1976.

American Crow (*Corvus brachyrhynchos*)

Identification: This familiar bird scarcely needs description, but might be confused with the common raven if its smaller size, rounded rather than wedge-shaped tail, and "caw" call rather than croaking voice are not noted. It also has a relatively smaller and weaker bill than do ravens, and does less gliding or soaring.

LATILONG STATUS

s	R	s	s
s	s	R	S
R	R	R	S

R	S	S	R
s	R	M	R
R	R	M	R

M	r	r	M
M	R	r	R
R	R	R	R

Status: A summer or year-round resident in wooded habitats throughout the region, but more common at lower altitudes, and varying from common to rare in the montane parks, breeding in several.

Habitats and Ecology: Forested areas, wooded river bottoms, orchards, woodlots, large parks, and suburban areas are all used by this species; it is often replaced by ravens in rocky canyons and higher montane areas.

Seasonality: Present throughout the year in the southern parts of the region, but distinctly migratory. Migration peaks are evident in Wyoming during March and April, and in Montana the birds are present mainly from late March to October, with a few overwintering in most years. Wyoming egg records are from May 5 to July 1, and Alberta records extend from May 2 to June 9, with a peak from May 13 to 24.

Comments: The American crow is a good deal smaller than the raven and in some areas, such as Jackson Hole, it seems to be competitively excluded from habitats that it might well occupy if the raven were not already there.

Suggested Reading: Chamberlain & Cornwell, 1961; Emlen, 1942.

Common Raven (*Corvus corax*)

Identification: A very large blackish bird with a rather wedge-shaped tail, a heavy bill, a low, croaking call, and a somewhat shaggy throat. Only the American crow can be mistaken for it, and any of the above-mentioned characteristics should separate the two species fairly easily.

Status: A permanent resident throughout much of the area, mainly in mountainous regions. Present in all of the montane parks, and breeding in all of them.

Habitats and Ecology: Generally associated with wilderness areas of mountains and forests, especially where bluffs or cliffs are present for nesting. Where these are unavailable, tall coniferous trees are used for nesting, as in Jackson Hole, where over 90 percent of the nests are in trees. Often found all the way to timberline in late summer, or even in alpine tundra areas. They also extend out into sage and grassland areas, scavenging for road-killed mammals and birds.

Seasonality: A resident wherever it occurs in the region. Colorado egg records are from April 13 to June 22, and this range encompasses the available records from farther north. In Colorado nestlings have been seen as early as mid-April. In Jackson Hole the egg period is from early April to mid-June.

Comments: This intelligent and adaptable bird has generally shunned human contact, in contrast to most of the jays and crows, and has remained associated with wilderness areas. They are effective predators of eggs and nestlings, and also share the carcasses of large mammals such as elk with eagles and coyotes in the Jackson Hole area.

Suggested Reading: Shiehl, 1978; Harlow, 1922; Tyrell, 1945; Dorn, 1972.

LATILONG STATUS

R	R	R	r
R	r	r	R
R	R	R	R

R	R	r	r
r	R	R	R
s	R	R	R

R	M	M	R
M	M	r	r
R	R	R	R

Black-capped Chickadee (*Parus atricapillus*)

Identification: This titmouse closely resembles the mountain chickadee, but has its black "cap" extending from the crown down through the eyes in an unbroken manner, so that the white also appears as a continuous patch. Its call is a whistled "phoe-be," with both notes clear and loud, and the second on a lower pitch than the first. Its song consists of a "chick-a-dee-dee-dee . . ."

LATILONG STATUS

R	R	R	r
r	R	R	R
R	R	R	R

R	R	R	R
r	R	R	R
r	R	r	R

r	M	R	r
	R	r	R
W	R	R	R

Habitats and Ecology: Associated with a wide variety of wooded habitats, both of coniferous and hardwood types, and breeding wherever suitable nesting cavities exist. These typically consist of old woodpecker holes, but sometimes the birds excavate their own nest cavities in the rotted wood of dead stumps. Bird houses are also occasionally used. Aspen groves and riparian woodlands are favored nesting areas in the Rocky Mountain region.

Seasonality: This species is a permanent resident throughout the area, although some seasonal shifts in distribution do occur. In Colorado, egg records extend from June 23 to July 4, although nest-building in Rocky Mountain N.P. has been observed as early as March 23. In Wyoming, nest-building has been observed in early May, and eggs found as early as May 20. Egg records in Montana extend from May 12 to July 28.

Comments: This is the most widespread and familiar of the North American titmice, and one that is often attracted to winter bird feeders, especially if suet is available. During the winter pair bonds appear to dissolve, and a certain degree of flocking often occurs.

Suggested Reading: Odum, 1941–42; Brewer, 1963; Sturman, 1968a, 1968b; Smith, 1974.

Mountain Chickadee (*Parus gambelii*)

Identification: Resembles the preceding species, but the white facial markings are interrupted by a black line extending through the eye region from behind. The song is nearly the same as that of the black-capped chickadee, but the call is usually a three- or four-noted descending whistle.

Status: A resident in mountain forests of the region, occurring in all the montane parks and probably breeding in all. More montane in distribution than the previous species.

Habitats and Ecology: Largely limited to montane coniferous forests, and usually absent from deciduous stands, although aspens are frequently used for nesting. Prefers open coniferous forests, especially pines, including both ponderosa pines and also pinyons. Woodpecker holes or self-excavated cavities in rotted wood are used for nesting.

Seasonality: A permanent resident throughout the region. Egg records in Colorado extend from April 5 to June 30, while Wyoming records are from May 31 to July 1. In Montana nests have been found from June until late July.

Comments: Like many other species, these birds tend to move upwards in elevation after the breeding season, and at times may winter close to timberline, although more commonly the birds retreat toward the plains at that time.

Suggested Reading: Dixon & Gilbert, 1964; Minoch, 1971; Bent, 1946.

LATILONG STATUS

R	R	R	?
r	R	R	R
R	R	R	R

R	R	R	r
r	R	R	R
r	R	R	R

r	r	r	r
	r	r	r
R	R	R	R

Boreal Chickadee (*Parus hudsonicus*)

Identification: This is the only North American chickadee with a brown rather than a black cap, and its back and flanks are also distinctly brownish, although not nearly such a bright brown as occurs in the chestnut-backed chickadee. Its call consists of wheezy and hoarse notes, without clear whistles.

LATILONG STATUS

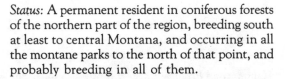

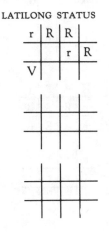

Status: A permanent resident in coniferous forests of the northern part of the region, breeding south at least to central Montana, and occurring in all the montane parks to the north of that point, and probably breeding in all of them.

Habitats and Ecology: Associated with coniferous forests, including swampy areas and muskegs on lower elevations, as well as those of montane areas. Nesting occurs in trees and stumps, either in old woodpecker holes or in self-excavated holes. Rabbit hair is often used as a nest-lining material.

Seasonality: There are relatively few nest records, but in Alberta the egg records extend from April 18 to June 13. Dependent young in Montana have been noted as late as August 1.

Comments: Boreal chickadees tend to winter in flocks, with pair formation occurring in late winter, then aggressive activity breaks up the flocks and pairs begin to form. Nest-inspection behavior may be a part of pair-forming behavior in these birds, and although they may pair for life they evidently do not use the same nest site in subsequent years.

Suggested Reading: McLaren, 1975; McClelland, 1977; Dixon, 1961.

Chestnut-backed Chickadee (*Parus rufescens*)

Identification: This is the only North American chickadee with bright rufous flank and back coloration. It has a sooty-black cap, and its call is a harsh and sibilant *shik-zee-zee*. It lacks clear, whistled call-notes.

Status: A permanent resident in the wetter coniferous forests of the western slopes of the region, south to about central Montana. A common breeder only in Glacier N.P.; not yet reported for Yoho or Kootenay.

LATILONG STATUS

r	R	R	
r	R	R	
R	r	M	?

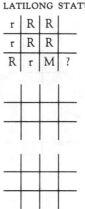

Habitats and Ecology: In this region essentially limited to the western hemlock, western redcedar association, but farther west having a broader ecological range extending into some deciduous woodlands adjacent to coniferous forests. Areas of stream courses and other forest margins are favored for foraging, and nesting occurs in woodpecker holes or self-excavated cavities.

Seasonality: Nesting records for the region are few, but in Washington the egg records extend from April 7 to June 10, with the majority for the period May 9 to 24. In Montana incubation has been observed on June 5, and feeding of young on July 15 and 16.

Comments: This species, perhaps more than any other bird of the region, has a range that conforms to that of the moist Pacific Northwest coniferous forest, which is best developed along the coast of Washington and British Columbia, but is locally extended inland on the wettest west-facing slopes of the Rockies in Montana and adjoining Canada.

Suggested Reading: McClelland, 1977; Sturman, 1968a, 1968b; Dixon, 1961.

Plain Titmouse (*Parus inornatus*)

Identification: The only typical titmouse of the region, with a chickadee-like shape but also with a small, bushy crest. In this species the plumage is almost uniformly gray, and no black or white evident anywhere. The species has a harsh *tsick-a-dee-dee* . . . song that is very much like that of a typical chickadee's.

LATILONG STATUS

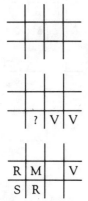

Status: A summer or permanent resident in the southwestern portions of the region, from the Snake River southward in Idaho, and in the southwestern and northwestern portions of Wyoming and Colorado, respectively. Common and probably breeding at Dinosaur N.M.

Habitats and Ecology: In this region associated almost exclusively with the pinyon–juniper association; in some other areas also extending into oak woodlands. Nesting is usually done in cavities of partially decayed and split-open trunks of junipers. They sometimes also use woodpecker holes, and occasionally will nest in birdhouses.

Seasonality: There are few regional records, but Colorado egg records extend from early May to May 21.

Comments: At least in California these birds tend to maintain permanent pair bonds, and very rarely mate with a new individual for as long as their original mate remains alive. They are relatively sedentary, and often nest in the same site in succeeding years.

Suggested Reading: Bent, 1946; Price, 1936.

Bushtit (*Psaltriparus minimus*)

Identification: This bird resembles a small chickadee in shape, but has a very long tail (about as long as the body) and is rather uniformly dull gray throughout, with a short black bill and pale yellow eyes, at least in our region. It usually is to be found in small flocks, feeding in low woodland trees and bushes and uttering sharp single-noted contact calls.

LATILONG STATUS

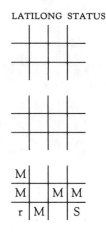

Status: Limited to the southern edge of our area, where it is most likely to be found in pinyon pine–juniper woodlands, and similar scrub oak–mountain mahogany habitats. Within the geographic limits of this book it is evidently mainly a nonbreeding vagrant, since the birds often move into ponderosa pines following breeding.

Habitats and Ecology: Commonest in open woodlands such as pinyon pine and juniper habitats, these birds also at times occur in sagebrush or even aspen-covered hillsides. Nests are usually in pinyon pines or junipers, and are beautiful soft, woven hanging structures woven of mosses, spider webs and hair or feathers, with lateral entrances and usually are less than ten feet above ground.

Seasonality: A permanent resident in most of Colorado and northern Utah, but seasonal or erratic at the extreme northern limits of its range in our region. In northern Utah the nesting records extend from April 25 to July 15, while in Colorado there are egg records for late April and early May.

Suggested Reading: Bent, 1946; Bailey & Niedrach, 1965.

Red-breasted Nuthatch (*Sitta canadensis*)

Identification: This small nuthatch (4 inches long) is the only North American species of the group that has a distinct black line extending back from the eye in the same manner as found in the mountain chickadee. It also has reddish brown flanks, but these are rather obscure in females. Its call is a nasal *nyack*, often repeated.

LATILONG STATUS

R	R	R	
r	R	R	R
R	R	R	r

R	r	R	r
r	R	R	
S	R	R	R

r	r	r	r
M	R	r	M
R	R	r	R

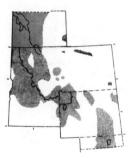

Status: Found in montane coniferous forests throughout the entire region, and a variably common breeding species in all the montane parks.

Habitats and Ecology: Limited largely but not entirely to coniferous forests, primarily those of relatively tall firs, and much of the foraging occurs at rather high portions of the trees. To a much more limited degree aspens and riparian woodlands are sometimes also used. Breeding occurs in the trunks of dead trees or the rotting portions of live trees, with the birds typically excavating their own nesting holes.

Seasonality: A permanent resident throughout the region. Egg records from Colorado are from May 17 to June 7, and from May 15 to July 25 in Montana. In Alberta, nest-excavation behavior has been observed as late as June 13, and in Montana nestlings have been noted into late July.

Comments: This species is unusual in that after the nest has been nearly excavated, both members of the pair bring resin in their bills and spread it above and below the nest hole. The function of this behavior is still somewhat obscure, but it probably helps deter other animals from entering the hole.

Suggested Reading: Kilham, 1972, 1973; McClelland, 1977; Anderson, 1976.

White-breasted Nuthatch (*Sitta carolinensis*)

Identification: This nuthatch is rather easily identified by its black "cap" that does not extend down to the eyes, which are thus entirely surrounded by white. Like all nuthatches it often forages in an upside-down manner on tree trunks, and has a rather nasal voice, which in this species' case includes a *yank-yank* call reminiscent of a toy trumpet.

Status: A permanent resident of deciduous forests and woodlands nearly throughout the region; rather rare in the Canadian montane parks, and known to breed only from Glacier N.P. southward.

LATILONG STATUS

R	R	r	
M	r	r	r
R	R	r	r

r	R	R	R
M	R	R	
	R	R	R

r		r	r
M	M	M	M
R	R	R	R

Habitats and Ecology: Largely confined to deciduous forests, but also extending into riparian woods and at least the lower zones of coniferous forest, especially the ponderosa pine zone and also locally into the pinyon zone. Nesting occurs in old woodpecker holes or in self-excavated holes in rotted wood of dead trees or partially dead ones.

Seasonality: A permanent resident throughout the region. Egg records in Colorado extend from May 13 to June 25, and in Montana active nests have been reported on June 17 and 25, the latter just hatched. Dependent young have been seen there in July.

Comments: These birds maintain their pair bonds for much of the year, and perhaps permanently, although in winter the members of the pair have little contact with one another and roost in different areas. This is the most widely distributed of the North American nuthatches, and the most adaptable ecologically.

Suggested Reading: Kilham, 1968, 1971a, 1972; McClelland, 1977.

Pygmy Nuthatch (*Sitta pygmaea*)

Identification: This is the smallest of the nuthatch species, and the only one in the region that does not have a black cap, although the grayish brown crown becomes darker near the eye. The call-notes include a high-pitched *peep* and a piping *wee-bee.*

LATILONG STATUS

R	R		
M	R	?	
R	R	R	M

M		r	r
M	M		
	M	M	M

	r	M	M
			R
V	V	R	R

Status: A local resident in some portions of the region, mainly in the drier areas of southern Idaho and western Colorado, but also local in western Montana. Generally rare or absent in the montane parks, but common in Rocky Mountain N.P.

Habitats and Ecology: Primarily associated with the ponderosa pine zone, but also occurring locally in the pinyon–juniper zone. It generally forages fairly high in tall pines, but nests closer to the ground in snags or stubs that have rotted trunks providing excavation opportunities.

Seasonality: A permanent resident in the region. Egg records in Colorado extend from May 5 to June 18, and those from Rocky Mountain N.P. are from June 5 to 18.

Comments: This species has a range that fairly closely matches that of the ponderosa pine in the western states, although during summer the birds sometimes range as high as the Engelmann spruce zone and in winter they spread out into riparian woods lining grassland streams. During the fall months communal roosting sometimes occurs, with as many as 150 or more birds sharing a single roosting tree cavity.

Suggested Reading: Bock, 1969; Stallcup, 1968; Norris, 1958.

Brown Creeper (*Certhia familiaris*)

Identification: This well-camouflaged bird is aptly named; it creeps upward on tree trunks in woodpecker fashion, but has a long, narrow and slightly decurved bill unlike that of any woodpecker. Its vocalizations are high-pitched notes that are easily overlooked.

Status: Resident in forested areas almost throughout the region, and present in all the montane parks, probably breeding in all. Breeding status in eastern Wyoming is uncertain but probable, since nesting is known to occur in the adjacent Black Hills of South Dakota.

Habitats and Ecology: Associated with forests throughout the year, including both deciduous and coniferous forests. Virtually all foraging is done on the trunks of fairly large trees, where the birds forage for insects in bark crevices and grooves.

Seasonality: A permanent resident throughout. There are few definite nesting records for the area, but in Rocky Mountain N.P. nesting usually occurs in June and July. The estimated nesting period for the Black Hills is from mid-May to mid-July.

Comments: Nests of this species are very hard to find, as they are hidden behind a piece of fairly loose bark, and thus very difficult to discover. One nest in Rocky Mountain N.P. was found at an elevation of 11,000 feet, only about 200 feet below timberline, but they have also been found as low as 7,000 feet in that state.

Suggested Reading: Davis, 1979; Braaten, 1975.

LATILONG STATUS

R	r	R	
	M	r	R
R	r	r	r

r	R		R
r	R	R	
s	R	r	M

r	M	r	M
	M	M	M
	V	R	R

Rock Wren (*Salpinctes obsoletus*)

Identification: This is a large and rather grayish wren, with conspicuous buffy tips on its outer tail feathers, a pale eye-stripe, and faint breast-streaking. It is always associated with rocky outcrops, and its loud and distinctive song consists of a complex mixture of buzzes and trilled notes that usually opens with two or three loud, challenging notes.

LATILONG STATUS

Status: Locally present throughout the region in rocky areas, especially in dry sagebrush dominated localities. Present in all the U.S. montane parks and possibly breeding in all; abundant in Dinosaur N.M.

Habitats and Ecology: Closely associated with eroded slopes, badlands, rocky outcrops, cliff walls, talus slopes, and similar rock-dominated habitats at generally rather low elevations, but sometimes occurring to 12,000 feet in Colorado. Crannies in cliffs are favorite nesting sites.

Seasonality: Records in Colorado are from March 3 to November 13. In Wyoming the records extend from April 2 to December 30, but with migration peaks in early May and September. In Montana the birds usually arrive in late April and depart in late September. Colorado egg records are from May 25 to June 28, while from Wyoming to Alberta there are egg records from May 27 to July 15, and nestling records from June to July 23.

Comments: One of the interesting features of this species is that the entrance to its nest is often "paved" with small, flat pebbles, for reasons still somewhat obscure.

Suggested Reading: Kroodsma, 1975; Tramontano, 1964.

Canyon Wren (*Caltherpes mexicanus*)

Identification: This wren occurs in similar habitats as the preceding species, but is especially associated with canyons. Unlike the rock wren it has a clear white breast, an almost uniformly rust-colored tail, and dark brown underparts. Its song consists of a series of descending and decelerating liquid notes sounding like *tee-you.*

Status: Largely limited to arid canyons south of the Snake River in Idaho, adjacent areas of southern Wyoming, and also western Colorado, with apparent areas of local distribution farther north and east. Absent from the montane parks except for Rocky Mountain N.P., where it is a common breeder; also common at Dinosaur N.M.

LATILONG STATUS

	s	S	
s	s		
S		s	S
	?	M	M
		S	S
M		M	s
S	s	M	M
S	R		R

Habitats and Ecology: Rocky canyons, river bluffs, cliffs, rock-slides, and similar topographic sites are favored, especially those offering shady crevices. Often found in canyons with streams at the bottom, but sometimes well away from water. Nesting usually occurs in rocky crevices, but at times buildings are also used.

Seasonality: Resident in parts of Colorado, but migratory farther north, and in Wyoming the records extend from April 17 to December 29. In Montana the birds usually arrive in early April and leave in late August. Colorado egg records extend from May 8 to July 10.

Comments: This species has an overlapping niche with the rock wren, but one study indicated that the canyon wren favors more secluded or covered habitats than the rock wren, and also tends to forage on cliff or canyon walls rather than on slopes of loose rocks.

Suggested Reading: Tramontano, 1964; Bent, 1948.

Bewick's Wren (*Thyromanes bewickii*)

Identification: This dark brown wren has a distinct white eye-stripe, a white-edged and long tail that is often jerked from side to side, and a song that usually consists of a variable number of introductory notes followed by a trill. Often found in brushy areas.

LATILONG STATUS

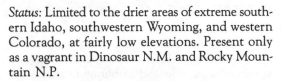

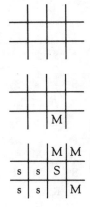

Status: Limited to the drier areas of extreme southern Idaho, southwestern Wyoming, and western Colorado, at fairly low elevations. Present only as a vagrant in Dinosaur N.M. and Rocky Mountain N.P.

Habitats and Ecology: Broken and rather low brushy areas, especially where heavier cover is present overhead, seems to be this species' favorite habitat. It occurs from riparian areas through sagebrush to pinyon–juniper and oak–mountain mahogany habitats in the region, but is perhaps most common in pinyon–juniper woodlands. Nesting is often done in natural tree cavities, but old woodpecker holes are sometimes also used, as are cavities in manmade structures.

Seasonality: Colorado records extend from February 12 to November 17. In that state eggs have been observed from May 13 to June 10, and in Washington egg records extend from March 29 to June 27.

Comments: This species is rarely common where the house wren also occurs, as it seems to be at a competitive disadvantage relative to that species, and apparently is prevented from breeding in areas that are used by house wrens. In general it has a much more southerly range than does the house wren, and is less migratory than that species.

Suggested Reading: Kroodsma, 1973; Cogswell, 1962; Miller, 1941; Bibbee, 1947.

House Wren (*Troglodytes aedon*)

Identification: This most widespread of wrens has a rather uniformly brown color, lacking the white eye-stripe and white underparts of the Bewick's wren, and a longer tail as well as a larger size than the winter wren. It has a loud and bubbling song of whistled notes that is often repeated many times per minute.

Status: Present almost throughout the region, although more common southwardly in the Rocky Mountain region, and rare or lacking in the montane parks of Alberta.

Habitats and Ecology: Generally most common in the lower elevation forests, but occasionally reaching timberline. In this region the birds favor riparian woodlands, aspen groves, and the lower and more open coniferous forest zones, as well as areas of human habitations. Nesting occurs in natural tree cavities, old woodpecker holes, artificial cavities such as birdhouses, and the like.

Seasonality: In Colorado these birds arrive about the first of May, and usually depart by the end of October. A similar schedule occurs in Wyoming and Montana, with migration peaks in May and September. Egg records in Colorado range from June 3 to July 17, and in Wyoming from June 1 to July 15. Double-brooding is common in many parts of this species' range.

Comments: This is one of the most familiar species of songbirds, and also one of the most conspicuous. Males are highly territorial, and at least in some areas tend to be non-monogamous, with frequent mate-changing prior to the second brood of the season.

Suggested Reading: Kroodsma, 1973; Kendeigh, 1941; Bent, 1948.

LATILONG STATUS

S	S	S	s
s	S	S	s
S	S	S	S

S	S	S	S
S	S	S	S
s	S	S	S

S	s	S	s
M	S	S	S
S	S	S	S

Winter Wren (*Troglodytes troglodytes*)

Identification: This tiny wren is only about 4 inches long, and has a very short tail as well as extensive flank and belly striping. Its song is very high pitched and greatly prolonged, often lasting more than 5 seconds. Usually found in very heavy cover, often near water.

LATILONG STATUS

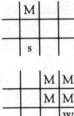

Status: A summer resident in the northern parts of the region, south to about central Montana, and rarely farther south (one Wyoming breeding record). Generally a common breeder in the montane parks south to Glacier, but only a vagrant farther south.

Habitats and Ecology: This species is typically found in heavy forests, usually coniferous, and often occurs in moist and shady canyons where brush-piles and tangles of vegetation cover the ground. Root-tangles or cavities in or under logs are favorite nesting locations, especially where there are undercut banks.

Seasonality: A winter resident in Colorado, present from late September to late May. In Wyoming a migrant, and in Montana a summer visitor, reported from April 23 to October 9. There are nesting records from Montana from early June to early July, and recently fledged young observed in late June and July. In a single Wyoming nesting (Freezeout Hills) young ready to fly were observed on July 15.

Comments: This rather mouse-like bird might well be overlooked if it weren't for the marvelous song of the male. Johnston Creek is one of the areas where it can be readily seen in Banff N.P., while in Glacier N.P. it is common along McDonald Creek.

Suggested Reading: Armstrong, 1955, 1956; Bent, 1948.

Marsh Wren (*Cistothorus palustris*)

Identification: Closely associated with marshy wetlands, this species' presence can often be detected by looking for its football-shaped nests in cattails or similar vegetation, and listening for its rattling and reedy song. If observed, the strong white eye-stripe and white striping on the back serve to identify it.

Status: Locally present, mostly at lower altitudes, throughout the region. Rare or absent in the montane parks except Grand Teton N.P., where occasional.

Habitats and Ecology: Restricted to marshy or swampy areas having an abundance of emergent plants such as reeds and cattails. Slow-moving waters, such as the inlets of reservoirs, are sometimes also used. Nesting is always done over water, usually from 3 to 5 feet above the substrate.

Seasonality: Wyoming records extend from March 30 to October 15, although the birds have also been known to winter rarely in the state. In Alberta they usually arrive by the third week of April and remain until about mid-October. Eggs have been found in Idaho as early as May 30, and in Montana nest construction has been observed as late as July 26. Probably two broods are normally raised in this region.

Comments: Males of this species build a number of "courting nests" that serve in part to advertise their territories, and provide females with a choice of nesting sites. After obtaining a mate, the male continues to court females, and may acquire up to three mates that nest within his territorial boundaries.

Suggested Reading: Verner, 1965, 1975a, 1975b; Kale, 1965, Welter, 1935.

LATILONG STATUS

S	S	M	
	S	s	S
S	s		M

S		s	s
S	S		M
S	S	M	B

M	s	M	
M	s	s	S
S	s	s	R

American Dipper (*Cinclus mexicanus*)

Identification: Easily identified, this species is confined to mountain streams and resembles an overgrown gray wren, with a cocked tail and a melodious, bubbling song. Foraging is done underwater, usually in rushing streams.

LATILONG STATUS

R	R	R	
r	R	R	R
R	R	R	R

R	R	R	r
R	R	R	M
s	R	R	M

R		R	M
		R	R
S	M	R	R

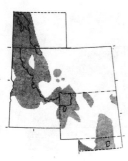

Status: Found in suitable habitats throughout the region, and a relatively common breeder in all the montane parks.

Habitats and Ecology: Rapidly flowing mountain streams, often with waterfalls or cascades present, are this species' prime habitat. Nesting is sometimes done on rock walls or overhangs near or even sometimes behind waterfalls, but more often the nests are constructed under bridges that cross suitable creeks or rivers. The birds are highly territorial, and pairs tend to be well separated.

Seasonality: A permanent resident in the region, although the birds tend to move to lower elevations in winter as their foraging areas freeze over. Eggs in Colorado have been observed from April 4 to June 10. There are few Wyoming records, but in the adjacent Black Hills of South Dakota eggs have been noted as late as July 5. In Glacier N.P. nest-building has been observed as early as April 21.

Comments: These birds often share their habitat with harlequin ducks, and both species feed on similar insect foods captured at the bottom of fast-moving streams by probing in the cobble. Unlike the harlequin, this species uses its wings to remain under water and to propel itself against the current as it searches for food.

Suggested Reading: Hann, 1950; Bakus, 1959a, 1959b.

Golden-crowned Kinglet (*Regulus satrapa*)

Identification: This tiny and inconspicuous bird is identifiable by its very short and sharply pointed beak, its strong white stripe above the eye, and a black-bordered crown that is yellow (in females) to yellow and orange (in males). The calls and songs are extremely high-pitched and inaudible to many.

Status: Present in coniferous forests throughout the region, mainly at higher elevations. An abundant to rare breeder in all the montane parks.

Habitats and Ecology: During the breeding season primarily associated with spruce–fir forests, but otherwise generally present in the coniferous zones and sometimes extending out into riparian woodlands. Nesting occurs in dense and fairly tall coniferous trees, usually spruces, and nests usually are placed rather high in the tree.

Seasonality: Resident in Colorado, and some wintering occurs as far north as Alberta, although seasonal movements are evident throughout the region. In Colorado nestlings have been seen from June 25 to August 26, and in Montana nest-building has been seen as early as late March.

Comments: This and the following species are members of an Old World group of birds sometimes called "Old World warblers." The kinglets are closely associated with rather dense coniferous forests throughout their range, and seem to compete very little with the New World warbler group.

Suggested Reading: Bent, 1949.

LATILONG STATUS

R	R	R	
r	r	r	s
R	r	R	r

r	s		s
s	R	R	
	r	M	M

			r
r	r	M	M
V	W	R	R

Ruby-crowned Kinglet (*Regulus calendula*)

Identification: Similar in general appearance to the preceding species, but lacking a white eye-stripe, and with a white eye-ring instead, and usually with no evident crest (males have a small red crown). The calls and songs are very high-pitched and surprisingly loud for such a small-sized bird.

LATILONG STATUS

S	S	S	
s	S	s	S
S	s	S	s

S	s	s	s
s	S	s	
s	S	S	S

s	s	s	s
	S	S	S
M	R	S	S

Status: Present in coniferous forests throughout the region, more widespread and generally more numerous than the golden-crowned kinglet. Present in all the montane parks, and probably breeding in all of them.

Habitats and Ecology: Breeding occurs in coniferous forests from the lower zones almost to timberline in the subalpine zone, but is usually in taller and denser forests of medium altitude. During winter the birds often move toward lower elevations, including prairie stream bottoms and sometimes into cities.

Seasonality: More migratory than the golden-crowned kinglet, but some wintering occurs as far north as Colorado. Wyoming records extend from April 14 to November 12, with peaks in May and October. In Alberta they are usually present from early May to September or early October. Egg records in Colorado range from June 3 to July 11, and in Montana eggs have been noted from June 18 to the latter part of July.

Comments: Surprisingly little fieldwork has been done on the kinglets in North America, but they typically breed well up in coniferous trees nearly out of sight from the ground, and are often almost completely hidden from view.

Suggested Reading: Rea, 1970; Bent, 1949.

Blue-gray Gnatcatcher (*Polioptila caerulea*)

Identification: This is a slim-bodied, long-tailed bird, with a generally gray color but with a white-tipped, black tail and a pale eye-ring. The song is a series of thin, insect-like notes, and the call is also thin and buzzy.

Status: Limited to the southernmost part of the region, mainly western Colorado and adjacent southwestern Wyoming (breeding known only for Green River latilong). Common at Dinosaur N.M. and probably breeding there; absent or a vagrant in the montane parks.

LATILONG STATUS

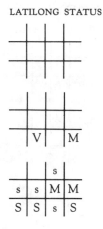

Habitats and Ecology: Breeding in the region occurs in pinyon–juniper and perhaps also adjacent oak woodland or sagebrush areas, up to about 7,000 feet elevation. Arid and park-like areas, with only scattered thickets, seem to be preferred for foraging, and nests are usually placed in low junipers.

Seasonality: Colorado records extend from April 27 to November 14. Nesting occurs from late May to late June.

Comments: These birds forage for insects among the branches of trees and shrubs, and build tiny felt-like nests on the branches of dead pines or junipers that blend almost perfectly with the bark of the tree. The outside is lined hummingbird-like with lichens and plant down, held together with spider webs.

Suggested Reading: Root, 1967, 1969; Fehon, 1955.

Eastern Bluebird (*Sialia sialis*)

Identification: Male eastern bluebirds are very similar to western blue-birds, but lack brown on the back, and have reddish brown rather than blue throats. Females of the two species are very similar, but female eastern bluebirds have paler, whitish throats, while those of westerns (and mountains) have much more dusky throats.

LATILONG STATUS

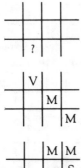

Status: Limited to deciduous woodlands along rivers on the plains of eastern Montana, Wyoming, and Colorado. Generally absent from the montane parks, but rare during migration in Rocky Mountain N.P.

Habitats and Ecology: Generally associated with open deciduous woods that are close to grass-lands, such as riparian forests, shelterbelts, farm-steads, and city parks. Nesting occurs in old woodpecker holes, natural cavities of dead trees, dead limbs, or sometimes utility poles. Birdhouses are also frequently used, especially where natural cavities are relatively rare.

Seasonality: Colorado records extend from March 19 to November 29. There are few records from farther north, but probably are comparable to those of the mountain bluebird. Egg records for Colorado and Montana are for June.

Comments: Eastern and western bluebirds have virtually complementary ranges, although in eastern Montana and eastern Wyoming a limited degree of geographic overlap does occur. In such areas the western bluebird is likely to utilize conif-erous woods and the eastern deciduous woods, although both use aspens.

Suggested Reading: Hartshorne, 1962; Thomas, 1946; Peakall, 1970.

Western Bluebird (*Sialia mexicana*)

Identification: Similar to the eastern bluebird, but males have a blue throat and a variable amount of chestnut on the upper back. Females are very similar to those of the eastern bluebird, but their throats are dusky gray rather than reddish buff. They also resemble females of the mountain bluebird, but are more rusty on the flanks and breast area.

Status: A local summer resident in northwestern and southwestern parts of the region, primarily in western Colorado and the northern parts of Idaho. Generally rare or absent from the montane parks, but an uncommon breeder in Rocky Mountain N.P.

LATILONG STATUS

S	S	s	
s	s		
S	S	S	?

M		?	
?	V	M	M
	M	M	M

		M	M
		s	M
S	s	S	S

Habitats and Ecology: Rather open timberlands, either of deciduous or coniferous trees, seem to be this species' favored habitats. It breeds in both aspens and ponderosa pine woodlands, and during the nonbreeding season extends out into the woodlands of pinyon–juniper, oak–mountain mahogany, and some agricultural or desert scrub habitats. A combination of timberlands having dead trees with natural cavities or living trees with woodpecker holes and nearby open grassy areas for foraging seems to provide optimum habitats. Breeding also extends to the level of mountain meadows, sometimes about 10,000 feet above sea level.

Seasonality: Colorado records extend from February 23 to November 5, and in Montana the birds are typically present from mid-March to mid-September. Egg records in Colorado are from May 6 to July 1.

Comments: In contrast to the eastern bluebird, this species is relatively silent, and its song is far less melodious. During migration the birds often are seen in company with mountain bluebirds.

Suggested Reading: Grinnell & Storer, 1924; Myers, 1912.

Mountain Bluebird (*Sialia currucoides*)

Identification: Male mountain bluebirds are the only birds of the region that are virtually entirely sky blue, except for their whitish underparts. Females are mostly grayish brown above and below, but exhibit blue coloration on the wing and tail surfaces. Like the western bluebird it is a relatively quiet species, and its song is soft and warbling.

LATILONG STATUS

S	S	S	s
S	S	S	S
S	S	S	S

S	S	S	S
S	S	S	S
s	S	S	S

S	S	S	S
S	S	S	S
S	S	S	S

Status: The most common and widespread of the bluebirds of the region, especially in rather open woodlands. Fairly common in all of the montane parks, and breeding in all.

Habitats and Ecology: Breeding occurs in open woodlands and forest-edge habitats from mountain meadows downward through the ponderosa pine zone, the aspen zone, and into the pinyon–juniper zone. Typically the birds favor nesting where either dead trees are available for nest cavities or where rock crevices or other suitable sites are present.

Seasonality: Generally resident in Wyoming, and present in Wyoming nearly throughout the year, with records from February 21 to November 2, with migration peaks in March and October. In Montana and Alberta the birds are usually present from mid-March to mid-October. Colorado egg records are from May 2 to June 12, while those from Wyoming are from May 10 to July 25. Montana egg records are from May 5 to July 9.

Comments: At least over much of their range, mountain bluebirds are double-brooded, and after the female begins her second clutch her mate typically remains with the first brood for about 10 days after they fledge. Rarely have the young of the first brood been observed helping to feed the second one.

Suggested Reading: Criddle, 1927; Power, 1966; Haecker, 1948.

Townsend's Solitaire (*Myadestes townsendi*)

Identification: This unusual member of the thrush family somewhat resembles a large flycatcher, and has a mostly grayish body with a white-edged tail, a pale eye-ring, and pale orange wing-patches that are evident only in flight. The song of the male is an extended melodious warbling that is one of the most beautiful of western bird songs.

Status: Widespread in wooded mountainous areas of the region during the breeding season, and extending into lower pinyon–juniper woodlands during winter. Present and probably breeding in all the montane parks.

Habitats and Ecology: Forested mountain slopes that provide snow-free areas for nesting on or near the ground, and which also offer sources of berries for food, are favored for nesting. In the winter the birds feed almost entirely on juniper or similar kinds of berries, but while breeding the usual thrush diet of insects is the most important source of food.

Seasonality: Essentially resident throughout the region, although there are marked seasonal movements into lower canyons or prairie areas where winter foods are available. Colorado egg records are from May 16 to July 10, and in Montana active nest records extend from June 2 to July 23.

Comments: This is one of the most unusual of thrushes, partly because of its flycatcher-like manner of catching insects, and partly because of its close association with berries (juniper berries in this region) during the winter months. It is also of interest for its beautiful territorial song, which is sometimes uttered in flight, and is followed by a plunging flight back to earth.

Suggested Reading: Dawson, 1919; Lederer, 1977; Salomonson & Balda, 1977.

LATILONG STATUS

R	R	S	
s	r	S	S
R	R	R	r

R	R	r	r
s	R	R	M
s	R	R	R

r	r	r	r
M	r	R	R
M	s	S	R

Veery (*Catharus fuscescens*)

Identification: This forest-dwelling thrush is rather uniformly rusty brown on the underparts, mostly whitish below, with indistinct breast spotting and with a buffy eye-ring. It is quite similar to the hermit thrush, but is less distinctly spotted on the breast and is less olive-colored on the back. Its song is a distinctive series of fluty *veery* notes that gradually descend the scale.

LATILONG STATUS

s	S	s	s
s	s	S	s
S	S	S	s

S	S	S	s
s	S	S	M
s	s	M	

M	s	s	M
M	s	S	s
	S	S	s

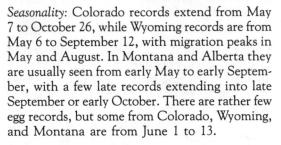

Status: Widespread in wooded areas of the region, especially near water. Variably common in the montane parks, becoming less common in the northern ones, and also rather rare in Rocky Mountain N.P.

Habitats and Ecology: In this region the favored habitats consist of wooded river valleys and canyons that range from deciduous gallery forests along prairie areas of Alberta, through aspen forests of the foothills, and willow-lined mountain streams up to about 8,000 feet at the southern end of the region. Areas with heavy and thickety undergrowth that are difficult for humans to penetrate are this species' favorite habitats, and most of its foraging is done on the ground.

Seasonality: Colorado records extend from May 7 to October 26, while Wyoming records are from May 6 to September 12, with migration peaks in May and August. In Montana and Alberta they are usually seen from early May to early September, with a few late records extending into late September or early October. There are rather few egg records, but some from Colorado, Wyoming, and Montana are from June 1 to 13.

Comments: This is one of the finest singers of the western thrushes, and its liquid song reminds some of a ball bearing rolling down a funnel. Their nests are usually well hidden at or near ground level, but in spite of this are often found by and parasitized by brown-headed cowbirds.

Suggested Reading: Bertin, 1977; Day, 1953; Dilger, 1956.

Gray-cheeked Thrush (*Catharus minimus*)

Identification: This forest-dwelling thrush is dull-colored, without any rusty tones, and with a head that is mostly grayish brown, without definite pale eye-ring markings. The breast is heavily spotted, very much like that of the Swainson's thrush, and its song is somewhat similar as well, but is a descending series of *wee-a* notes that sometimes rises sharply at the end.

Status: A rare migrant in the region, primarily east of the mountains. Reported as a summer vagrant at Banff N.P.; the nearest known breeding areas are extreme northern Saskatchewan or possibly northern Alberta.

LATILONG STATUS

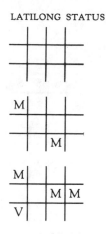

Habitats and Ecology: In our region these birds are likely to be seen foraging on or near the ground in almost any dense woodland, but their breeding habitats are typically scrub willows, alders, and dwarf birches near arctic timberline. While on migration the birds are often seen in association with Swainson's thrushes or other forest thrushes; at this time they are relatively quiet and elusive, as they move about through the shaded woodland floor searching for ground-dwelling insects and worms.

Seasonality: There are few records for this region, but during migration the birds are likely to be present at the same time as such commoner species as the Swainson's thrush and hermit thrush.

RARE MIGRANT

Comments: Although this species breeds all the way to the Bering coast of western Alaska, its migration route is almost exclusively east of the continental divide, and thus it is of only accidental occurrence in Idaho and western Colorado.

Suggested Reading: Bent, 1968; Wallace, 1939.

Swainson's Thrush (*Catharus ustulatus*)

Identification: This forest-adapted thrush is another ground-foraging species similar to the preceding one, but has a more buffy-colored face, with a definite buffy eye-ring, and with a rather buffy-toned breast with darker brown spotting. At least in this region the back coloration lacks any rusty tones such as occur in the veery and hermit thrush. The usual song is a spiralling series of whistles ascending in pitch, just the opposite of a veery's.

LATILONG STATUS

S	S	S	
s	S	S	S
S	S	S	s

S	S	s	s
s	S	S	
S	S	s	M

s		s	s
	s	M	M
M	s	s	S

Status: Widespread during summer through the forested areas of the region, and relatively common in the montane parks, probably breeding in all of them.

Habitats and Ecology: On migration these birds are likely to be found in almost any fairly dense woodlands, but during the breeding season the birds are likely to be found at higher and cooler elevations, where shaded canyons occur but where there are also fairly large areas of tangled brushy undergrowth, permitting ground-level foraging. Riparian thickets, often of willows or alders, and moist mountain slopes supporting aspens, are usually used for nesting in this region.

Seasonality: Colorado records extend from April 29 to October 29, with a few winter records. Wyoming records are from May 9 to October 20, with peaks in May and September. In Alberta the birds usually arrive the second week of May, and are generally gone by the end of October. Colorado egg records are from June 22 to August 1, and in Montana and Alberta there are egg records from mid-June onward, and recently fledged young observed from mid- to late July.

Comments: This is one of the commonest of the forest thrushes of the region, and it is somewhat more arboreal than the other *Catharus* thrushes. Likewise the nests are usually elevated from 2 to 20 feet above ground, and are often placed in rather small trees near water.

Suggested Reading: Morse, 1972; Sealy, 1974; Dilger, 1956.

Hermit Thrush (*Catharus guttatus*)

Identification: Hermit thrushes are perhaps best identified by their relatively rusty tails, which contrast with relatively non-rusty back coloration. They also have heavily spotted breasts, and fairly conspicuous pale buffy eye-rings. Their song is complex and varied; a fluty series of phrases that are often repeated; typically it consists of three phrases with long pauses between each phrase and with each phrase considerably higher in pitch than the one before.

Status: A summer resident in wooded areas almost throughout the region; present in all the montane parks and probably breeding in all of them.

Habitats and Ecology: Moist woodlands, especially of coniferous or mixed hardwoods and conifers, are preferred for breeding. Spruces, ponderosa pines, and higher zones of coniferous forests almost all the way to timberline are sometimes used. Shady and leaf-littered forest floors are favored for foraging, and in some areas the altitudinal range of breeding spans several thousands of feet.

LATILONG STATUS

s	S	S	
	s	s	M
S	s	S	s

s	S	s	s
s	S	s	
	S	S	S

s	s	S	s
S	s	S	S
S	S	s	S

Seasonality: Colorado records extend from May 2 to November 11, and in Wyoming the range is from April 19 to November 27, with migration peaks in May and October. In Montana and southern Alberta the birds arrive late April or early May, and they sometimes remain until early October. Colorado egg records are from May 14 to July 11, while in Montana egg records extend to July 16.

Comments: The hermit thrush is much more often heard than seen, but nevertheless the species' magnificent song is likely to be more memorable than a fleeting sight of a fleeing bird. Nests are typically on the ground, nearly sunk out of sight in deep mosses, but sometimes also are placed on the lower limbs of trees.

Suggested Reading: Bent, 1965; Pettingill, 1930; Morse, 1972; Sealy, 1974.

306

American Robin (*Turdus migratorius*)

Identification: This is such a familiar bird that description of the adult seems superfluous, but juveniles are heavily spotted on the breast and somewhat resemble other thrushes, and western birds are generally duller than those from more eastern parts of the continent. The song, a repeated *cheerio* of about three or four phrases that may be repeated several times, is fairly distinctive.

LATILONG STATUS

R	R	S	s
S	R	S	S
R	R	R	S

R	R	S	R
S	R	R	R
R	R	R	R

R	R	R	R
M	R	R	R
R	R	R	R

Status: Widespread throughout the entire region, especially in open woodland areas. Fairly common to abundant in all the montane parks, and breeding in all.

Habitats and Ecology: Open woodlands, whether natural or artificial, such as suburbs, city parks, farmsteads, etc., are typical habitats, but the birds tend to occur almost anywhere there are at least scattered trees and soft ground suitable for probing for insects and worms, and where mud can be gathered for the nest. Nesting on human-made structures seems to be preferred over natural nest sites such as trees, at least in protected areas.

Seasonality: Resident as far north as Colorado, but with seasonal movements evident. In Wyoming migrations are evident in April and again in September or October, while in Montana the range of records is from February to November. In Alberta April and September are also major months of migration, with a few birds sometimes attempting to overwinter. Colorado egg records are from May 15 to July 5, and in Wyoming there are egg records from May 10 to mid-July. In Montana nestlings have been noted from late May to mid-August.

Suggested Reading: Howell, 1942; Bent, 1949.

Varied Thrush (*Ixoreus naevius*)

Identification: This thrush is nearly the same size and shape as the American robin, and the adult male has a somewhat similar rusty red breast, but the rusty area is divided by a black band, and there is also a rusty eye-stripe and rusty wing-bars. Females are much duller in appearance, but show the same general patterning as the male, except for a much fainter dark breast band. The song is an unusual series of phrases on different pitches, with rapid trills that tend to be loudest in the middle, and with definite pauses between the phrases.

Status: A permanent or summer resident in northern and western portions of the region, south to about west-central Montana. A common breeder in the montane parks as far south as Glacier N.P., but only a vagrant farther south.

LATILONG STATUS

S	S	R	
s	s	s	M
S	s	s	M

M			M
	V		
	V		

		V	
			V
V			W

Habitats and Ecology: In this region the varied thrush is associated with mature coniferous forests, especially rather wet forests that have completely shaded floors and a relatively open understory vegetation that permits ground foraging. In the winter the birds turn to berries and fruits such as those of mountain ash and Russian olive, as well as frozen apples.

Seasonality: A permanent resident as far north as southern Idaho, and an early migrant elsewhere, with birds usually arriving in mid- or northern Idaho in mid-March, and remaining until late October. Hatched young have been seen in Idaho as early as May 26, and eggs reported as late as August 7. Active nests have been noted in Montana as early as mid-June, and fledged young as early as June 25.

Suggested Reading: Martin, 1970; Bent, 1968.

Gray Catbird (*Dumatella carolinensis*)

Identification: This mostly grayish bird has a blackish cap and dark reddish brown under tail coverts. It has a distinctive cat-like *meow* call, and often flicks its tail. It is usually found close to heavy cover. Its song is a highly variable mixture of squeaky notes, nasal sounds, and more melodious phrases.

LATILONG STATUS

S	S	s	s
s	S	s	s
S	S	S	S

S	S	s	S
s	S	S	S
s	s	S	S

s	s	S	M
s	s	M	S
M	S	S	S

Status: Present in wooded habitats nearly throughout the entire region, but rarer to the north, and absent or only a vagrant in the Alberta montane parks. Breeding probably occurs in all the U.S. montane parks.

Habitats and Ecology: Dense thickets, ranging from riverine forests or prairie coulees, city parks and suburbs, orchards, woodland edges, shrubby marsh borders, and similar overgrown areas that provide a combination of dense vegetation and "edge" situations are the ideal habitats of this species. Coniferous forests are avoided, although aspen groves are used, as are other natural vegetational habitats that offer rich sources of insects and berries.

Seasonality: Colorado records extend from May 8 to November 13, while Wyoming records are from May 11 to October 2. In Montana and southern Alberta the birds usually arrive in late May and leave in September. Colorado egg records are from June 9 to July 3, and those from Wyoming are from June 5 to 30. In Montana active nests have been seen as late as the end of July.

Comments: A member of the "mimic-thrush" group, this species sometimes incorporates the songs of other species into its songs, although not to the extent true of the mockingbird.

Suggested Reading: Harcus, 1973; Laskey, 1962; Adkisson, 1966.

Northern Mockingbird (*Mimus polyglottos*)

Identification: This species is mostly grayish above and white below, with white outer tail feathers and white wing patches that flash in flight. Flying birds sometimes closely resemble shrikes, but shrikes always have black facial masks that mockingbirds lack. In spring, the complex song of the male, that usually imitates those of other species, and usually consists of phrases that are repeated several times, is fairly distinctive.

Status: A local permanent resident in northern Colorado and a highly local and infrequent breeder in eastern Alberta; in other areas a migrant or vagrant, primarily east of the mountains. Absent from the montane parks except for Rocky Mountain N.P., where it is a vagrant. Local and irregular in southern Alberta, which is substantially north of its primary breeding range.

LATILONG STATUS

	M	?	
V	M		
M		M	M
M		M	
	M		M
		M	M
M			M
s	s	s	R

Seasonality: A local permanent resident in Colorado, and farther north variably migratory, with Wyoming records from April 30 to November 17, and migration peaks in May and September. Too irregular farther north to judge seasonally.

Comments: This is essentially a southern species which has extended its range considerably in recent decades, perhaps in part because of its attraction to human habitations and associated protection and feeding. The species is highly territorial, and its loud and persistent singing during the breeding season is a reflection of this fact.

Suggested Reading: Michener & Michener, 1935; Laskey, 1962; Adkisson, 1966.

310

Sage Thrasher (*Oreoscoptes montanus*)

Identification: This species is rarely seen far from sagebrush habitats, and in that environment it is unique in its heavily spotted white breast, its pale wing-bars, and its yellowish eyes. In shape and color it is rather thrush-like, but lacks rusty tones, and has a white-cornered tail. The song is a prolonged series of warbling phrases.

LATILONG STATUS

M			
	M		
V	M	M	s

S	s	s	s
s	S	S	S
S	S	s	S

s	S	S	S
S	S	S	S
S	S	S	S

Status: A local summer resident in sagebrush habitats almost throughout the region, north as far as north-central Montana. Largely limited to lower altitudes, and absent or rare in the montane parks except Rocky Mountain N.P., where uncommon and not known to breed, and Yellowstone N.P., where a rare breeder.

Habitats and Ecology: This species is closely associated with sage-dominated grasslands and to a much lesser extent other shrublands dominated by shrubs of similar growth-forms such as rabbitbrush and greasewood. Most foraging is done on the ground, but nesting is done in the shrubs. A greater array of shrublands is used in other seasons.

Seasonality: Colorado records are from April 1 to October 24, while Wyoming records extend from March 25 to October 3, with peaks in April and September. Wyoming egg records are from May 17 to mid-July, and in Colorado nestlings have been observed as early as May 13.

Comments: The little that has been written on this species suggests that it is much like the other North American thrashers, with pairs sometimes remaining mated during successive years, and showing a high level of territorial activities.

Suggested Reading: Reynolds & Rich, 1978; Killpack, 1970; Bent, 1948.

Brown Thrasher (*Toxostoma rufum*)

Identification: This fox-colored bird resembles a long-tailed and slim-bodied thrush, which is heavily streaked with dark brown below. It is usually found in woodland edge situations, in similar habitats to those of gray catbirds, and its song is somewhat like that of the catbird, but is more melodious and varied, the phrases usually being repeated several times in mockingbird-fashion.

Status: A local summer resident east of the mountains throughout the entire region; present in the montane parks only as rare migrants or vagrants.

Habitats and Ecology: Associated with open, brushy woodlands, scattered clumps of woodland in open environments, shelterbelts, woodlots, and shrubby residential areas. In prairie areas the birds are mostly confined to shrubby coulees or to riparian forests that provide sources of berries as well as foraging locations in open grasslands.

Seasonality: Winters regularly as far north as Colorado, and occasionally to Wyoming, where most records are from February 3 to September. In Montana and Alberta the birds typically arrive in May, and most are gone by September. Colorado egg dates are from May 13 to July 31, while egg and nest records for Wyoming and Colorado are for June.

Comments: Catbirds and brown thrashers have very similar territorial requirements, and these birds often fight over territories, with the larger thrashers usually evicting the catbirds. At least in some areas the birds are regularly double-brooded, with the birds sometimes changing mates between broods.

Suggested Reading: Erwin, 1935; Bent, 1948.

LATILONG STATUS

		?
V		s
V		

	s	S
		S
		S

	s	S
M	M	S
s	M	S

Water Pipit (*Anthus spinoletta*)

Identification: Like the following species, this is a ground-dwelling, brownish bird that somewhat resembles a large sparrow, but has a much sharper and weaker bill. Both species have white-edged tails and somewhat spotted breasts, but the water pipit has a more uniformly dark back plumage and an unstreaked crown. A *pip-it* call is often uttered in flight.

LATILONG STATUS

s	s	S	M
M	M	M	M
S	s	S	M

S	S	S	M
S	S	S	
M	S	S	M

s		M	M
M	M	S	M
M	S	S	S

Status: A breeding summer resident in high montane areas over much of the region, and present on open grasslands or beaches during migration. Present in all the montane parks and probably breeding in all.

Habitats and Ecology: During the breeding season this species is found on alpine tundra and high meadows, while at other seasons it occurs on similar very open terrain, usually with only sparse vegetation, and often a moist substrate. Shorelines, flooded fields, river edges, and similar habitats are commonly used by migrants or wintering birds.

Seasonality: In Colorado some birds regularly winter on the plains, while Wyoming records are from March 22 to November 6, with peaks in April or May and October. By mid-April the birds have arrived in Montana and southern Alberta, and they usually remain in those regions until early October. Colorado egg records are from June 22 to July 23, while Wyoming egg records are from June 15 to July 13. Active nests in Montana have been noted from July 4 to August 18.

Comments: In Colorado, nests of water pipits have been noted at elevations of from 11,500 to 14,000 feet, making it one of the highest nesters of all of the region's birds. At these elevations the breeding season is very short, and the families begin to move to lower elevations shortly after the young have fledged.

Suggested Reading: Pickwell, 1947; Verbeck, 1970.

Sprague's Pipit (*Anthus spraguei*)

Identification: Very similar to the preceding species, but closely associated with prairie grasslands, and with a distinctly striped crown and a buff-streaked back that appears rather scaly in pattern. A flight song is uttered on the breeding ground that is highly musical and descending in pitch, and is frequently repeated.

Status: A summer resident on native grasslands east of the mountains in Alberta and Montana, and a rather rare migrant in plains areas farther south. Absent or at most a rare vagrant in the montane parks.

LATILONG STATUS

Habitats and Ecology: Associated during the breeding season with native grasslands of only short to moderate stature. Breeding also occurs in alkaline meadows and around the edges of alkaline lakes. Outside the breeding season the birds are also associated with grassy habitats.

Seasonality: Wyoming records are from April 17 to September 20, and Montana records extend from early May to late August. Nests in Montana have been seen from May 29 to July, and in North Dakota egg dates range from June 7 to 30, but nestlings have been observed as late as August 2.

Comments: This is one of the grassland endemics of central North America, and like many grassland species it has a spectacular song-flight display that helps to compensate for a generally inconspicuous plumage pattern. During aerial display the white outer tail feathers are spread conspicuously, and the bird flies in a circular manner around its territory in full song.

Suggested Reading: Bent, 1950.

Bohemian Waxwing (*Bombycilla garrulus*)

Identification: Like the following species, this waxwing has a distinctive olive-yellow general plumage color, a short crest, a black mask, and a yellow-tipped tail. Unlike the cedar waxwing, this species has rusty brown under tail coverts, yellow and white markings on the wings, and a somewhat larger body size. Both species utter almost constant high-pitched hissing or trilling sounds, but that of the Bohemian waxwing is lower-pitched and more rasping.

LATILONG STATUS

r	W	R	
	W	W	W
S	W	W	W

W	W		W
W	W	W	W
W	W	W	W

W	W	W	W
		W	W
W	W	W	W

Status: A breeding summer resident in montane forests as far south as Glacier N.P., and a migrant or winter resident farther south, including the other montane parks.

Habitats and Ecology: During the breeding season this species is associated with coniferous and mixed forests, often nesting in loosely associated groups in conifer groves. Outside the breeding season the birds move about opportunistically, seeking out sources of berries and small fruits in trees and hedges, such as mountain ash (rowan), crabapples, pyracantha, and the like.

Seasonality: In Alberta the birds are present year-around, and egg records extend from May 24 to June 13, with the majority of records between May 29 and June 6. Except where they breed in northern Montana the birds are mainly winter residents, present from October to April. In Wyoming and Colorado they are typically present from early November to April.

Comments: The winter distribution of this species is quite variable, and is probably determined by the severity of winter weather and especially the availability of berries in more northerly areas.

Suggested Reading: Bent, 1965.

Cedar Waxwing (*Bombycilla cedrorum*)

Identification: Very similar in appearance to the preceding species, but without white and yellow wing markings, and lacking the distinctively rusty under tail covert coloration. The birds are distinctly smaller than Bohemian waxwings, and their voices are higher-pitched.

Status: Widespread over the wooded areas of the region, present in all the montane parks and probably breeding in all of them except Rocky Mountain N.P., where apparently only a rare summer visitor.

Habitats and Ecology: Somewhat open woodlands, primarily of broadleaved species, are used for nesting, including riparian forests, farmsteads, parks, cedar groves, shelterbelts, and brushy edges of forests. Areas that have abundant growths of berry-bearing bushes are especially favored, although insects, buds, and other food sources are also consumed.

Seasonality: Locally resident as far north as Colorado, although more common in winter than in summer. In Wyoming the records are from early May to November, and they usually do not arrive in central Alberta until early June. Most have left by the end of September, but some occasionally winter with flocks of Bohemian waxwings. Montana egg records are from June 11 to August 19, and similar dates seem to apply to Alberta.

Comments: Like the Bohemian waxwing, this is a highly social species, and nesting occurs in somewhat clumped patterns, with breeding seemingly timed to coincide with the period of maximum berry and fruit availability.

Suggested Reading: Lea, 1942; Bent, 1950; Putnam, 1949.

LATILONG STATUS

S	S	S	s
s	S	s	S.
S	S	S	S

S	S	s	S
M	s	S	S
M	r	s	S

s	M	M	M
M	M	M	M
r	s	s	R

Northern Shrike (*Lanius excubitor*)

Identification: Like the following species, this predatory songbird has a mostly grayish body color, with a black mask, a white-edged black tail, and black wings with a large white patch at the base of the primary flight feathers. Young birds of both species are less distinctly patterned. Separation of the northern from the loggerhead shrike is difficult, but this species is generally paler, especially on the head and back, and has a narrower black mask that does not extend forward above the beak, which is longer and heavier than that of the loggerhead. Mainly present during winter months.

LATILONG STATUS

W	W	M	
W	W	W	M
W	W	W	W

W	W	M	W
M	W	W	W
	W	W	W

W	W	W	W
	W	W	W
W	W	M	W

WINTERING MIGRANT

Status: A winter resident or migrant throughout the region, mainly at lower elevations. Reported from all the montane parks, but generally uncommon to rare in all.

Habitats and Ecology: Invariably associated with open landscapes such as agricultural lands or grasslands that have scattered observation points such as fence posts or small trees. Reported as high as 9,500 feet in Colorado, but primarily occurring at lower elevations. Usually solitary, and sometimes seen where small birds such as sparrows are likely to gather. The nearest breeding area is in northeastern Alberta near Lake Athabaska.

Seasonality: Colorado records are from October 11 to June 1, and Wyoming records extend from October to April. Likewise, Montana records span a similar period, while in southern Alberta the birds typically migrate through in October and again in March and April, rarely wintering as far north as Edmonton.

Comments: Like the loggerhead shrike, this species is an effective predator, but tends to concentrate on small prey such as insects, small rodents, and sparrow-sized birds. Summer foods are primarily insects; only in winter are birds and small mammals regularly killed.

Suggested Reading: Cade, 1962, 1967; Miller, 1931.

Loggerhead Shrike (*Lanius ludovicianus*)

Identification: Very similar to the preceding species, but slightly smaller, considerably darker on the head and upperparts, and with a broader dark mask through the eye that extends forward above the bill and is often only slightly margined with white on the upper edge. Both species are found in open country, but this one is most likely to be seen from spring through fall, while the northern is more prone to be seen during the worst winter months.

Status: A summer resident and migrant nearly throughout the region, except for the mountainous portions of Alberta and northern Idaho, where rare or accidental. Probably breeds in most or all of the U.S. montane parks of the region, although specific records are lacking.

LATILONG STATUS

M	M	s	s
	s	M	s
s	M	s	?

S	?	S	S
M	s	S	S
s	s	s	S

s	S	s	S
S	s	S	S
S	S	s	S

Habitats and Ecology: Like the northern shrike, this species is associated with open habitats having scattered perching sites, and ranges altitudinally from agricultural lands on the prairies to montane meadows. Sagebrush areas, desert scrub, and pinyon–juniper woodlands offer ideal nesting and foraging areas, but some nesting also occurs in woodland edge situations, farmlands, and similar sites.

Seasonality: Colorado records are from March 8 to November 20, with birds occasionally wintering. Wyoming records are from mid-April to mid-October, with rare wintering. Montana and Alberta records are from late April to about September, generally overlapping only very slightly with periods of northern shrike occurrence. Wyoming egg records are from May 29 to June 12, and in Montana and Alberta active nests have been noted from mid-June to late July.

Suggested Reading: Linsdale, 1938; Smith, 1973; Wemmer, 1969; Porter et al., 1975.

318

European Starling (*Sturnus vulgaris*)

Identification: Starlings might be easily confused with other "blackbirds," but have much slimmer and more pointed beaks, which are variably yellow in adults. They also have much more pointed wings, a feature very evident in flying birds, and a very rapid wing-beat. The plumage varies greatly with age and season, but during the summer months is mostly iridescent over most of the head and body, and during the winter the body plumage is variably tipped with white spotting.

LATILONG STATUS

R	R	R	s
S	R	s	S
R	R	R	R

R	R	S	R
S	R	R	R
R	R	R	R

R	M	R	R
r	R	R	R
R	R	R	R

Status: Present throughout the entire region, usually as a year-round resident, although substantial migration does occur. A fairly common breeder in all the montane parks.

Habitats and Ecology: Largely associated with humans, and most abundant in cities, farm areas, and suburbs, but also utilizing natural woodlands with woodpecker holes or other nest sites, such as aspen groves, where it competes effectively with native hole-nesting birds.

Seasonality: Present year-around throughout the region, although in Alberta only relatively few birds actually overwinter. Colorado egg records are for May, as are available records from Montana, but in adjacent North Dakota breeding records extend from mid-April to early July. Probably double-brooded in this region.

Comments: This introduced species has caused a great deal of damage to native species such as bluebirds, which it manages to dominate and exclude from their nesting holes. It also has caused a great deal of damage to agricultural interests, and as such is one of the least desirable of all songbirds.

Suggested Reading: Planck, 1967; Dunnett, 1955; Kessel, 1957.

Solitary Vireo (*Vireo solitarius*)

Identification: This species is the only vireo of the region that has distinctive white wing-bars and a white eye-ring (rather than an eye-stripe), and furthermore is relatively large, with a bluish gray upperpart coloration and a white throat. Its song consists of a series of slurred, deliberate phrases similar to those of an American robin, usually with 2 or 3 notes per phrase.

Status: A summer resident in forested areas over most of the region, occurring in all the montane parks and known to breed in several, but apparently rare in Yellowstone and Grand Teton N.P.

Habitats and Ecology: Open, coniferous or mixed forest with considerable undergrowth seem to be this species' favored habitat, especially those that offer open branches for foraging at low to medium tree levels. Fairly dry and warm forests are favored over moist and cool ones, and breeding extends from the open oak or aspen and ponderosa pine zones upward through the lower coniferous communities, to about as high as 8,000 feet in Colorado. The nests are usually in the lower branches of pines or oaks, often only a few feet above ground level.

Seasonality: Colorado records are from April 21 to October 29, while those from Wyoming are from May 7 to October 16, with peaks in May and September. In Montana and southern Alberta the birds arrive in early to mid-May, and are usually gone by mid-September. Colorado egg records are from May 30 to June 28, while in Wyoming and Montana active nests have been observed from mid-June to mid-July.

Suggested Reading: Barlow & Williams, 1970; Barclay, 1977; James, 1973.

LATILONG STATUS

S	S	s	
s	S	s	S
S	s	S	s

s	s		?
M	s	s	
	M		

M		M	s
	s	s	M
S	s	S	S

Warbling Vireo (*Vireo gilvus*)

Identification: This inconspicuous vireo is more often heard than seen, for it usually forages at considerable height in tall trees. It is mostly grayish green above, without wing-bars and with only an indistinct pale eye-stripe, which is not outlined in black. Its song is distinctively cadenced, a long warble that is regularly accented and typically ends on a strong emphatic note.

LATILONG STATUS

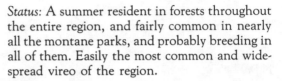

S	S	S	s
s	S	S	S
S	S	S	S

S	S	s	s
s	S	s	M
s	S	S	S

S	s	S	s
	S	S	S
s	S	s	S

Status: A summer resident in forests throughout the entire region, and fairly common in nearly all the montane parks, and probably breeding in all of them. Easily the most common and widespread vireo of the region.

Habitats and Ecology: Fairly open woodlands, especially of deciduous trees, are favored by this species. It is probably most common along riparian forests supporting tall trees, but also occurs in aspen groves, well wooded residential or park areas, especially where tall cottonwoods are present. In coniferous forest areas the birds favor areas where single or clumped broadleaved trees such as aspens or birches occur, and foraging is done near the crowns of fairly densely leaved trees, and nests are sometimes located as high as 90 feet above ground in very tall forests.

Seasonality: Colorado records extend from May 8 to October 1, and Wyoming records are from May 12 to October 28, with peaks in late May and September. In Montana and southern Alberta the birds usually arrive in mid-May, and depart in early to mid-September. Nest records in Colorado are from June 15 to July 29, while in Wyoming and Montana egg records are from June 15 to July 13.

Suggested Reading: James, 1976; Sutton, 1949; Dunham, 1964; Grinnell & Storer, 1924.

Philadelphia Vireo (*Vireo philadelphicus*)

Identification: A rather rare vireo of the region, limited to the area east of the mountains during the breeding season, and somewhat resembling a warbling vireo, but much more yellowish on the breast and underparts, and with a distinct black lower border to the white eye-stripe. The song is similar to that of the red-eyed vireo but is slower, higher-pitched, and consists of a series of mostly two-noted phrases with rather long pauses between phrases.

Status: A local summer resident in central Alberta south to about Cold Lake and Sundre. Otherwise a spring and fall migrant east of the mountains; reported as a vagrant in Banff, Jasper, and Yoho N.P.

LATILONG STATUS

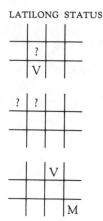

Habitats and Ecology: Breeding habitats of this species include open deciduous woodlands such as regrowth areas of aspens and poplars, or secondary birch–poplar communities that have developed following logging of coniferous forests. The birds also occur in muskeg areas having willow or alder thickets around their edges.

Seasonality: There are few migration records for the area south of Canada, but probably the migration schedule is similar to that of the other vireos. In Alberta the birds arrive on their breeding grounds in late May and begin to move south the second half of August, with only stragglers remaining after the first week of September.

Comments: Like all vireos, this species is protectively colored to match the green to yellow-green half-shaded environment in which it is found, and in spite of the male's loud songs it may prove a frustrating job to actually see the bird among the foliage around it. Its small size and distinctly greenish yellow color makes this species even more difficult to see than the other vireos of the region, and its nests are usually very well camouflaged with lichens, birch bark, or similar materials.

Suggested Reading: Barlow & Rice, 1977; Bent, 1950.

Red-eyed Vireo (*Vireo olivaceous*)

Identification: This is a fairly large but still inconspicuous vireo, without white wing-bars or eye-ring, but with a definite white eye-stripe bounded above and below with black. Its song is distinctive, a nearly continuous series of robin-like phrases given in a querulous manner, usually of three-noted series ("Got any eggs? Don't have any eggs . . ."), and endlessly repeated.

LATILONG STATUS

S	S	S	
s	S	S	s
S	S	S	s

S	s	s	s
M	s	M	S
	s		

M		M	M
	s	M	M
V	V		S

Status: A summer resident over the northern half of the region, and local farther south as a breeder or migrant. Largely limited to deciduous areas, and thus rare or occasional in most of the montane parks except Glacier N.P., where a common breeder.

Habitats and Ecology: This species is primarily associated with deciduous forests, especially those with semiopen canopies, and in the Rocky Mountain region is largely limited to broadleaved riparian forests in prairie areas, or to planted areas such as city parks and farmsteads, as well as aspen groves or poplars growing among conifers.

Seasonality: Colorado records are from May 8 to September 12, while in Wyoming the records extend from late May to November 2, and in Montana from mid-May to early September. In southern Alberta the birds usually arrive in late May, and are normally gone again by early September. Colorado egg dates are from June 22 to July 7, while Montana dates for active nests are from mid-June to early August.

Comments: This is one of the most widespread of the North American vireos and certainly one of the most frequently encountered in deciduous forests. It often nests fairly conspicuously in the horizontal forks of large trees, suspending its nest in the usual distinctive manner of vireos, and typically camouflaging it with lichens, bits of wasp nests, and similar items that are attached to spider webbing around its exterior.

Suggested Reading: Rice, 1978; Williamson, 1971; Southern, 1958; Lawrence, 1953a.

Golden-winged Warbler (*Vermivora chrysoptera*)

Identification: Males in spring plumage are easily identified by their bright yellow crown and yellow wing-patch, and their black cheek- and throat-patches. Females also have a rather large area of yellow on the wings, and a dark gray throat and facial patch. Their buzzy song consists of one *zee* note followed by three to five more on a distinctly lower pitch.

Status: An accidental vagrant over most of the region, but reported from 9 latilongs (and at least 18 times as of the late 1970s) in Colorado, including Rocky Mountain N.P., where reported in July of 1974.

LATILONG STATUS

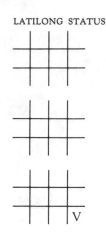

Habitats and Ecology: During the migration period, this species is likely to be found in various open habitats, but on the breeding ground it is primarily associated with forest-edge habitats having a dense undergrowth of ferns and other moisture-loving plants. It occurs in both upland and lowland habitats, including hillside thickets, overgrown pastures, brushy fields, and aspen-lined edges of forests.

Seasonality: The Colorado records are mainly for the second half of May, and again in mid-September. There are no other regional records.

Comments: This eastern species of warbler ranges from the southeastern states to southern Canada, and its nearest breeding area to the Rocky Mountains is in north-central Minnesota. In some areas the golden-winged and blue-winged warblers overlap and occasionally hybridize, producing a variety of hybrid types that are fully fertile.

ACCIDENTAL VAGRANT

Suggested Reading: Ficken & Ficken, 1962, 1968a; Murray & Gill, 1976; Eyer, 1963.

Tennessee Warbler (*Vermivora peregrina*)

Identification: This warbler is an inconspicuous, vireo-like species that is mostly yellowish green above and below, with faint pale wing-bars and a strong white eye-stripe, lined with black below. During the breeding season the underparts are mostly whitish, but they are more yellowish in the fall. The usual song is a series of loud and staccato *tsseet* notes, gradually speeding up and often ending in a trill.

LATILONG STATUS

M	S	M	
	s		
?	M		M

M			M
	M	M	M
	M	M	

		M	
	M	M	M
			M

Status: A summer resident and local breeder in the northwestern part of the region, and a variably common spring and fall migrant east of the mountains elsewhere in the region. An occasional breeder in Banff and Jasper N.P., and perhaps some of the other Canadian montane parks. Reported as summering in Clearwater National Forest (Idaho), but not proven to breed there.

Habitats and Ecology: The usual breeding habitat consists of coniferous boggy areas such as those of spruce and tamarack or white cedar, usually where sphagnum mosses are abundant. It also occurs on brushy hillsides, along forest clearings, and deciduous forests, and in Alberta favors deciduous or mixed woods that have poplars or aspens present. Foraging is done rather high up in the crown foliage, although nesting is on the ground, usually in sphagnum-covered hummocks.

Seasonality: Colorado records are from May 2 to 27, and from September 25 to October 7, while in Wyoming they are from May 12 to June 13, and from August 28 to October 5. In their Alberta breeding grounds the birds are usually present from mid-May to mid-September. Egg records for Alberta range from June 1 to 16.

Suggested Reading: Bowdish & Philipp, 1916; Bent, 1953.

Orange-crowned Warbler (*Vermivora celata*)

Identification: A dingy and nondescript warbler, with no bright markings anywhere; its generally olive-green color is marked only with a brighter yellow eye-stripe and yellow undertail coverts. Its song is a series of rather weak staccato trilled notes that typically becomes lower and slower toward the end, and somewhat similar to that of a chipping sparrow.

Status: A summer resident in most wooded areas throughout the region, at least at lower elevations. Present and probably breeding in all the montane parks, but common only in the more northerly ones.

LATILONG STATUS

s	s	S	
	S	s	s
S	s	S	M

s	s	s	M
	S	M	s
s	s	s	S

M	s	S	M
	S	M	M
V	s	s	S

Habitats and Ecology: A variety of woodland and brushy habitats are used by this species for breeding, ranging from riparian woodlands, pinyon–juniper habitats, and aspen groves. In montane areas they favor willow or alder thickets near streams, or willow thickets at treeline, while at lower elevations they tend to breed along riverine woods or in brushy vegetation surrounding beaver ponds in northern coniferous woodlands. On migration the birds are found in a wide variety of brushy or wooded habitats, but favor the brushy areas of river bottoms.

Seasonality: Colorado records are from April 19 to November 10, with a few later stragglers, while in Wyoming the species has been recorded from early May to October 25. In Montana and southern Alberta the birds arrive in early May and remain until late September or early October. There are few specific nesting records for the region, but in Colorado the birds nest from June to late July, while in Montana active nests have been seen as late as August 17.

Comments: This is a rather widespread and adaptable warbler that because of its drab color and rather weak song is much more likely to be overlooked than many of the more spectacularly plumaged birds.

Suggested Reading: Bent, 1963.

Nashville Warbler (*Vermivora ruficapilla*)

Identification: Like the other *Vermivora* species, this is a rather dull-colored species, with a rather bright yellow belly, breast, and throat, and a conspicuous white eye-ring. Females are less colorful than males, but show the same general pattern. The male's song is a series of slow and high-pitched *see-it* notes followed by a trill of rapid *ti* notes.

LATILONG STATUS

S	S	M	
s	s		
S	s	s	

M	M		M
	V		
	M		

	M	M	
V	V	V	M

Status: A local summer resident in the northwestern part of the region, south to about west-central Montana and adjacent northern Idaho. A vagrant or rare migrant in the montane parks. Reported as present in summer but of uncertain breeding status in the Cypress Hills.

Habitats and Ecology: Moderately open deciduous woods, or the deciduous portions of mixed woods, are the primary breeding habitats of this species. It seems limited to those woodlands sufficiently open to allow for the growth of shrubbery under which nesting occurs. Forest areas that allow for feeding at heights of 25–40 feet seem preferred, although some foraging in shrubbery also occurs. On migration a wider array of habitats are used, but riparian woodlands are apparently favored.

Seasonality: Colorado migration records are from April 24 to May 23, and from September 1 to November 21, while in Idaho the records extend from April 21 to September 17. In Alberta there are records from April 30 to September 3. There are few regional breeding records, but fledged young have been observed in Montana in July, while in the Cypress Hills of Alberta this species has been observed feeding a fledged cowbird in late August, suggesting local nesting.

Comments: Nesting occurs on the ground in this species, often in clumps of sphagnum mosses concealed from above by overhanging vegetation, making the nests extremely hard to locate.

Suggested Reading: Johnson, 1976; Lawrence, 1948; Bent, 1953.

Virginia's Warbler (*Vermivora virginiae*)

Identification: Another dingy *Vermivora* species, which is quite similar to an orange-crowned warbler, but has a definite white eye-ring and a yellow-tinged breast and undertail area. Its song is a rapid series of accelerating weak notes that may end on some lower notes.

Status: A local summer resident in the southernmost part of the region, including extreme southern Idaho and northwestern Colorado. A common breeder in Rocky Mountain N.P., and a probable breeder in Dinosaur N.M., but absent from the other montane parks. In Rocky Mountain N.P. nesting occurs at lower elevations of the eastern slope, such as Moraine Park.

LATILONG STATUS

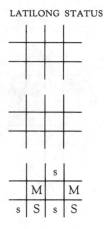

Habitats and Ecology: In Idaho this species is essentially limited to mountainsides that are covered with dense thickets of mountain mahogany; more generally it is associated with scrubby oak, open pinyon–juniper woodlands, and similar semi-arid and brush-dominated habitats. In Colorado the typical nesting habitats are where scrub oaks meet the ponderosa pine zone, at 5,000 and 7,000 feet elevation. The trees and taller shrubs provide singing and foraging posts, while nests are located at ground-level under bushes.

Seasonality: Colorado records are from April 29 to October 20, and probably a similar phenology applies to Idaho. Colorado egg records are from June 1 to 26, although nestlings have been observed as early as June 5.

Comments: The Virginia's warbler is an extremely close relative (some would say only a subspecies) of the Nashville and Colima warbler complex, and these three have non-overlapping breeding distributions.

Suggested Reading: Jackson, 1976; Bent, 1963.

Northern Parula (*Parula americana*)

Identification: Males of this species have a unique breast-band of black and red surrounded by a bright yellow throat and breast. Females also have a fairly bright yellow throat and a bluish back, and both sexes have broad white wing-bars and white eye-rings. The song is a rising, buzzy trill that ends abruptly on a lower note.

LATILONG STATUS

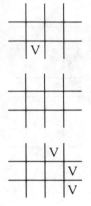

Status: A rare migrant east of the mountains in the region; accidental farther west. Absent from the montane parks except Rocky Mountain N.P., where an accidental vagrant (one record). Reported three times from Montana, from 8 Wyoming latilongs, and from 12 Colorado latilongs. The nearest breeding areas are in southern Manitoba and northern Minnesota.

Habitats and Ecology: On migration these birds are likely to be seen in riverine forests or other deciduous forest areas, but on the breeding grounds the birds are closely associated with swampy woodlands, especially those with mosslike lichens (*Usnea*) or "Spanish moss" (*Tillandsia*).

Seasonality: There are relatively few regional records, but in Colorado the species has been observed from April 3 to May 27, and from September 22 to October 24.

Comments: The nest of this species is especially interesting, and consists of a hanging mass of inconspicuous lichens, which is often further concealed by a curtain of *Usnea* lichens.

Suggested Reading: Graber et al., 1983; Bent, 1953.

RARE MIGRANT

Yellow Warbler (*Dendroica petechia*)

Identification: This is the most uniformly yellow of all the warblers, without any white present in the plumage. Males have a series of reddish brown breast streaks, while females are generally duller and more olive-yellow throughout, with only faint brownish streaking. The song is a distinctive "Tseet-tseet-tseet-sitta-sitta-see" (or "Sweet, sweet, sweet; summer's sweet").

Status: A local summer resident throughout the region in suitable habitats, including the montane parks, where abundant to uncommon, and probably breeding in all.

Habitats and Ecology: Generally moist habitats, such as riparian woodlands and brush, the brushy edges of marshes, swamps, or beaver ponds, and also drier areas including roadside thickets, hedgerows, orchards, and forest edges. A combination of open areas and dense shrubbery seem to be important for breeding, although migrant birds are rather more widely distributed.

Seasonality: Colorado records are from April 30 to October 7, and Wyoming records are from May 2 to September 14. In Montana and southern Alberta the birds normally arrive before the middle of May and leave by mid-September. Egg records in Colorado are from June 18 to July 6. In Wyoming eggs or unfledged young have been noted from June 1 to July 25.

Comments: This is one of the most widespread and abundant of North American warblers, and is often called "the wild canary" by laymen. This species is very often the victim of parasitic egg-laying by cowbirds, although it typically deals with such alien eggs by simply building a new level of nest above the old clutch and begins again.

Suggested Reading: Schrantz, 1943; Frydendall, 1967.

LATILONG STATUS

S	S	S	s
s	S	S	S
S	S	S	S

S	S	S	S
s	S	S	S
s	S	S	S

S	s	S	S
s	S	S	S
S	S	S	S

Chestnut-sided Warbler (*Dendroica pensylvanica*)

Identification: Breeding males have a bright chestnut-colored breast and a yellow crown, with otherwise white underparts and white cheeks. Females have the same general patterning but are much less colorful. The song is a distinctive whistled "Pleased, pleased, pleased to meecha," in a similar but more rapid manner than the yellow warbler.

LATILONG STATUS

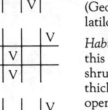

Status: A rare migrant in the region east of the mountains, and a vagrant farther west. The nearest breeding areas are in east-central Alberta and also in the front range area of Colorado (Georgetown latilong, possibly also Fort Collins latilong).

Habitats and Ecology: During the breeding season this species is generally associated with low shrubbery, forest edges and clearings, briar thickets, overgrown pastures, and similar rather open and dry areas having scattered trees and shrubs. In Alberta the birds inhabit fairly open but mature deciduous woodlands with loose understories of cranberries and dogwoods.

Seasonality: Migration records are few, but in Colorado the birds have been reported from April 29 to August 25. In central Alberta it normally arrives in mid-May, and probably normally has left by the end of August, although actual departure dates are lacking.

Comments: This is an open-country species of warbler that is generally much easier to observe than many of the other brightly colored species. It is highly active, and sometimes flies out to capture flying insects in a flycatcher-like manner.

Suggested Reading: Ficken & Ficken, 1962; Tate, 1970; Cripps, 1966.

Magnolia Warbler (*Dendroica magnolia*)

Identification: This eastern warbler is the only species that has the combination of a yellow breast and throat and white tail-patches. Males have blackish streaking on the breast, large white wing-bars, and a song that is similar to that of the yellow warbler but has fewer syllables and sounds like "Wee-o, wee-o, wee-chy."

Status: A local summer resident in the northwestern corner of the region, south to the vicinity of Banff N.P., although not yet proven to breed there. A rare migrant farther south to the east of the mountains, and a vagrant in Rocky Mountain N.P. Reported from 8 Montana latilongs, 7 Wyoming latilongs, and 11 Colorado latilongs.

LATILONG STATUS

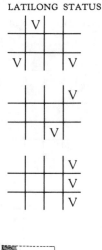

Habitats and Ecology: In Alberta breeding occurs in open coniferous and mixed forests, especially areas of young spruce and pines only about 6 to 8 feet high and which are not too dense. In some areas open coniferous bogs that are dominated by white cedars or other species are preferred, and likewise coniferous forest edges, second-growth following logging, and other habitats dominated by bush and saplings are also used.

Seasonality: Colorado records are from April 25 to May 22, and from November 13 to December 3. In Alberta the birds arrive shortly after the middle of May, and leave early in September. There are no nesting records for the region, but records from farther east suggest June nesting.

LOCAL MIGRANT

Comments: Magnolia warblers build their nests in low trees, typically small conifers, from 1 to 8 feet above ground, and they are usually well concealed and placed near the tip of a horizontal branch. Wing- and tail-spreading displays, which exhibit the white wing markings, are used for advertising the territories by males and for aggressively posturing toward other birds.

Suggested Reading: Kendeigh, 1945; Bent, 1953.

Cape May Warbler (*Dendroica tigrina*)

Identification: Males of this rare species have white wing-patches, a heavily streaked and tiger-like black-and-yellow breast, and generally a yellowish face that sets off a chestnut ear-patch. Females also have a strongly streaked breast and a somewhat yellowish face, but have little white on the wings and no chestnut on the cheeks. The male's song is a weak, thin and repeated *seet* note.

LATILONG STATUS

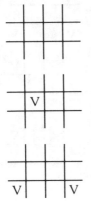

Status: A rare migrant in the easternmost parts of the region; accidental farther west, and a vagrant in the montane parks. Reported at least once in Montana and Wyoming, and from 6 Colorado latilongs. The nearest breeding area is in north-central Alberta.

Habitats and Ecology: In Alberta the nesting habitat consists of mature spruce stands in coniferous or mixed woods. Fairly open stands of tall conifers, or the edges of coniferous forests, especially if birches or hemlocks are present. On migration a much broader array of habitats are used, and they are often found in deciduous trees.

RARE MIGRANT

Seasonality: Colorado records are from May 10 to June 16. In Alberta the birds usually arrive about the middle of May, and are generally gone by mid-September. There are no specific breeding records for the region.

Comments: Cape May warblers are prone to nest very high in tall coniferous trees, and thus very little has been learned of their breeding biology and behavior. Their nests are almost invisible from the ground, and are built of mosses such as sphagnum.

Suggested Reading: Walley, 1973; Bent, 1953.

Yellow-rumped Warbler (*Dendroica coronata*)

Identification: Males of this species have a distinctive combination of a yellow rump, a yellow flank patch, a yellow crown and (usually) a yellow throat, but otherwise are mostly bluish gray above and white below. Females also have yellow rumps and yellow flank markings, but otherwise are rather dull-colored. The male's song is a slow, trilling whistle that may rise or fall at the end.

Status: A common and widespread species in wooded areas throughout the region; perhaps the commonest breeding warbler. A common to abundant breeder in all the montane parks.

LATILONG STATUS

S	S	S	
s	S	S	S
S	S	S	S

S	s	s	s
s	S	S	S
s	S	S	S

s	s	S	s
M	S	s	S
S	S	S	S

Habitats and Ecology: This species breeds in a wide array of coniferous forests, from the ponderosa pine zone upwards, and also breeds in riparian forests with conifers present. Habitats range from open, park-like ponderosa pine communities through dense montane forests to timberline species, foraging from low branches to the highest crown levels. During winter the habitats used are more varied, and include berry-eating and nectar-drinking to aerial flycatching.

Seasonality: One of the most hardy warblers, with Colorado records extending occasionally over winter. Wyoming records are from April 20 to mid-October, and in Montana and southern Alberta the birds are usually present from late April to early October. Colorado egg records are from June 19 to July 6, while in Montana and Wyoming egg records extend from June 10 to 27, with nestlings reported as early as June 5.

Comments: This species occurs as an eastern form ("myrtle warbler") that has a white throat, and a yellow-throated race ("Audubon's warbler"), a form that breeds in the Rocky Mountain region.

Suggested Reading: Morse, 1980; Hubbard, 1969; Ficken & Ficken, 1966.

334

Black-throated Gray Warbler (*Dendroica nigrescens*)

Identification: The head of this species is strongly marked with black and white (white eye-stripes and white chin-stripes), the breast is black, and the underparts are white with black flank striping. Females are similar but have mostly whitish breasts and more grayish heads; both sexes have small but distinctive yellow spots in front of the eye. The song is a *weezy-weezy-weezy-weezy-weet.*

LATILONG STATUS

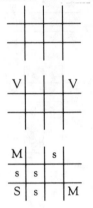

Status: Limited to the southernmost portions of the region, including southern Idaho, southwestern Wyoming (*American Birds* 35:964), and northwestern Colorado. Common only in Dinosaur N.M., and a probable breeder there.

Habitats and Ecology: During the breeding season this species is closely associated with pinyon-juniper woodlands, and in Colorado they have been found nesting at about 7,000 feet elevation. In Idaho they nest on low ridges covered by large, gnarled junipers. Elsewhere they have been found breeding in oak woodlands, and in general they seem to prefer trees with dense and stiff foliage, of relatively low stature, and in dry environments. The birds forage in dense terminal foliage, and nest at medium heights.

Seasonality: In Colorado they have been reported from April 25 to September 7. Eggs have been reported in Colorado and northern Utah from June 11 to July 19, and nestlings seen during the latter part of June in Idaho.

Comments: This is another of the southwestern chapparal species that barely reaches the region covered by this book, and is more likely to be seen at Dinosaur N.M. than in any of the montane parks.

Suggested Reading: Grinnell & Storer, 1924; Stein, 1962.

Townsend's Warbler (*Dendroica townsendi*)

Identification: This beautiful western warbler is perhaps best identified by the black (males) or gray (females) cheek patch surrounded by yellow, and with black spots extending down a yellow-tinted breast. Males have a large black throat-patch and are a brighter yellow throughout than females. The song is usually six to eight repeated notes of one pitch, followed by three or four very rapidly repeated notes on a higher or lower pitch.

Status: A summer resident in the northwestern part of the region, south at least to west-central Montana and possibly farther. Reported as an occasional breeder in Yellowstone N.P. by some early observers, but currently believed to be only a migrant. Also reported to be a confirmed breeder in Clearwater N.F., Idaho, possibly representing its southernmost limits.

LATILONG STATUS

S	S	S	
s	S	S	S
S	s	S	

M	M		
	M	M	
	M	M	M

		M	M
	M	M	M
M	M		M

Habitats and Ecology: This is a crown-level forager in tall conifers, favoring dense and mature montane forests. This general adaptation seems to be true during the non-breeding season as well as when nesting. Nesting appears to be in spruces and firs, sometimes within 15 feet of the ground, but rather few nests of this species have been described. In Alberta the nesting habitat consists of dense stands of spruce or fir, often with a stream or a willow-lined swamp nearby.

Seasonality: Colorado migration records are from May 7 to 26, and again from August 19 to October 29. Idaho records are from May 1 to September 16, and fledged young have been observed in early July. In Alberta the birds typically arrive during the last week of May, and leave by early September. Wyoming records extend from May 11 to November 13. Eggs have been reported in Montana as early as June 2, and in Washington egg records extend from May 24 to June 24.

Suggested Reading: Stein, 1962; Bent, 1953.

Black-throated Green Warbler (*Dendroica virens*)

Identification: Quite similar in appearance to the closely related Townsend's warbler, but both sexes lack dark cheek markings, and have no yellowish tinges on the underparts below the breast region. The golden cheeks and gray to black breast area provide a distinctive combination of colors. The song is a series of about four *see* notes that end with two final notes that are accented and clearer whistled notes, sounding something like "Zee-zee-zoo'-zee."

LATILONG STATUS

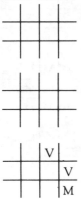

Status: A rare migrant in the eastern part of the region; accidental elsewhere, including the montane parks, where an irregular vagrant. The nearest breeding area is in northern Alberta (Athabaska drainage). Reported from 2 Montana latilongs, 3 Wyoming latilongs, and 10 Colorado latilongs.

Habitats and Ecology: Associated during the breeding season with mature and rather open coniferous or mixed forests, especially those with pines present. It also has been reported from some tall but second-growth timber and also with scattered trees in pastures or hillsides. In Alberta it seems to be associated with mature spruce stands.

Seasonality: Colorado records are from May 4 to 24, and from September 16 to November 24. In Alberta the birds usually arrive about the middle of May, and are present until about mid-September.

RARE MIGRANT

Comments: This is the ecological equivalent of the Townsend's warbler in eastern areas, and the two also are somewhat similar in voice and appearance, suggesting a common ancestry. The ranges of the two species approach one another in Alberta, but are not known to be in contact.

Suggested Reading: Mores, 1980; Stein, 1962; Pitelka, 1940; Nice & Nice, 1932.

Blackburnian Warbler (*Dendroica fusca*)

Identification: Adult males are distinctive, with brilliant orange-red throat color, and with patches of bright orange on the crown, above the eye, and behind a small black cheek-patch. There is also a large white wing-patch. Females have a yellowish throat and upper breast, a yellow eye-stripe, white wing-bars, and streaked flanks. The male's song is extremely high-pitched, and is terminated by a very high-pitched trill.

Status: A rare migrant in the eastern parts of the region; a vagrant elsewhere, including the montane parks. The nearest breeding area is probably in central Alberta or central Saskatchewan. There are records for 5 Montana latilongs, 6 Wyoming latilongs, and 9 Colorado latilongs.

LATILONG STATUS

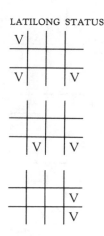

Habitats and Ecology: In Alberta breeding birds are associated with heavy stands of spruce and fir in mixed forests, while farther east the birds occur in a variety of coniferous and deciduous forest habitats. However, mature coniferous forests, especially those in swampy areas and with *Usnea* lichens present, seem to be the favored breeding habitats. In any habitat, the birds tend to forage high in the trees, and nests are typically placed very high in tall trees as well.

Seasonality: There are few regional records south of Canada, but in Colorado the birds have been reported from May 9 to June 12. In Alberta they arrive about the middle of May, and remain for an uncertain period, but probably no later than the end of September.

RARE MIGRANT

Comments: This is one of the most beautiful of warblers and one of the hardest to see at close range, since it always forages at considerable height, much like the Townsend's and black-throated green warblers.

Suggested Reading: Bent, 1953; Lawrence, 1953b.

338

Grace's Warbler (*Dendroica graciae*)

Identification: This is a rather modestly colored warbler with only a yellow to whitish eye-stripe and a yellow throat and breast to give it color; it also has white wing-bars and white underparts but is otherwise generally grayish above, with blackish spotting on the back. Its song is a rapid, staccato, musical trill, speeding up toward the end.

LATILONG STATUS

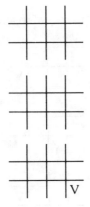

Status: An accidental vagrant in the southern part of the region; the nearest breeding areas are in southwestern Colorado. Reported during July and September in Rocky Mountain N.P.

Habitats and Ecology: Associated with the ponderosa pine and to a lesser extent the juniper zones in southern Colorado, up to about 8,500 feet. Extends into desert scrub and riparian habitats during the non-breeding season, as well as into other coniferous forest types. Elsewhere it is generally limited to sparse stands of small pines, from 15 to 30 feet tall.

Seasonality: Reported in Colorado from April 25 to September 4, and with egg records for May, but actively singing males noted as late as June.

Comments: This species is a bird of the open pine forests, where it forages in the upper levels of trees, and occasionally flies out to catch insects in a flycatcher-like manner.

Suggested Reading: Webster, 1961; Bent, 1953.

ACCIDENTAL VAGRANT

Palm Warbler (*Dendroica palmarum*)

Identification: This is a distinctly brownish warbler, with a chestnut crown, a whitish eye-stripe, yellow under tail coverts, and a somewhat yellowish throat that is streaked with brown. The birds wag their tails almost constantly, and their song is a buzzy trill much like that of a chipping sparrow, usually of six or seven notes.

Status: A rare migrant over the eastern part of the region; accidental in the western areas and the montane parks. There are at least 10 Montana sightings, records from five Wyoming latilongs, and from 14 Colorado latilongs. The nearest breeding area is in central Alberta, south to about Grande Prairie, Elk Island N.P., and Cold Lake.

LATILONG STATUS

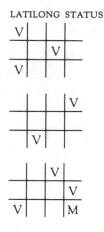

Habitats and Ecology: During the breeding season these birds are associated with dense boggy areas dominated by larch, spruce, and white cedar, and with alders, willows, and cranberry thickets present. Nesting is done on the ground, in fairly dry sphagnum mosses. Outside of the breeding season and on migration the birds are seen in various habitats such as roadside shrubbery and deciduous trees.

Seasonality: Colorado records are from May 12 to 20, and from September 16 to December 17. In Alberta the birds usually arrive during the first week of May, and remain until at least the middle of October. Egg records in Alberta are from May 30 to June 16.

RARE MIGRANT

Comments: This warbler is obviously poorly named, at least insofar as its breeding habitats are concerned, and might better have been called the "muskeg warbler" or some such name. It is also not very warbler-like in appearance, and instead has a head pattern and song somewhat reminiscent of a chipping sparrow.

Suggested Reading: Bent, 1953; Welsh, 1971; Graber et al., 1983.

Bay-breasted Warbler (*Dendroica castanea*)

Identification: Males of this species are easily identified by their chestnut crown, throat, breast, and flanks, with a contrasting buffy white neck-patch and white wing-bars. Females are much more difficult to identify, and have a small amount of chestnut on the crown, throat, and sides, but otherwise are generally dull-colored. The male's song is a *seetzy-seetzy-seetzy-see*, given in a high and weak voice.

LATILONG STATUS

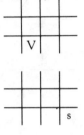

Status: A rare migrant in the eastern parts of the region; accidental in western areas and in the montane parks. The nearest breeding areas are in central Alberta, south to about Cold Lake and perhaps to Jasper N.P. (where rare).

Habitats and Ecology: During the breeding season this species is associated with coniferous forests, especially those in rather swampy areas and with birches or maples present, and also with mixed forests having clearings or edge areas. In Alberta it is found in extensive stands of mature spruce or mixed spruce, larch, and pine. It may also occur in mixed forests, but only those that are dominated by conifers.

Seasonality: Colorado records are from May 11 to 25, and from August 22 to October 1. In Alberta the birds arrive about the third week of May, and remain until about mid-September. There are apparently no regional egg records.

RARE MIGRANT

Comments: This species is a low-level forager, generally searching for insects on branches below the main tree foliage, and its nests are likewise typically placed less than 25 feet above the ground level.

Suggested Reading: Mendall, 1937; Bent, 1953; Graber et al., 1983.

Plate 27. Mountain chickadee, adult at nest. Photo by author.

Plate 28. Chestnut-backed chickadee, adult. Photo by author.

Plate 29. American dipper, juvenile. Photo by author.

Plate 30. American robin, adult at nest. Photo by Scott Johnsgard.

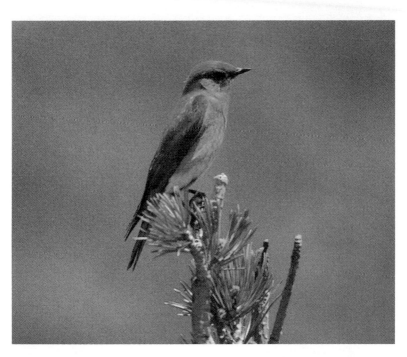

Plate 31. Mountain bluebird, adult male. Photo by author.

Plate 32. Swainson's thrush, adult at nest. Photo by Alan G. Nelson.

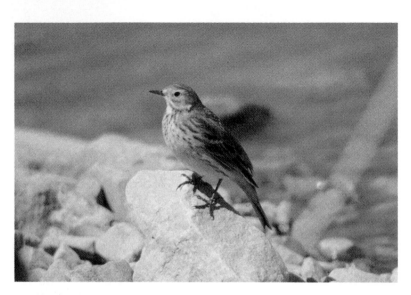

Plate 33. Water pipit, adult. Photo by author.

Plate 34. Yellow-rumped (Audubon's) warbler, adult male at nest. Photo by Kenneth Fink.

Plate 35. Western tanager, adult male. Photo by author.

Plate 36. Lazuli bunting, adult male. Photo by author.

Plate 37. Dark-eyed (Oregon) junco, adult. Photo by Kenneth Fink.

Plate 38. White-crowned sparrow, adult. Photo by author.

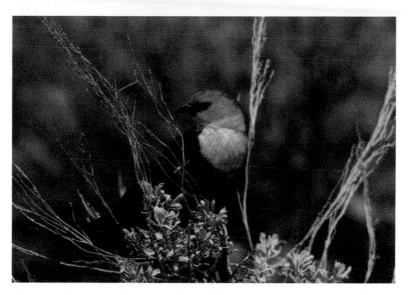

Plate 39. Yellow-headed blackbird, adult male. Photo by author.

Plate 40. Red crossbill, immature male. Photo by author.

Plate 41. Cassin's finch, adult male. Photo by Alan G. Nelson.

Plate 42. Gray-crowned rosy finch, winter male. Photo by Alan G. Nelson.

Blackpoll Warbler (*Dendroica striata*)

Identification: Males of this species are distinctively patterned in black and white, with no yellow or other colors present. The crown is black, the cheeks are white, and the flanks are heavily spotted with black and white. Females are much more drab, but have a spotted throat and flanks, white wing-bars, and white tail spots. The male's song is a very high-pitched series of notes on a single pitch and loudest in the middle.

Status: A local summer resident in the montane forest of the northwestern corner of the region, south at least through Banff N.P. and Bragg Creek; a migrant in the eastern parts of the region east of the mountains.

LATILONG STATUS

M	M		
			M
V	M	M	

M		M	M
	V		

M		M	M
		M	M
			M

Habitats and Ecology: In Alberta the favored breeding habitat consists of mixed woodland, especially spruce and aspen or alder regrowth on previously burned areas. In northern Alberta they inhabit deciduous shrubs as frequently as young coniferous growth. Outside the breeding season they occur over a much broader array of wooded habitats.

Seasonality: Migration records in Colorado are from April 29 to May 27; fall records are apparently lacking. In Alberta the birds usually arrive about the middle of May, and are gone by the end of September. Egg records seem to be lacking for the region.

Comments: Although not very colorful, this is one of the more attractive warblers, and one that can usually be seen very easily during migration, as it tends to forage low in the trees and in rather open situations, where the male's contrasting plumage can be easily observed.

Suggested Reading: Morse, 1979; Bent, 1953; Graber et al., 1983.

Black-and-white Warbler (*Mniotilta varia*)

Identification: This distinctly plumaged warbler can be readily identified by its strongly striped black and white plumage, with a strongly striped crown, black and white striping on the back, and heavy black spotting on the white breast and flanks. The male's song is a series of six to eight high, double whistling notes, with the first syllable of each doublet stressed and the second syllable lower. No other warbler crawls up and down the trees in a nuthatch-like manner.

LATILONG STATUS

Status: A local summer resident in the northeastern parts of the region, and a rare migrant east of the mountains throughout the region. An accidental vagrant in the montane parks, except for Banff, where occasional but breeding is unproven.

Habitats and Ecology: During the breeding season this species inhabits deciduous or mixed woods bordering lakes and streams, or in shrubbery around muskeg areas. It also breeds in immature or scrubby trees on hillsides or ravines, and in riverside forests in grassland areas.

Seasonality: Records in Colorado mostly extend from April 9 to November 30, with rare records to December 24, while in Wyoming they are from May 5 to 25, and from August 20 to September 23. In Montana and southern Alberta the birds are usually present from the first half of May to early September. There are no egg records for the region, but probably nesting occurs mainly during June.

Comments: Because this species forages in a nuthatch- or creeper-like manner, it is not so dependent on flying insects, and so tends to be an earlier spring and later fall migrant than the more typical warblers.

Suggested Reading: Bent, 1953; Smith, 1934; Graber et al., 1983.

American Redstart (*Setophaga ruticilla*)

Identification: Males are readily identified by their bright orange patches on their flanks, wings, and tail, and an otherwise mostly black and white color. Females exhibit yellow patches on the flank, wings, and tail where the males are orange, and both sexes often can be seen flitting about in the trees, where their wing and tail markings flash brightly. The song of the male is usually a series of 5 or 6 rapid notes with the last or last two notes strongly accented and downslurred.

Status: A relatively common summer resident in woods over most of the region, including montane forests. Present and variably common in all the montane parks, but rarer southwardly, and apparently absent from Rocky Mountain N.P. during the summer (but breeding in the nearby foothills).

LATILONG STATUS

S	S	S	
s	S	S	s
S	S	S	S

s	s	S	s
s	S	s	
M	M	s	M

M		M	S
	s	M	M
M	S		S

Habitats and Ecology: Breeding habitats of this species include moist bottomland woodlands, the margins or openings of mature forests, young or second-growth stands of various types of forests, and especially deciduous forests. The presence of nearby water and of a brush layer seem to be important habitat components.

Seasonality: Records in Colorado extend from April 30 to September 14, plus a late record of November 14. In Wyoming the records extend from April 30 to September 12, and in Montana and southern Alberta the birds are present from about the middle of May until the latter part of September. In Wyoming, Colorado, and Montana there are records of eggs or active nests from June 5 to 30, and records of nestlings for the first half of August.

Comments: Redstarts are among the most visible and active of the warblers, with almost constant singing and flying about capturing insects on the wing as frequently as foraging in the foliage.

Suggested Reading: Ficken, 1962, 1963; Bent, 1953.

Prothonotary Warbler (*Protonotaria citrea*)

Identification: Males of this species are mostly a bright golden yellow on the head and underparts, and almost uniformly dark grayish blue on the back, wings, and tail, except for white tail patches. Females are similar, but much less colorful on the head and underparts. The male's song is a series of loud, clear, ascending notes.

LATILONG STATUS

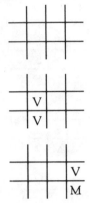

Status: An accidental vagrant or rare migrant in the region, with no records for Montana, 2 reports for Wyoming, and reports for 8 Colorado latilongs. Reported once for Yellowstone N.P. (*Auk* 49:91–2) and once for the Jackson Hole area. The nearest breeding area is in easternmost Kansas.

Habitats and Ecology: Breeding habitats consist of moist bottomland forests and wooded swamps or periodically flooded woodlands in the vicinity of running water or pools. Nesting occurs in old woodpecker holes or other natural cavities, usually over water.

Seasonality: The relatively few Colorado records are from May 13 to August 25. Egg records from Kansas are from May 11 to July 10.

Comments: This species of warbler is centered in the southeastern United States, and is the only warbler of this region that nests in tree cavities. Often these nest holes are only five or six feet above water, and frequently are old downy woodpecker holes that are lined with mosses, grasses, or other vegetation to make them suitable for nesting.

Suggested Reading: Walkinshaw, 1953; Bent, 1953; Graber et al., 1983.

Worm-eating Warbler (*Helmintheros vermivorus*)

Identification: This is the only warbler that has a strongly striped black and orange-yellow head, with clear buffy-orange underparts. There are no white wing-bars, tail-patches, or other bright colors present. The birds forage on the ground in dense vegetation. The male's song is a chipping sparrow-like series of rapid and sharp notes.

Status: A rare migrant or vagrant in the region, with no Montana records, only one for Wyoming, and 11 latilong records for Colorado. The nearest breeding region is in eastern Oklahoma.

LATILONG STATUS

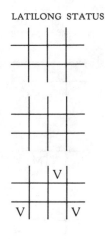

Habitats and Ecology: During the breeding season this species inhabits wooded hillsides with medium-sized deciduous trees and an undergrowth of small shrubs and saplings, particularly where there are streams or swampy areas nearby. Nesting occurs on the ground, under shrubs or sometimes under a canopy of dead leaves at the base of a tree.

Seasonality: Colorado records are from April 22 to May 7, and again for late October. There are no regional egg records.

Comments: This is an extremely poorly studied species, which is relatively elusive and often overlooked because of its song's similarity to that of a chipping sparrow.

Suggested Reading: Bent, 1953; Graber et al., 1983.

RARE MIGRANT

Ovenbird (*Seiurus aurocapillus*)

Identification: This ground-foraging warbler has a chestnut brown crown bordered with black, a white eye-ring, and with bold black spotting on the white breast and flanks. It is similar to the northern waterthrush, but that species lacks the white eye-ring and black-and-brown crown. Its song is a very loud, repeated *teacher, teacher, teacher* . . . of about 10 or 12 notes that gradually rise in volume.

LATILONG STATUS

Status: A local summer resident east of the mountains in Alberta, Montana, Wyoming, and Colorado, and a rare migrant or vagrant in the mountains, including the montane parks.

Habitats and Ecology: During the breeding season this species occupies well-drained, bottomland deciduous forests, and well-shaded and mature upland forests, especially on north-facing slopes or shady ravines. In Alberta the birds favor deciduous or mixed woods in which the undergrowth is not too dense for ground foraging, and avoid the darkest coniferous forests.

Seasonality: Colorado records extend from May 6 to October 14. In Montana and southern Alberta the birds usually arrive during the second or third week of May, and remain until the latter half of September. Colorado egg records are from June 1 to 23, and in Montana nesting has been noted in June and young observed in July.

Comments: The ovenbird is one of the most abundant warblers in eastern deciduous forests of North America, but in this region it is at the edges of its range. In many ways it acts more like a small thrush than a warbler, and indeed has the general plumage characteristics of the forest-dwelling thrushes. Like them too, it has a much louder song than do most warblers.

Suggested Reading: Hahn, 1937; Bent, 1953; Graber et al., 1983.

Northern Waterthrush (*Seiurus noveboracensis*)

Identification: This species rather closely resembles the previous one, but instead of having a white eye-ring it has a white eye-stripe, and it is uniformly dull brown on the crown. Its underparts are white with black striping, and the species forages on the ground, walking in the same manner as the ovenbird. However, it teeters and bobs its tail almost constantly while walking, something in the manner of a spotted sandpiper. Its song is loud and robust, with many notes that usually end in a downslur, sounding like "Twit-twit-twit-chee-chee."

Status: A local summer resident in the northwestern portions of the region, south to about the Wyoming border. Breeding occurs in the montane parks south at least to Glacier N.P. and perhaps to Yellowstone, although the latter is still unproven.

LATILONG STATUS

S	S	S	s
	s	s	
S	S	s	s

S	s	s	M
s	M		
	M	M	M

	M		
	M		M
V	M		M

Habitats and Ecology: During the breeding season this species inhabits woodlands with ponds, lakes and streams, especially those with brushy bogs and swampy areas of forest. Standing-water habitats are favored over those with moving streams, and in Alberta the birds are found in deciduous forests that have heavy underbrush and are often partially or recently flooded.

Seasonality: Colorado records are from April 15 to May 28, and from August 13 to October 28. Wyoming records extend from May 10 to September 13, and Montana records from mid-May to late September. The birds usually arrive in central Alberta during the third week of May, and some occasionally remain as late as early October. There are rather few regional breeding records; egg records from British Columbia are from June 5 to 25, and in North Dakota eggs have been found in mid-July.

Suggested Reading: Eaton, 1957; Bent, 1953.

348

MacGillivray's Warbler (*Oporornis tolmei*)

Identification: This is the only gray-headed warbler, and one which in addition has an incomplete white eye-ring and an underpart coloration that is mostly yellow. Foraging is done close to the ground, and the song of males is a series of *chu-weet* notes that ends in a downslurred buzzy trill.

LATILONG STATUS

S	S	S	s
s	S	S	S
S	S	S	s

S	S		s
s	S	S	S
s	S	S	S

s	s	s	s
S	S	S	s
S	S	s	S

Status: Widespread in woodlands and brushy areas throughout the region, including montane areas; relatively common and a probable breeder in all the montane parks.

Habitats and Ecology: Generally associated with brushy thickets, especially riparian woodlands. Less often it occurs in dense deciduous woods or mixed woodland on upland slopes, or in mature riverbottom forests. In Alberta the birds are usually found close to water in thick brushy growth, in prairie coulees, mountain slopes with dense shrubbery, or along forest clearings.

Seasonality: Colorado records extend from April 24 to October 8, and in Wyoming from May 7 to October 13. In Alberta they arrive on breeding areas in late May, and are mostly gone by the middle of September. Colorado egg records are from June 1 to July 5, and early June appears to be the nesting period for Wyoming. In Washington there are egg records from May 29 to June 22.

Comments: This is a close relative of and an ecological replacement form of the mourning warbler of eastern North America, and which has a breeding range reaching central Alberta, close to the eastern limits of the MacGillivray's warbler.

Suggested Reading: Griscom et al., 1957; Bent, 1953.

Common Yellowthroat (*Geothlypis trichas*)

Identification: Males of this water-loving species are uniquely patterned with a black mask through the eyes, and otherwise are mostly yellowish on the underparts. Females are quite dull, and have no trace of the mask, but do have a yellowish throat and breast. The male's song, a loud, repeated *witchity*, usually allows a person to identify the species' presence long before it is observed.

Status: A summer resident throughout the region, including montane areas, and a common to uncommon breeder in all of the montane parks except Rocky Mountain N.P., where rare.

Habitats and Ecology: Moist to wet ground, with associated vegetation such as tall grasses, shrubs, and small trees, are the primary breeding habitat, although at times the birds extend to upland thickets of shrubbery and low trees. Willow thickets around beaver ponds, the edges of muskegs, and scrub alders are among its favorite nesting areas.

Seasonality: Colorado records are from April 18 to October 4, and Wyoming records are from April 16 to October 16. In Montana and southern Alberta they are usually present from mid-May to mid- or late September. Colorado and Wyoming egg records are from June 9 to 25, and active nests in Montana have been noted from May 28 to June 29.

Comments: This species and the yellow warbler may well be the most widespread warblers of the region, but whereas the yellow warbler prefers dry thickets, this species is mostly limited to wet ones.

Suggested Reading: Stewart, 1953; Hofslund, 1959.

LATILONG STATUS

s	S	S	s
s	S	S	s
S	S	S	s

S	S	s	S
s	S	M	S
S	S	s	S

M	M	S	S
M	s	s	s
S	S	s	S

Hooded Warbler (*Wilsonia citrina*)

Identification: This is the only warbler in which the male has a yellow face bounded above by a black crown and below by a black breast, the two black areas connected by a narrow black band. The underparts are bright yellow, and the tail has white spotting. Females are similarly yellow, with white tail markings, but lack the black "hood" and breast markings. The male's call is loud and clear, and usually of repeated *weet-a* notes.

LATILONG STATUS

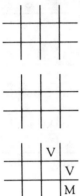

RARE MIGRANT

Status: A rare migrant or vagrant east of the mountains; accidental in the montane areas. There are no Montana records, only 3 for Wyoming, and records from 9 Colorado latilongs. The nearest breeding area is in eastern Oklahoma.

Habitats and Ecology: On its breeding grounds this species is associated with thick bottomland woods and wet, open woods, especially those of a swampy nature. Extralimital sightings in this region have been of vagrant birds usually seen in gardens, hedges, and similar suburban sites.

Seasonality: Colorado records are from April 8 to May 18. Breeding in the central Great Plains apparently occurs in late May and June.

Comments: This is clearly a species well out of its range and habitat in the Rocky Mountains, and reflects the abilities of birds to move well away from their expected localities. Most often these are immature birds, but at least some of the Colorado records are of adult males in breeding plumage.

Suggested Reading: Griscom et al., 1957; Bent, 1953; Graber et al., 1983.

Wilson's Warbler (*Wilsonia pusilla*)

Identification: This little yellow warbler is distinctively marked with a black "skullcap" in males; there is no white on the wings or tail. Females are similarly yellowish, but have only a dusky cap, and somewhat resemble a female yellow warbler. The male's song is a series of short descending, slurred and staccato *chi* notes that drop in pitch at the end.

Status: A widespread summer resident in woodlands throughout most of the region, including the montane parks, where generally common and probably breeding in all.

Habitats and Ecology: On their breeding grounds these birds inhabit willow, alder thickets along rivers or beaver ponds, brushy edges of lakeshores, the edges of mountain meadows, timberline areas of low shrubby vegetation, and sometimes aspen thickets. In Colorado they regularly breed at altitudes of more than 10,000 feet, near timberline, especially in willow thickets around high mountain lakes.

Seasonality: Colorado records are from April 15 to November 6, and Wyoming records extend from May 4 to November 10. In Montana and southern Alberta the birds are usually present from about the middle of May until the end of September or early October. Colorado egg records are from June 1 to July 3, and in Montana and Wyoming the records extend from June 6 to 27. Dependent young have been seen in late July in Banff N.P.

Comments: This is the most widespread species of the genus *Wilsonia* in North America, breeding from coast to coast, and from Alaska to northern Mexico, in a wide variety of climates and plant habitats.

Suggested Reading: Stewart, 1973; Harrison, 1971; Stewart et al., 1977.

LATILONG STATUS

s	S	S	M
s	s	S	M
S	s	s	M

s	s	M	M
s	S	S	
s	S	s	M

s	M	M	M
M	M	s	M
S	S	S	S

352

Canada Warbler (*Wilsonia canadensis*)

Identification: Males of this attractive warbler have yellow "spectacles," a bright yellow throat, and yellow underparts, the two yellow areas separated by a black "necklace." Females also have spectacle-like markings around the eyes and bright yellow underparts, but the "necklace" markings are barely apparent. The song of the male is rapid and variable, usually of repeated *ditchety* phrases, and ending with a single *chip*.

LATILONG STATUS

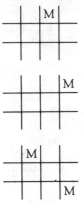

Status: A local and rare migrant east of the mountains, with two Montana records, one Wyoming record, and five Colorado records. There are June records for Banff and Jasper, but the nearest known breeding area is in the general vicinity of Lesser Slave Lake.

Habitats and Ecology: On its breeding grounds this species seeks out thick stands of willow and alder along streamsides, or dense shrubs in swampy forest areas. It also breeds in mature mixed or deciduous forests, and in the heavy undergrowth of regenerating deciduous forests. Nesting is done on the ground, often under tree roots, in a bank cavity, or in a rotted and moss-covered stump.

Seasonality: Most of the regional records are for the second half of May or the first half of June, but there are also records for the first half of September. The birds arrive on their central Alberta breeding areas in late May or early June, and are gone by early September. There are no regional egg records.

LOCAL MIGRANT

Comments: This is a ground- and low-level foraging warbler, using similar habitats to those of the MacGillivray's warbler, in considerably wetter situations.

Suggested Reading: Krause, 1965; Bent, 1953.

Yellow-breasted Chat (*Icteria virens*)

Identification: This very large warbler is nearly 8 inches long, and has conspicuous "spectacles" as well as bright yellow underparts and gray to olive-green upperparts. Its voice is extremely unusual for a warbler, consisting of a strange medley of loud and clear whistles intermixed with squeals, squawks, and other unusual notes, in seemingly random order. The birds inhabit thick underbrush, and thus are more often heard than seen.

Status: A summer resident at lower altitudes over most of the drier portions of the region, generally under 7,000 feet. A vagrant only in the montane parks south of Canada, and a possible occasional breeder at Dinosaur N.M.

LATILONG STATUS

		s	?	
		s	?	
S	M	s		

s		s	S
s			S
	M	M	S

M		S	s
	s		s
S	s	s	S

Habitats and Ecology: During the breeding season this species occurs along the shrubby coulee areas of the plains, the oak and mountain mahogany woodlands of the foothills, along alder and willow-lined creeks of the prairies, brushy forest edges, and in shrubby overgrown pasturelands.

Seasonality: Colorado records extend from May 6 to October 3, and those from Wyoming are from May 10 to September 5. Montana and Idaho records are from May 10 to early September. Egg records in Colorado are from June 6 to 28, and in Montana and Wyoming extend from June 11 to July 23.

Comments: This species is perhaps the most unusual of all the New World warblers, both in terms of its anatomy and its behavior, and perhaps should be removed from the group altogether. In its ecology it approaches the catbird and other "mimic thrushes," with a high degree of territoriality.

Suggested Reading: Dennis, 1958; Thompson & Nolan, 1973; Bent, 1953.

Hepatic Tanager (*Piranga flava*)

Identification: Males of this species are the only almost uniformly brick-red birds of the region; the male otherwise has a blackish bill and dark cheeks. The female is extremely similar to the female scarlet tanager and probably cannot be separated from it in the field. The male's song is much like that of the American robin, and is not as hoarse as that of other tanagers.

LATILONG STATUS

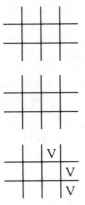

Status: A very rare migrant or accidental vagrant in the southern parts of the region; there are no records from Montana, records from only 3 southeastern Wyoming latilongs, and from 4 Colorado latilongs. The nearest breeding area is in northeastern New Mexico or possibly rarely in adjacent southeastern Colorado (La Junta latilong).

Habitats and Ecology: In Arizona and New Mexico this species is primarily associated with the ponderosa pine zone, especially between 5,000 and 7,500 feet. They also occur in pine–oak and oak woodlands, as well as in pinyon–juniper woodlands, especially near streams.

Seasonality: There are very few regional records, but several are for May and presumably reflect vagrant birds.

ACCIDENTAL VAGRANT

Comments: This species is probably a very close relative of the summer tanager, another generally southern and southeastern species that closely resembles it and in some areas occurs in the same general habitats. However, the hepatic tanager tends to occur in lower valleys, and remains closer to water, than does the summer tanager, which favors cottonwood groves in the southwestern states.

Suggested Reading: Bent, 1958.

Scarlet Tanager (*Piranga olivacea*)

Identification: Males of this species are unmistakable; they are a brilliant red except for black wings and tail. Females are mostly yellow below and yellow-green above, without the wing-bars typical of female western tanagers. They cannot be separated readily from female hepatic tanagers. The male's song is a series of about 6 or 7 hoarse and robin-like phrases.

Status: An accidental vagrant or rare migrant in the region; there is 1 Montana record, 4 from Wyoming, and records from 12 Colorado latilongs. The nearest breeding areas are in the western Dakotas and Nebraska.

Habitats and Ecology: Breeding typically occurs in mature hardwood forests growing in river valleys, slopes, and bottomlands. Less often it occurs in coniferous forests and in city parks or orchards.

Seasonality: The available regional sight or specimen records extend from May 8 to October 15. There are no regional breeding records, but in North Dakota active nests have been found between mid-June and mid-July.

Comments: Although this is one of the most spectacularly colorful of all North American birds, it rarely can be observed close to ground level, as it is a canopy-zone forager that nests very high in tall trees, sometimes as high as 75 feet above ground. By comparison the western tanager is much easier to observe closely.

Suggested Reading: Bent, 1958; Prescott, 1965.

LATILONG STATUS

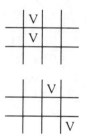

ACCIDENTAL VAGRANT

Western Tanager (*Piranga ludoviciana*)

Identification: The male of this species is the only regional bird that is mostly lemon yellow, with black wings and tail, and with a reddish head, at least during the spring and summer. The female is much duller, and mostly yellow below, with a greenish yellow back and a black tail and wings, the latter crossed by broad white wing-bars. The male's song is robin-like but more hoarse, consisting of two- and three-syllable notes.

LATILONG STATUS

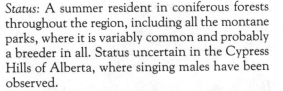

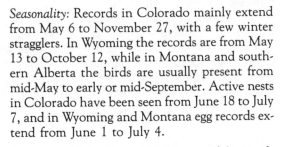

Status: A summer resident in coniferous forests throughout the region, including all the montane parks, where it is variably common and probably a breeder in all. Status uncertain in the Cypress Hills of Alberta, where singing males have been observed.

Habitats and Ecology: Breeding occurs in various habitats, including riparian woodlands, aspen groves, ponderosa pine forests, and occasionally in Douglas fir forests and pinyon–juniper or oak–mountain mahogany woodlands. It is usually found in areas having a predominance of coniferous trees, preferably those that are fairly open, but occasionally extending into fairly dense forests.

Seasonality: Records in Colorado mainly extend from May 6 to November 27, with a few winter stragglers. In Wyoming the records are from May 13 to October 12, while in Montana and southern Alberta the birds are usually present from mid-May to early or mid-September. Active nests in Colorado have been seen from June 18 to July 7, and in Wyoming and Montana egg records extend from June 1 to July 4.

Comments: This splendid bird is one of the jewels of the Rocky Mountain forest region, and outside of the breeding season can often be found in suburbs or other areas well away from their breeding sites, at times even appearing at bird feeders.

Suggested Reading: Bent, 1958.

Rose-breasted Grosbeak (*Pheuticus ludovicianus*)

Identification: Males are easily recognized by their black upperparts and white underparts, the two areas separated by a rose-red breast. There is also a large white wing-patch and pink under wing linings, both evident in flying males. Females have a strongly spotted and striped brown and white pattern, much like a large sparrow or female red-winged blackbird. The male's song is robin-like, but uttered more rapidly.

Status: A local summer resident in the extreme northern portion of the region (Red Deer and Rocky Mountain House area, possibly south to Bottrel and the Porcupine Hills), and a rare to occasional migrant farther southeast of the mountains, with vagrants occurring in the montane parks.

LATILONG STATUS

		M
M		M

M	M		M
		M	
	M	M	

	M		
M			M
M		M	M

Habitats and Ecology: During the breeding season this species is found in relatively deciduous areas or the deciduous portions of mixed forests on floodplains, slopes, and bluffs. Forests where the undergrowth is tall but not too dense are apparently preferred, although a variety of undergrowth conditions are utilized.

Seasonality: Colorado records are from April 27 to October 3. Alberta records extend from May 8 to September 1. There are no regional egg records, but in North Dakota egg records are from May 31 to June 27.

Comments: This eastern North American species replaces the black-headed grosbeak east of the Rocky Mountains, and overlaps to some degree with it on the western plains, where a limited degree of hybridization occurs. The ecology and behavior of these two species are very similar, and probably the male plumage pattern differences are important in reducing the rate of hybridization.

Suggested Reading: Dunham, 1966; West, 1962; Kroodsma, 1970, 1974.

Black-headed Grosbeak (*Pheucticus melanocephalus*)

Identification: Males of this species are distinctively patterned with orange-yellow underparts, and with a black head, wings, and tail; the wings and tail are variously spotted or striped with white. It might be confused with the evening grosbeak, but that species is more brightly yellow on the back and underparts, and has a bright yellow forehead. The female is a chunky sparrow-like bird, with heavy dark cheek markings and a yellowish tinge on the underwing area and belly. The male's song is very similar to that of the rare rose-breasted grosbeak, sounding much like an American robin in having a series of fluty, slurred whistles.

LATILONG STATUS

s	S	s	s
s	S	s	S
s	S	S	s

s	S	s	S
s	s	M	S
M	S	s	S

S	s	s	s
	S	s	M
S	s	M	S

Status: A summer resident over most of the region in wooded areas excepting the northernmost parts in Alberta, where only a vagrant. A variably common breeder in all the U.S. montane parks.

Habitats and Ecology: During the breeding season this species is associated with open deciduous woodlands having fairly well developed shrubby understories, and usually on floodplains or upland areas. It extends into wooded coulees and riparian forests of cottonwoods and similar vegetation in the plains, and sometimes also nests in orchards, oak–mountain mahogany woodlands, and aspen groves.

Seasonality: Colorado records extend from April 26 to October 10, and Wyoming records are from May 12 to October 2. In Montana and southern Alberta the birds are usually present from mid- or late May to the end of August or early September. Colorado egg records are from May 24 to June 28, and Wyoming and Montana active nests have been noted from June 5 until the latter part of July.

Comments: These are highly territorial and vocal birds, and like the rose-breasted grosbeak, males sometimes even sing while sitting on eggs.

Suggested Reading: Kroodsman, 1970, 1974; Weston, 1947.

Blue Grosbeak (*Guiraca caerulea*)

Identification: Males of this species are the only dark blue birds with a short and sparrow-like beak and brownish wing-bars. Male indigo buntings are smaller, much rarer, and lack brown wing-bars, and male mountain bluebirds are paler, especially on the underparts. Females look very much like large house sparrows, but have distinctive buffy-brown wing bars. The male's song is an extended series of warbling phrases that variably rise and fall in pitch.

Status: A local summer resident in the southeastern corner of the region, reaching almost to Rocky Mountain N.P. and possibly breeding in Dinosaur N.M. Breeding has been reported from one Wyoming latilong, and there is local breeding in Elmore County, Idaho (Glenns Ferry).

Habitats and Ecology: During the breeding season these birds are found in brushy and weedy pastures, old fields with scattered saplings, forest edges, hedgerows, and streamside thickets. The presence of large seeds, such as sunflowers, seems to favor its occurrence, and the birds are also often found near water.

Seasonality: Colorado records are from May 17 to September 18, and active nests in that state have been reported from June 5 to August 14. In Idaho the birds have been observed with juveniles in mid-August.

Comments: Although somewhat larger, this species is a fairly close relative of the indigo and lazuli buntings, and like these species is associated with edge habitats rather than forests. Double-brooding is apparently common in this species.

Suggested Reading: Stabler, 1959; Bent, 1968.

LATILONG STATUS

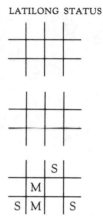

Lazuli Bunting (*Passerina amoena*)

Identification: Males of this species have bright blue heads and upper-parts, an orange-brown breast, and white to buffy wing-bars. The females are rather nondescript, with unspotted tan breasts, darker brown upper-parts, and two conspicuous whitish wing-bars. The male's song is a loud warbling that toward the end includes many rapid rising and falling *tree-tree* or *trit-a-tree* phrases and has a few scratchy or buzzy notes.

LATILONG STATUS

s	S	s	s
s	S	s	s
S	S	S	s

s	S	S	s
s	S	S	S
M	S	s	S

s	s	s	s
M	s	M	s
s	s	s	S

Status: A summer resident in suitable habitats nearly throughout the region, but becoming rarer eastwardly and northwardly; reported in all the montane parks except Banff/Jasper, and probably breeding in most.

Habitats and Ecology: In the mountain areas these birds breed along the edges of deciduous forests on gentle valley slopes, such as aspen groves, or thickets of willow or alder, while on the foothills and plains the birds are usually found in riparian woodlands supporting a mixture of shrubs, low trees, and herbaceous vegetation. Plant diversity and discontinuity of cover seem to be important habitat characteristics for this species.

Seasonality: Colorado records extend from April 25 to October 3, and Wyoming records are from May 5 to September 15. In Montana and southern Alberta the birds are usually present from mid-May to early September. Colorado egg records are from June 6 to July 9, and Wyoming egg records extend from June 1 to August 2, a span that also encompasses available Montana records.

Comments: This species is the western replacement form of the indigo bunting, and has very similar habitat characteristics and behavior patterns to that species.

Suggested Reading: Emlen et al., 1975; Bent, 1968.

Indigo Bunting (*Passerina cyanea*)

Identification: Males of this species are the only small and sparrow-like birds that are almost entirely dark blue; the more common lazuli bunting has a brownish breast and the larger blue grosbeak has brown wing-bars. The mountain bluebird is much paler underneath and has a much more pointed bill. Females cannot be separated easily from female lazuli buntings but have less conspicuous wing-bars and more definite breast streaking. The male's song is a long series of whistled phrases that consists mostly of paired phrases.

Status: A local summer resident primarily east of the major mountain ranges, but extending west locally to the upper Missouri drainage, the Bighorn Mountains of Wyoming, and central Colorado. A vagrant in the southern montane parks, and with the nearest known breeding in the Fort Colorado latilong.

Habitats and Ecology: This species typically breeds in relatively open hardwood forests on floodplains or uplands. Open woodlands, with a high density of shrubs and an open canopy, are favored, and thus forest edges, second-growth areas, orchards, overgrown pastures, and similar habitats are typically utilized.

Seasonality: Colorado records are from May 7 to September 14, with nesting records extending from July 26 to August 8.

Comments: This species rather often hybridizes with the lazuli bunting at the western edge of the Great Plains, where the two forms overlap considerably. Male song types may be important in such areas for proper species recognition and maintenance of reproductive isolation.

Suggested Reading: Bradley, 1948; Emlen et al., 1975.

LATILONG STATUS

M		
		S
V	M	

M	M		S
	M		
	M	M	

M		M	M
	M	M	
M			S

Dickcissel (*Spiza americana*)

Identification: Males resemble miniature meadowlarks, with yellow breasts and a black "bib." Females look much like typical sparrows, but have rufous markings on the anterior wings, a yellowish stripe above the eye, and a yellow tinge on the breast. The male's song is a distinctive *Dick, dick, dick, sissssssss-sss-sss.*

LATILONG STATUS

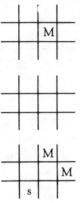

Status: A local summer resident in the plains area east of the mountains, breeding in extreme eastern Montana and very locally in eastern Wyoming (Fort Laramie and Buffalo latilongs), and more extensively in eastern Colorado. A vagrant in montane areas, and not yet reported from any of the montane parks, but possibly breeds near Dinosaur N.M. in the Craig latilong.

Habitats and Ecology: This is a prairie-adapted species that breeds in grasslands having a combination of tall forbs, grasses, and shrubs, or in grassy meadows having nearby hedges or brushy fencerows.

Seasonality: Colorado records extend from May 1 to October 4, and active nests have been noted there from July 3 (eggs) to August 15 (nestlings). The breeding season is generally long in the Great Plains, and suggestive of double-brooding.

Comments: This is one of the Great Plains endemic species that reaches the western edge of its range in the Rocky Mountain region, much like the lark bunting and several other grassland sparrows.

Suggested Reading: Harmeson, 1974; Zimmerman, 1966.

Green-tailed Towhee (*Pipilo chlorurus*)

Identification: Adults of this species are mostly grayish to greenish gray, with a contrasting chestnut crown, and with a black-bordered white throat. Young birds also have a black-bordered whitish throat, but are otherwise mostly a streaked brownish. The male's song consists of two or three whistled *sweet-to* notes, followed by a series of buzzy trills, and the call is a distinctive cat-like mewing note. Usually seen on or near the ground in shrubby vegetation.

Status: A local summer resident in the southern parts of the region, breeding north to central Montana (White Sulfur Springs latilong, possibly to Lewiston), and a variably common summer resident in the montane parks north to Yellowstone N.P.

LATILONG STATUS

		V	
		?	s

S	S	s	s
s	S	S	s
S	S	S	S

s	S	S	s
s	S	S	S
S	S	S	S

Habitats and Ecology: During the breeding season this species occurs in brushy foothills areas dominated by sagebrush, scrub oaks, saltbush, and greasewood flats, scrubby riparian woodlands, and similar open and semi-arid habitats. Forested areas are avoided, but scattered trees in brushlands are used as singing posts. Spreading shrubs that allow for easy movement and foraging on the ground surface below are favored vegetation types.

Seasonality: Colorado records primarily extend from April 10 to November 24, with occasional wintering individuals. Wyoming records are from May 2 to October 12. In Montana the birds are usually present from mid-May to early September. Colorado egg records are from May 19 to June 30, and in Wyoming and Montana there are egg records from May 27 to June 18.

Comments: This is one of the most arid-adapted of the North American towhees, but like the others it is a ground-forager, which typically retreats into heavy cover rather than flushing when faced with danger.

Suggested Reading: Grinnell & Storer, 1924; Bent, 1968.

Rufous-sided Towhee (*Pipilo erythropthalma*)

Identification: Males of this species are easily identified by their black heads and breasts, white-spotted black backs, and white-cornered black tails, together with chestnut flanks. In the female, the black areas are replaced by brown, but both sexes have bright red eyes. The male's song is a distinctive *Drink your teeeee*, and his call in the western states is a nasal and cat-like *wheeee* note.

LATILONG STATUS

s	s	s	s
S	S	s	M
S	S	S	s

S	s	S	s
s	S	M	s
	M	M	S

s		s	
	s	M	S
S	S	s	R

Status: A summer resident over most of the region, becoming rarer in the Canadian mountains and also in the more arid southwestern areas. Common in Dinosaur N.M., but otherwise uncommon to rare in the montane parks, but probably breeding in several.

Habitats and Ecology: Breeding occurs in brushy fields, thickets, woodland openings or edges, second-growth forests, city parks, and well-planted suburbs. Habitats that have a good accumulation of litter and humus, and a protective screen of shrubby foliage above the ground, are highly favored by these birds.

Seasonality: In parts of southern Colorado these birds are resident, but in Wyoming the records extend from April 9 to October 29, with migration peaks in May and September. In Montana and southern Alberta the birds arrive in late April or early May, and usually are gone by the end of September or early October. Egg records in Colorado are from May 17 to July, and in Wyoming from May 25 to July 24. Alberta egg records extend from June 10 to 22. In at least some areas double-brooding is apparently common.

Comments: Rufous-sided towhees in the Rocky Mountain area may appear unusual to persons from eastern North America, as males are heavily spotted with white on their upperparts. These types were once considered different species, but are now known to be only racially separated.

Suggested Reading: Sibley & West, 1959; Baumann, 1959; Davis, 1960.

Brown Towhee (*Pipilo fuscus*)

Identification: This is the dullest of the towhees in the region, and appears almost uniformly brownish, with paler underparts and a buffy striped throat. Young birds are more heavily striped and distinctly sparrow-like, but the ground-foraging feeding method is characteristic. The song consists of a series of two-syllable *chili* notes followed by a variable ending. The usual call is a loud *shut-up*.

Status: Probably only an accidental vagrant in the region, with no Montana records, only one (Jackson latilong) for Wyoming, and only a few for northern Colorado. The nearest known breeding area is in southern Colorado (north to the Canon City latilong), but reported as "rare" at Dinosaur N.M. and a potential breeder there.

LATILONG STATUS

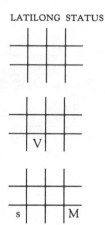

Habitats and Ecology: Breeding occurs in dry shrubby areas such as sagebrush, desert scrub with cholla cactus, and pinyon–juniper habitats; Colorado nestings most often occur in junipers. Flat areas in the vicinity of dense scrubby thickets, and open ground that is closely adjacent to brushy cover, and suitable foraging areas of ground under such shrubbery, seem to be important habitat components.

Seasonality: A resident in southern Colorado, and with egg records extending from May 14 to 31. The extralimital Wyoming record is for early June.

ACCIDENTAL VAGRANT

Comments: This species should be looked for in Dinosaur National Monument, where breeding might well occur rarely, even though that is north of the species' known northern range limits in Colorado.

Suggested Reading: Tvrdik, 1978; Davis, 1961; Marshall, 1960.

American Tree Sparrow (*Spizella arborea*)

Identification: This winter sparrow is readily recognized by its clear grayish breast with a black "stickpin" spot in the middle, as well as a rufous crown and somewhat rufous wing markings set off by two white wing-bars. During winter the male's song is unlikely to be heard, but the musical call-note, a *teedle-deet* is quite distinctive. Generally found in small groups rather than single birds while in this region.

LATILONG STATUS

M	M	M	
	M	M	M
W	W	W	W

W	W		W
M	W	W	W
	W	W	

W		W	W
	W	W	W
W	W	W	W

WINTERING MIGRANT

Status: An overwintering migrant virtually throughout the entire region, including both montane areas and plains, but rarer in mountains farther north, and unreported from Kootenay N.P. The nearest breeding areas are in northeastern Saskatchewan.

Habitats and Ecology: While in the Rocky Mountain region this species occupies brushy prairie areas, roadside thickets, farmsteads, old orchards, overgrown and weedy pastures, and similar relatively open habitats. They often occur in company with juncos and other gregarious and hardy sparrows, and feed about on the ground or snow surface, industriously searching out small seeds. During the breeding season they are associated with arctic timberline habitats.

Seasonality: In Colorado the records extend from October 26 to May 21, and in Wyoming from September 28 to May 1, with migration peaks in October and April. In Montana they are present from about mid-September to mid-April, and in southern or central Alberta they typically pass through in September, wintering occasionally in the extreme southwest, and again move north in late March.

Comments: Tree sparrows are extremely hardy and charming birds that are likely to be seen on winter bird walks when few other birds are to be seen, and occasionally coming into feeding trays to supplement the foods that they somehow manage to locate in snow-covered areas.

Suggested Reading: Weedon, 1965; Heydweiller, 1935.

Chipping Sparrow (*Spizella passerina*)

Identification: Chipping sparrows are easily identified by their rusty brown crowns, which are bounded below first by a white line and then by a black eye-streak, and by their clear, unmarked gray breast. Tree sparrows are similar, but have blackish breast spots, and no white on the head. The male's song is an extended trill of notes all on the same pitch, sounding something like a sewing machine in its monotonous uniformity.

Status: A widespread and common summer resident throughout the region in all wooded areas; possibly the most common breeding sparrow in the montane parks, and very probably breeding in all of them.

LATILONG STATUS

S	S	S	s
S	S	S	S
S	S	S	S

S	S	S	S
s	S	S	S
S	S	S	S

s	s	S	s
s	S	s	S
S	S	s	S

Habitats and Ecology: Breeding in this species is done in open deciduous or mixed forests, the margins of forest clearings, the edges of muskegs, in timberline scrub, riparian woodlands, pinyon–juniper or oak–mountain mahogany woodlands, and similar diverse habitats. Generally scattered trees, an unshaded forest floor, and a sparse ground covering of herbaceous plants seem to be the kinds of habitat considerations that are important.

Seasonality: Resident in parts of Colorado, but in Wyoming the birds are present from late March to late October, with migration peaks in April and September. In Montana and southern Alberta they are usually present from mid- or late April to about the end of September. Colorado nest records are from June 19 (nestlings) to June 27, and in Wyoming there are egg records from June 3 to July 14. Alberta and Montana egg records are from May 30 to June 20.

Comments: This widely distributed sparrow survives well in disturbed, successional habitats, and as such probably benefits from human activities such as lumbering and other deforestation activities. It breeds virtually everywhere in North America except in tundra regions and extreme deserts.

Suggested Reading: Hebrand, 1974; Tate, 1973; Walkinshaw, 1944.

Clay-colored Sparrow (*Spizella pallida*)

Identification: This is a dull-colored sparrow that has a grayish nape, a brownish cheek patch that is outlined by gray or whitish, and a pale buffy stripe through the middle of the crown. The back is also heavily striped with buff and blackish streaks, but the breast is uniformly whitish gray. The male's song is a series of up to five low-pitched buzzing sounds sounding more like an insect than a bird.

LATILONG STATUS

	M	s	s
	S		s
V	s		s

s	s	M	s
	M		
?			

	M	s	
	M	M	
	s	M	

Status: A local summer resident in the northern part of the region, mainly east of the mountains and also north of Wyoming, but locally breeding in Banff and Jasper N.P., and probably in extreme northern Wyoming (Burgess Junction and Buffalo latilongs). Also reported as breeding in the Jackson latilong (Oakleaf et al., 1982), but without specific documentation.

Habitats and Ecology: Favored breeding habitats consist of brushy thickets in prairies, fenceline shrubbery along pastures or meadows, mixed-grass prairies with scattered shrubs or low trees, brushy woodland margins, early successional stages of forests following logging or fires, and retired croplands. Nesting sometimes also occurs in city parks or residential areas.

Seasonality: In Colorado the records extend from March 29 to October 21, and in Wyoming from April 29 to October 20. They usually arrive in Alberta by late April or early May, and remain until about the end of September. There are no available nesting dates for Wyoming or Montana, but egg records for Alberta extend from June 5 to July 7.

Comments: This is another prairie endemic, that is most common in the prairie provinces of Canada, and does not extend south into the central Great Plains. To the southwest it tends to be replaced by the Brewer's sparrow, and to the east by the field sparrow, all of which have somewhat similar ecological requirements.

Suggested Reading: Knapton, 1979; Salt, 1966.

Brewer's Sparrow (*Spizella breweri*)

Identification: This small sparrow is nearly identical in appeaance to its close relative the clay-colored sparrow, but lacks a buffy crown-stripe, has a less definite grayish nape, and its brown ear-patch is not so distinctly set off from the rest of the head and nape striping. The male's song is much more elaborate than that of the clay-colored sparrow, and consists of an extended series of complex and musical or buzzy trills. Usually found in semi-desert scrub habitats, but also at alpine timberline in the Rockies.

Status: A summer resident in nearly the entire region, except for the plains of Alberta, where replaced by the clay-colored sparrow. Variably common in most and probably all of the montane parks, and known to breed in several.

LATILONG STATUS

M	M	S	
	S		s.
S	s	s	s

Habitats and Ecology: In the Rocky Mountain region this species breeds in two very different habitats. The first is in short-grass prairies with sage or other semi-arid shrubs present in varying densities. In Colorado, mountain mahogany or currants growing in brushy hillsides or mesa edges are sometimes used, while in Idaho the birds have been found breeding on sagebrush flats as well as in serviceberry-covered slopes of mountain ridges. In southern Alberta the birds breed on short-grass plains with scattered sage and cacti, as well as along timberline in Banff and Jasper parks, in stunted spruces, firs, willows, and alders.

S	S	s	S
s	S	S	S
S	S	S	S

s	S	S	S
S	S	S	S
S	S	S	S

Seasonality: Colorado records are from April 25 to October 25, while Wyoming records are from April 22 to October 3. In Montana and Alberta the birds usually arrive in early May, and remain until late September. Colorado nesting records are from June 15 to July 13, and egg records for Wyoming and Montana are from May 24 to June 16, with nestlings observed into late July. Eggs in Jasper N.P. have been noted from June 25 to July 14.

Suggested Reading: Bent, 1968; Best, 1972.

Field Sparrow (*Spizella pusilla*)

Identification: Field sparrows rather closely resemble American tree sparrows, but lack the dark breast patch, and have pink beaks rather than black upper and yellow lower mandibles. The breast is clear gray, except in young birds which have somewhat streaked breasts. The male's song is easily recognized; it is a sweet series of whistled notes that speed toward the end, like a ping-pong ball coming to a stop on a tabletop.

LATILONG STATUS

Status: A very local summer resident in extreme eastern Montana (Glendive and Baker latilongs, possibly others), and a local migrant farther south. Only a vagrant in the montane parks; reported once in Glacier N.P. Possibly also breeds in extreme eastern Wyoming and the eastern counties of Colorado.

Habitats and Ecology: Breeding occurs in brushy, open woodlands, brushy ravines or coulees, sagebrush flats, abandoned hayfields, forest clearings, and similar habitats having a combination of low grassy areas and scattered shrubs or trees. Very similar habitats are used by the chipping sparrow, but that species tolerates a greater tree density and a later successional stage.

Seasonality: Colorado dates extend from March 29 to November 24; no specific breeding records are available. Montana records extend from May 11 to August 12, but there are no specific breeding dates available. Available egg dates for North Dakota are for June.

Comments: The field sparrow is extremely common and widespread from the Great Plains eastward, and it is interesting that its range terminates at about the point where that of the Brewer's sparrow begins, suggesting that competitive exclusion may be occurring.

Suggested Reading: Best, 1977, 1978; Crook, 1948.

Vesper Sparrow (*Pooecetes gramineus*)

Identification: This is a plain-colored grassland sparrow which has only white outer tail feathers as a distinctive fieldmark. Otherwise it has a chestnut patch at the bend of the wing (often invisible), a pale whitish eye-ring, and a weakly spotted whitish breast. Its song is musical and somewhat like that of a song sparrow, but usually with two pairs of preliminary slurred notes followed by a descending trill: "Here, here; where-where; all together down the hill."

Status: A summer resident throughout the region in grassland areas; variably common in all the montane parks and probably breeding in all.

Habitats and Ecology: During the breeding season this species is found in overgrown fields, prairie edges, grasslands with scattered shrubs and small trees, sagebrush areas where the plants are scattered and stunted, and similar open habitats, but not extending to mountain meadows or tundra zones.

Seasonality: Most Colorado records are from February 26 to October 5, with rare overwintering, while Wyoming records are from April 12 to October 9. In Montana and southern Alberta the birds are usually present from late April to late September. Colorado egg records are from May 21 to July 4, and Wyoming egg records extend from May 15 to August 4. In Alberta there are egg records from May 20 to June 7. Double-brooding is regular over most of the species' range.

Comments: This is one of the more widely distributed and abundant species of grassland sparrows, ranging from coast to coast and from Mexico almost to Alaska. Like several other grassland species (meadowlarks, longspurs, etc.) they have conspicuous white tail feathers that probably are effective visual signals in flight, but can be easily hidden when on the ground.

Suggested Reading: Best, 1972; Bent, 1968.

LATILONG STATUS

s	S	S	s
S	S	S	S
S	S	S	s

S	S	S	S
S	S	S	S
S	S	S	S

s	S	S	S
S	S	S	S
S	S	S	S

Lark Sparrow (*Chondestes grammacus*)

Identification: This grassland sparrow has conspicuous white corners on its tail feathers (most evident in flight) and a contrasting head pattern with bright chestnut ear-patches, a chestnut stripe over the eye, and various white and black markings, as well as a small blackish spot on the otherwise uniformly gray breast. The song is a complex mixture of buzzy notes and trills, but typically begins with two loud and clear introductory notes.

LATILONG STATUS

M	M	s	
	s	M	s
V	S	s	S

S	s	S	S
s	S	s	S
M	M	M	S

M	M	S	s
s	M	s	S
S	s	S	S

Status: A summer resident over most of the region in grassland habitats, but rarer northwardly and absent from the montane areas of Alberta. Rare to occasional in the montane parks farther south.

Habitats and Ecology: This species favors grasslands that have scattered trees, shrubs, large forbs, or adjoin such vegetation; thus weedy fencerows near grasslands, open brushland on slopes, sagebrush flats, scrubby and open oak woodlands, orchards, and similar habitats are all suitable. Mountain meadows and alpine areas are not used.

Seasonality: Colorado records extend from April 19 to September 23, and Wyoming records are from May 1 to September 25. In Montana and southern Alberta the birds are usually present from early May to mid-September. Colorado egg records are from June 2 to July 19, and Wyoming egg records are from May 20 to June 14. Montana egg records are from June 17 to July 10. Double brooding may occur in some more southern areas.

Comments: This is one of the most attractive of the grassland sparrows, and one of the easiest to watch in courtship and territorial display, during which the white tail feathers are exposed by tail-spreading.

Suggested Reading: Tramontano, 1971; Newman, 1970.

Black-throated Sparrow (*Amphispiza lineata*)

Identification: This sparrow is very easily identified by its generally grayish color, with a black bib, white "mustache" and eye-stripes, and an otherwise gray to blackish head. The outer tail feathers are also whitish. The male's song is mostly of high-pitched bell-like notes, usually opening with two clear notes and followed by a trill.

Status: A local summer resident in the southwestern part of the region, mainly in Idaho south of the Snake River, and possibly also in southwestern Wyoming and adjacent Colorado. Breeding status at Dinosaur N.M. is uncertain, but known to breed in the adjoining latilong to the south.

LATILONG STATUS

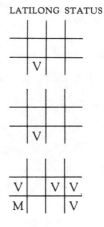

Habitats and Ecology: Breeding habitats consist of thinly grassed pastures with scattered cactus, yucca, or mesquite. Desert uplands with much exposed ground, but with hiding places in thick and woody twig growth or cactus plants, are especially favored. Nests are also often placed in cactus or dense shrub growth.

Seasonality: A few Idaho records are from May 6 to August 10, and a similar small number of Colorado records are from April 23 to late July. However, wintering occurs as far north as southern New Mexico, so the absence of fall dates is misleading. Eggs have been noted in Colorado on May 19, and young observed in late July.

Comments: This is perhaps the most arid-adapted of the North American sparrows, and one that is notable for its ability to survive in the absence of drinking water, obtaining fluids from the foods that it eats.

Suggested Reading: Heckenlively, 1970; Bent, 1968.

Sage Sparrow (*Amphispiza belli*)

Identification: This rather pallid-colored sparrow is somewhat grayish brown, with white underparts that are marked with a dark central breast spot and some dark spotting on the sides of the breast and anterior flanks. The face is grayish, with a whitish eye-ring, a white "mustache" stripe, and a white spot between the eye and beak. The song consists of 4 to 7 thin and high notes, with the third note the highest and loudest.

LATILONG STATUS

V		
V		

s	s	
s	M	S
M	s	S

s	s	s
S	S	M	s
S	S	S	S

Status: A summer resident in sage areas in the southern parts of the region, north to the Snake River in Idaho and north-central Wyoming. Generally rare or absent from the montane parks, but occasional at Dinosaur N.M., and a probable breeder there.

Habitats and Ecology: The species is closely associated with fairly dense to sparse and scrubby sagebrush vegetation during the breeding season, but also breeds at times in similar semi-desert vegetation types, such as in saltbush. Foraging is done on rather bare ground areas of gravel or alkali soil around the bushes, and escapes are made by fleeing into the shrubbery.

Seasonality: Colorado records are from February 11 to September 30, and Wyoming records are from April 2 to October 25. Colorado egg records are from May 20 to June 25, and there is an egg record from Idaho for July 7.

Comments: Nesting in this species is usually in sagebrush, but at times the birds also place their nests under the bushes. They are often associated with such sage-adapted species as sage grouse, sage thrashers, and Brewer's sparrows.

Suggested Reading: Sumner & Dixon, 1953; Moldenhauer & Wiens, 1970.

Lark Bunting (*Calamospiza melanocorys*)

Identification: Males are immediately identifiable by their mostly black plumage except for large white wing-patches, which are evident both in perching and especially in flying birds. Females are very similar to vesper sparrows but also have white wing markings and also white corners on their tails. Males utter a song-flight while hovering that is a varied mixture of harsher notes, whistles, and trills. Closely associated with grasslands, and usually seen in groups, even during the breeding season.

Status: A summer resident in the eastern half of the region, mainly on plains and foothill grasslands; a rare migrant or vagrant in the montane parks, probably not breeding in any.

Habitats and Ecology: This species favors mixed-grass prairies for nesting, but also can be found in short-grass and tall-grass prairies, as well as sage grasslands, retired croplands, alfalfa fields, and stubble fields. Areas with abundant shrubs are avoided, but fence posts or scattered trees may be used as song posts.

Seasonality: Most Colorado records are from April 19 to October 1, with a few later reports of stragglers; Wyoming records are from April 30 to September 11. In Montana and southern Alberta the birds are usually present from early May until late August. Egg records in Colorado are from May 22 to July 25, and in Wyoming from May 21 to July 8. Alberta and Montana records are from May 15 to August 10.

Comments: This species is of special interest because of the somewhat colonial nesting behavior it exhibits, with nests often placed only 10 to 15 yards apart, and males often singing from adjacent fenceposts.

Suggested Reading: Bent, 1968; Butterfield, 1969.

LATILONG STATUS

	M	M	s
M	M		S
V	M	M	M

S	M	s	S
M	M	s	S
s	M	S	S

S	s	S	S
s	S	s	S
s	s	S	S

Savannah Sparrow (*Passerculus sandwichensis*)

Identification: This is a drab grassland sparrow without white on the tail, bright colors on the head, or patterned wings. It has a variably yellowish and white (the yellow portion in front of the eye) stripe above the eye, a pale crown stripe, a fairly heavily streaked (but not blotched) breast, and a rather short tail. The male's song is usually begun with two or three *chip* notes, followed by two buzzy trills, and dropping at the end. The birds most resemble vesper sparrows, but lack white outer tail feathers, and also resemble pale song sparrows but have shorter tails and lack a central breast spot.

LATILONG STATUS

s	S	S	s
	S	S	S
S	s	S	s

S	s	s	s
S	S	s	s
S	S	M	S

	s	S	M
s	s	s	S
S	S	S	S

Status: A summer resident throughout the region, mainly at lower altitudes, but occurring commonly in all the montane parks and probably breeding in all.

Habitats and Ecology: During the breeding season this species is closely associated with moist but low-stature prairies, the wet meadow zones around marshes or other wetlands, and the moist and open areas of mountain meadows. A growth of dense ground cover, preferably only a few inches tall, with scattered bushes or clumps of taller vegetation for song perches, are typical aspects of nesting habitats. Nests are placed on the ground, under thick herbaceous cover of grasses or sedges, and are usually hidden from above by overhanging leaves.

Seasonality: Colorado records are from March 23 to November 11. In Montana and southern Alberta the birds are usually present from mid- or late April until late September or early October. Colorado egg records are from June 7 to July 12, and in Wyoming and Montana the records extend from June 1 to July 18, with hatched young observed as early as June 9.

Suggested Reading: Dixon, 1978; Wiens, 1973; Welsh, 1975; Potter, 1972.

Baird's Sparrow (*Ammodramus bairdii*)

Identification: This grassland sparrow rather closely resembles the savannah sparrow, especially in its short tail, spotted breast, and pale-striped crown, but the breast spotting is more restricted and necklace-like, the crown is distinctly tinted with golden buff, and the stripe above the eye is also distinctly buffy throughout rather than yellowish in front and white posteriorly. The male's song consists of about three preliminary notes that are followed by a lower pitched trill.

Status: A summer resident in grasslands of eastern Alberta and northeastern Montana, and a migrant east of the mountains farther south. Reported in the montane parks only as a vagrant.

Habitats and Ecology: Closely associated during the breeding season with native prairie areas, including ungrazed or lightly grazed mixed-grass prairies, wet meadows, and various disturbance habitats such as hayfields, stubble fields, and retired croplands. A dense but low vegetation over the soil and a few scattered shrubs for singing posts seem to be desirable aspects of the habitat.

Seasonality: Colorado records extend from April 29 to May 20, and from August 23 to December 26. In Montana and southern Alberta the birds are usually present from the middle of May until early September. Egg records from Montana and Alberta extend from June 9 to July 22. Single-brooded.

Comments: This is one of the prairie endemics associated with the northern grasslands of North America, with a somewhat similar breeding range as those of the LeConte's sparrow and sharp-tailed sparrow, both of which are more associated with wetter habitats.

Suggested Reading: Cartwright et al., 1937; Bent, 1968.

LATILONG STATUS

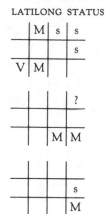

Grasshopper Sparrow (*Ammodramus savannarum*)

Identification: Grasshopper sparrows are among the smallest and plainest of the grassland sparrows, with plain buffy breasts (somewhat striped in juveniles), a pale crown stripe, a short and unmarked tail, and a pale, somewhat yellowish face with a contrasting and "beady" dark eye. The male's "song" is a grasshopper-like buzzy trill, sometimes preceded by two or more *tick* notes.

LATILONG STATUS

	M	s	
	S	?	M
S			

s		s	s
?	S	M	M
	M	M	S

		S	s
			M
V	s	M	M

Status: A local summer resident, mainly at lower altitudes, in grasslands throughout the region. Generally absent from the montane parks except as a vagrant, but reported as a rare breeder at Yellowstone N.P.

Habitats and Ecology: During the breeding season this species is mostly associated with mixed-grass prairies, but also occurs in short-grass and tall-grass areas, as well as on sage grasslands and disturbed grasslands such as hayfields, stubble fields, and retired croplands. Mountain meadows are not used, nor are grassland areas that have largely grown up to shrubs.

Seasonality: Colorado records extend from April 21 to November 30. In Wyoming the records are from May 18 to September 17, while in Montana and southern Alberta the birds are usually present from early May to early September. Records of active nests in Colorado are from late June to July 28, while in Montana there are nesting records from June 10 to July 27.

Comments: Except when males are actively singing, grasshopper sparrows are unlikely to be seen, as they are prone to hide in grassy cover rather than flush, and when flushed fly but a short distance before landing in the grasses again.

Suggested Reading: Tromantano, 1971; Wiens, 1973; Smith, 1963.

LeConte's Sparrow (*Ammodramus leconteii*)

Identification: This grassland sparrow has the same small size, short tail, and rather flat-headed shape of the grasshopper sparrow, but has a slightly streaked and buffy-yellow breast, a more nearly whitish central crown stripe, and especially a more golden-yellow head except from grayish ear coverts, making it the most distinctly yellow-faced of the regional sparrows. The male's song is an insect-like, buzzy two- or three-parted hissing similar to that of a grasshopper sparrow.

Status: A local summer resident in southeastern Alberta east of the mountains. Known breeding in Montana is limited to the Kalispell latilong (Camas Creek, Flathead County), but may also occur in the Browning latilong and perhaps others. Rare during summer in Glacier N.P. and breeding there is unproven, as is also true of Watertown Lakes N.P.

LATILONG STATUS

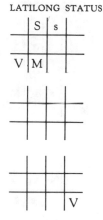

Habitats and Ecology: Breeding in this species is largely limited to hummocky bogs with alder or willows present, but it also nests in the wet meadows around prairie ponds or marshes, in moist tall-grass prairies, and in moist hayfields or retired croplands. A favorite nesting cover is cordgrass (*Spartina*); usually the nest is on a dry hummock surrounded by dense grass or a shrub.

Seasonality: In Montana these birds are usually present from at least early June (probably earlier) until early October. In Alberta they normally arrive about the middle of May, and migrate south in early September. Egg records for Montana and Alberta extend from June 6 to July 15, and in Saskatchewan and Manitoba there are egg records from June 4 to 21.

Comments: This attractive but elusive wetlands sparrow is best located by its "song," which is easily confused with that of a grasshopper sparrow but is even briefer, weaker, and more buzzy.

Suggested Reading: Murray, 1969; Bent, 1968.

380

Fox Sparrow (*Passerella iliaca*)

Identification: This is a fairly large and long-tailed sparrow that closely resembles a song sparrow, but has a more rusty-brown tint to its tail and wings, a less heavily streaked and more grayish head coloration, and (in our region at least) a more uniformly brownish to grayish upper back, without blackish streaking. The male's song is highly melodious, and usually begins with a single introductory whistled note, followed by a series of trills and slurs.

LATILONG STATUS

s	S	S	
?	s	S	s
S	s	s	s

s	s		M
s	S	s	M
S	S	M	

		M	
M	s	M	M
	s	s	S

Status: A summer resident in wooded areas almost throughout the region; present and variably common in all the montane parks, and probably breeding in all.

Habitats and Ecology: During the breeding season dense brushy thickets, and the brushy margins of thick forests, are the favored habitats. Riparian thickets of willows or alders, alder clumps on mountain slopes, and the twisted and stunted conifers near timberline all serve to attract this species. Thickets that provide sufficient space underneath for ground foraging, and have a carpet of leaves and litter for scratching towhee-like for food, are particularly favored.

Seasonality: In Colorado the species is locally resident throughout the year. In Montana the birds are usually present from early April to mid-October. In Idaho it has been reported from March 8 to October 8, and in Alberta it is usually present from mid-April until late October. In Colorado there are egg records from June 8 to 28, and in Idaho young out of the nest have been noted as early as mid-May.

Comments: The fox sparrows of the Rocky Mountains are a good deal less rusty brown than their eastern counterparts, and might easily be taken for a different species by persons unfamiliar with these variations.

Suggested Reading: Linsdale, 1928; Threlfall & Blacquiere, 1982.

Song Sparrow (*Melospiza melodia*)

Identification: This abundant sparrow is recognized by its heavily streaked brown and white breast; the streaks form a large central spot in the middle of the breast, and by the long tail that is pumped up and down while the bird is in flight. The head is marked with dark brown, grayish, and white, and the upper back is also streaked with blackish brown. Young birds are less heavily streaked on the breast and lack a definite breast spot. The male's song begins with three introductory whistles, which are followed by a musical trilled series of notes that tend to drop in pitch.

Status: A summer or permanent resident in suitable habitats throughout the region; present in all the montane parks and probably breeding in all.

Habitats and Ecology: Breeding habitats include such woodland edge types as the brushy margins of forest openings, the edges of ponds or lakes, shelterbelts, farmsteads, coulees on prairies, aspen groves, and the like. Foraging occurs mostly on the ground, both in open areas and leaf-covered ones, where the birds scratch to expose foods.

Seasonality: Resident in Colorado, and also largely resident in Wyoming, although migration peaks are evident in March and October. Some wintering also occurs in Montana, while in central Alberta the birds are seasonal, arriving in mid-April and usually not remaining past mid-October. Colorado egg records are from May 24 to July 2, and in Montana and Wyoming there are egg records from June 1 to 22. Alberta egg records extend from May 18 to July 28, with a peak in late May and early June.

Comments: Early studies by M. M. Nice on territoriality and life history of this species provided some of the best information available even today on the biology of any North American songbird; her classic studies are models of careful observation.

Suggested Reading: Nice, 1937, 1943; Knapton, 1974.

LATILONG STATUS

R	R	R	s
s	R	R	s
R	R	R	R

r	R	r	R
s	R	M	r
S	R	r	R

r	r	R	r
M	r	R	R
S	R	R	R

Lincoln's Sparrow (*Melospiza lincolnii*)

Identification: This sparrow somewhat resembles an immature song sparrow, with a lightly streaked breast, a long tail, and somewhat grayish cheeks. However, it is distinctive in that the upper breast is buffy rather than white, the central crown stripe, "eyebrow" stripe, and sides of the neck are more extensively and uniformly gray, and the entire plumage seems more grayish in tone. The birds are typically much more furtive than are song sparrows, and pump their tails in flight less strongly.

LATILONG STATUS

s	S	s	
	M	s	s
S	S	s	M

s	s	s	s
s	S	s	S
s	S	S	S

S	S	s	s
M	S	S	S
M	M	S	S

Status: A summer resident in wooded areas almost throughout the region, from lowland bogs to alpine timberline. Present in all the montane parks, and probably breeding in all of them.

Habitats and Ecology: In Alberta this species is mainly associated with marshes and bogs having extensive growths of willows and alders. Willow thickets along slow-moving streams are also utilized, as are the brushy borders of muskeg pools. In mountainous areas the birds favor boggy mountain meadows, especially those fringed with willow thickets and supporting a fairly tall growth of grasses, sedges, and herbs.

Seasonality: In Colorado this species is locally resident. In Wyoming the records extend from April 8 to October 20, with peaks in May and September. In Montana and southern Alberta the birds are usually present from early or mid-May to the end of September. Active nests in Colorado have been observed from June 15 to July 1, and in Montana and Wyoming there are egg records from June 16 to 22, and young reported from June 20 to July 15. Alberta egg records are from May 27 to June 28.

Suggested Reading: Grinnell & Storer, 1924; Bent, 1968.

Swamp Sparrow (*Melospiza georgiana*)

Identification: This is one of the few sparrows likely to be seen in swamps and marshy areas; it has a similar color pattern to the chipping sparrow, with a bright rusty crown patch and a dark eye-stripe, but the "eyebrow" is gray rather than white, and the relatively short tail and wings are distinctly chestnut-brown in tone. The male sings from cattails or other marsh vegetation, and its song is a loud series of metallic *weet* notes.

Status: A local summer resident in Alberta east of the mountains and south to the vicinity of Red Deer. Generally absent from the montane parks, but a rare visitor (May, June, September) to Banff N.P.

LATILONG STATUS

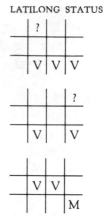

Habitats and Ecology: This species is strongly associated with wetlands during the breeding season, especially areas that are well grown with cattails, phragmites, shrubs, or small trees. In Alberta muskeg-like woodland swamps that have willows, alders, birches, and sometimes black spruces are favored areas, but the birds also nest in dense shrubbery along woodland streams or pools.

Seasonality: In Colorado this is a locally wintering migrant, with records from August 22 to May 19. In Montana it has been noted as an early spring (February) and fall (August to November) migrant. In Alberta it usually arrives in late Aril or early May, and remains until late September. Specific breeding records are not available for the region.

Comments: Swamp sparrows build their nests among the stalks of cattails or sometimes in flooded bushes, often in water up to two feet deep, and in company with marsh wrens and similar water-loving species.

Suggested Reading: Bent, 1968.

White-throated Sparrow (*Zonotrichia albicollis*)

Identification: These are rather large sparrows with white throats, white crown and "eyebrow" stripes that are separated by a black line, and a somewhat golden tinge to the area between the eye and the beak. Immature and fall-plumaged birds have far less defined head markings, but the throat is white even in fairly young birds. Males have a distinctive song that (in the Rocky Mountain area) can be transcribed as, "Oh Canada, sweet, sweet Canada."

LATILONG STATUS

M	W		
	M		?
V	M	W	M

W	W		W
	M		M
	M	M	

M		M	M
		M	M
M			W

Status: A local summer resident in northern and central Alberta south as far as Banff N.P. and the area west of Calgary. Elsewhere a migrant or overwintering visitor throughout the region, but rarer in the mountains and generally only a vagrant in the montane parks.

Habitats and Ecology: In Alberta this species nests in deciduous and mixed woodlands, particularly woodland edge habitats such as lake shores, river banks, old burned areas, logged areas, and roadsides. Outside the breeding season the birds are often observed foraging on the ground in somewhat brushy situations.

Seasonality: In Colorado these birds are wintering migrants, with records from September 23 to May 19. In Wyoming and Montana the records extend from May 6 to mid-June, and from August to mid-November. In Alberta the birds usually arrive in early May, and remain until well into October. Alberta egg records are from June 1 to 19.

Comments: This species often occurs in company with other "crowned sparrows" while on migration and on wintering areas, and the male's song can often be heard well before the birds arrive on their northern breeding grounds.

Suggested Reading: Fisher & Gills, 1946; Wythe, 1938; Rees, 1973.

Golden-crowned Sparrow (*Zonotrichia atricapilla*)

Identification: This large member of the sparrow group has a unique black and gold crown (the gold surrounded by black), and an otherwise rather uniformly gray face, throat, and breast. Immature birds have less colorful head patterns and closely resemble young white-throated sparrows, but lack white throats and never have breast streaking. The male's call is a clear whistled "Oh dear me," in a descending and rather mournful way.

Status: A local summer resident in the Alberta mountains south to Banff N.P., and a migrant from that area south to northern Idaho. Elsewhere in the region only a vagrant.

LATILONG STATUS

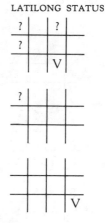

Habitats and Ecology: During the breeding season these birds seek out low coniferous or deciduous growth at or even above tree line. Thickets of stunted willows, alders, and conifers growing in high meadows or on scree slopes provide the nesting habitats; the nests are placed in low woody vegetation or on the ground. While on migration and on wintering areas the birds prefer interrupted brushlands, where leafy litter provides for ground-foraging opportunities.

Seasonality: Idaho records extend from May 6 to October 6. In southern Alberta the birds usually arrive the second week of May, and they have been recorded as late as October 13. In Banff and Jasper parks nestlings have been noted from June 26 to July 15, and newly fledged young seen as late as August 4. British Columbia egg records extend from June 19 to July 28.

Comments: These birds are the only "crowned" sparrows that are restricted to the western slope of the Rocky Mountains, with a migration to and from the Pacific coast, rather than directly southward as in the other species of this group.

Suggested Reading: Davis, 1973; Robertson, 1957.

White-crowned Sparrow (*Zonotrichia leucophrys*)

Identification: Breeding adults of this species have strongly striped black and white crowns, with a grayish throat. Unlike the similar white-throated sparrow there is no yellow in front of the eye, and the beak is mostly pink to yellowish, rather than blackish, the latter trait being useful for separating the immatures of the two species, which are more similar than adults. The male's song begins with notes similar to those of the white-throated sparrow, but ends in buzzy trills of varying pitch and loudness.

LATILONG STATUS

s	s	S	s
s	s	s	s
S	s	s	s

S	S	S	s
S	S	S	S
S	S	S	S

S	s	S	s
M	S	S	S
S	R	R	R

Status: A summer or permanent resident in suitable habitats almost throughout the region, including the montane parks, where it is a common to abundant breeder in all.

Habitats and Ecology: During the breeding season this species occurs in riparian brush, in coniferous forests with well developed wooded undergrowth, in aspen groves with a shrubby understory, in willow thickets around beaver ponds or marshes, and on mountain meadows with alders or similar low and thick shrubbery, often to timberline. Damp, grass-covered ground and nearby shrubbery seem to be important habitat components. On migration and during winter the birds are found in a variety of habitats that offer a combination of brushy cover and open ground for foraging.

Seasonality: In Colorado this species is locally a year-round resident. In Wyoming the records extend from April 18 to October 26, and in Montana and southern Alberta the birds are usually present from late April to early or mid-October. Wyoming and Colorado egg records are from June 1 to July 19, and in Montana active nests have been seen from June 24 to August 8. Alberta egg records are from June 11 to July 6.

Suggested Reading: Blanchard, 1941; Blanchard & Erickson, 1949; Morton et al., 1972.

Harris' Sparrow (*Zonotrichia querula*)

Identification: This is the largest of the sparrows, and the only sparrow that has an unbroken black breast, throat, anterior face, and crown, as well as a pinkish bill. During fall and winter the amount of black on the face, crown, and breast varies greatly, and at that time the pinkish bill, large size, and variably spotted or streaked sides serve to identify the species. The male's song is a series of from 2 to 4 long, spaced, and quavering whistles on the same or slightly different pitch.

Status: A seasonal or overwintering migrant over much of the region, but far more common on the plains east of the mountains, and rare or accidental in the montane parks. The nearest breeding areas are in eastern and northeastern Saskatchewan.

LATILONG STATUS

M	W	M	
	W	W	
V	W	W	M

W	W		W
M	W	W	
	W	W	

W		W	W
			W
W	W		W

Habitats and Ecology: During the breeding season this species is associated with the edges of the spruce forest adjoining arctic tundra, especially rather damp and open areas where the trees are low and scattered. Outside the breeding season the birds are much like the other "crowned" sparrows, foraging on the ground in areas close to thickets, which are used for protection. At that time hedgerows, orchards, farmsteads, riparian thickets, woodland edges, and even sagebrush and desert scrub habitats are often utilized.

Seasonality: Overwintering occurs in Colorado, with records extending from September 29 to May 23. In Wyoming the records extend from October 7 to May 14, with peaks in November and May. Montana and Alberta records are mainly for fall, from September to early November; the few spring records are mostly for May.

WINTERING MIGRANT

Comments: This is one of the last species of North American birds for which the nesting site was described; the first nests were discovered in 1931 near Churchill, Manitoba. The migration corridor is relatively narrow, and is centered on the Great Plains.

Suggested Reading: Semple & Sutton, 1932; Rowher et al., 1981.

Dark-eyed Junco (*Junco hyemalis*)

Identification: This rather variable sparrow ranges from being mostly grayish or "slate-colored" above to almost entirely black on the head and upper breast, with a more brownish back and sides, and generally white underparts. In all, however, the outer tail feathers are white and the central ones blackish. All of these forms occurring in the Rocky Mountains are now considered a single species. Males have a song that is a musical trill all on the same pitch, something like that of a chipping sparrow.

LATILONG STATUS

R	R	R	
r	R	R	S
R	R	R	R

R	R	r	s
S	R	R	M
M	R	R	R

r	R	R	R
M	R	R	R
R	r	R	R

Status: A seasonal or permanent resident in wooded habitats throughout the region, including all the montane parks, where the species is a common to abundant breeder. Winters at lower altitudes, including the entire plains region.

Habitats and Ecology: Breeding habitats include open coniferous forests, especially pinyon–juniper woodlands, ponderosa pine forests, mixed forests, aspen woods, forest clearings, the edges of muskegs or jackpine-covered ridges, and similar habitats that offer ground-foraging and ground-nesting opportunities as well as tree or brush cover for escape.

Seasonality: A year-round resident almost throughout the entire region, although some seasonal movements do occur. Colorado nest records are from May 29 to August 1, while Wyoming egg records are from June 11 to July 27. Montana egg records are from late April to July 19, and Alberta egg records are from May 16 to July 10.

Comments: The juncos of this region include "white-winged" forms in the Black Hills region and surrounding areas, the "Oregon juncos" breeding east to northwestern Montana, the "pink-sided" forms breeding in the interior from Alberta to Wyoming, and the "gray-headed" forms occurring north to northern Colorado and adjacent western Wyoming.

Suggested Reading: Hostetter, 1961; Sabine, 1959; White, 1973; Williams, 1942.

McCown's Longspur (*Calcarius mccownii*)

Identification: Males of this short-grass plains sparrow have a pale grayish head and underparts, except for a black breast-band and crown, and a tail that is mostly white, with a T-shaped black pattern produced by central and terminal banding. Females are much more sparrow-like, but also have a white tail with a black T-shaped pattern along the middle and tip. Males have a flight song uttered during a parachuting, slow descent. In the winter the male is less colorful and more female-like, with the black breast and crown markings partially obscured.

Status: A summer resident on the plains east of the mountains from southern Alberta southward through the region. A rare migrant in the mountains, and only a vagrant in the montane parks.

LATILONG STATUS

		s	s
			S

s		s	?
	M		s
	M		M

	S	s	M
M	M	s	S
M		M	M

Habitats and Ecology: During the breeding season this species is mostly limited to short-grass prairies and grazed mixed-grass prairies, but also breeds to some degree on stubble fields or newly sprouting grainfields. While on migration and during the winter period the birds occur on open grasslands, low sage prairies, mountain meadows, and similar open habitats.

Seasonality: Colorado records are primarily from early March to mid-October, although a few birds often remain through the winter. Wyoming records are from April 6 to October 27, and in Montana and southern Alberta the birds are usually present from early or mid-April to middle or late October. Colorado egg records are from May 23 to July 11, and Wyoming egg records extend from May 17 to August 8. Montana egg records are from May 9 to July 28.

Comments: This is one of the two plains-nesting longspurs, and both sometimes occur on the same areas.

Suggested Reading: Mickey, 1943; Bent, 1968.

Lapland Longspur (*Calcarius lapponicus*)

Identification: Males of this tundra- and grassland-adapted species are the only sparrows with a black breast, throat, and anterior face color, and with a chestnut nape. In the winter period, when the birds are most abundant in the region, males have much lighter underparts and only a grayish breastband similar to that of a McCown's longspur, and a less bright chestnut nape. However, both sexes have white only on their outer tail feathers, and this feature is most useful for recognizing females, which are nearly identical to those of other longspurs. A three-syllable rattling call is often uttered by migrating birds, especially in flight.

LATILONG STATUS

W	M	
		M
M	M	M

W	M	M	M
	W	W	W
W			W

		W	W
			W
	V	W	W

WINTERING MIGRANT

Status: A seasonal or overwintering migrant over much of the region, primarily at lower altitudes and in grassland habitats. Generally rare to accidental in the montane parks. The nearest breeding area is in northeastern Manitoba, along the coast of Hudson Bay.

Habitats and Ecology: Associated during the breeding season with arctic tundra. While on migration and on wintering areas they are typically found on open habitats, such as snow-covered grasslands, mud flats, and the like. Shortly after their arrival in fall the birds often are found in the lower mountain parks, but as the weather becomes more severe the birds move to the foothills and plains, where they usually occur in fairly large flocks.

Seasonality: Colorado records extend from September 18 to April 1. In Wyoming the records range from October 7 to April 3. In Montana the birds are usually present from September to April, with variable abundance during winter. They usually arrive in southern Alberta about mid-September, and have usually passed southward by the end of October. They return in early April, and remain until about the first of May.

Suggested Reading: Sutton & Parmelee, 1955; Sensteadt & MacLean, 1979.

Chestnut-collared Longspur (*Calcarius ornatus*)

Identification: During the breeding season males have a distinctive black breast and underparts, and a bright chestnut nape patch. At other seasons the black areas are obscured by gray, and the nape is only slightly chestnut-colored. At all seasons birds of both sexes can be identified by the mostly white tail with a black terminal triangle. Otherwise females are extremely similar to those of the other longspur species. The call is a two-syllable *kittle* repeated up to 5 times, and it also has a weak rattling call of 3 to 5 syllables. The flight song, given on the nesting grounds, is similar to that of a western meadowlark.

Status: A summer resident on the plains areas east of the mountains over most of the region, and a local migrant somewhat farther west of the breeding range, but generally uncommon to accidental in the montane parks.

Habitats and Ecology: Primary breeding habitats consist of grazed or hayed mixed-grass prairies, short-grass plains, the meadow zones of salt grass around alkaline ponds or lakes, mowed hayfields, heavily grazed pastures, and the like. Outside the breeding season the birds often are found in cultivated fields rich in weed seeds, especially of such species as amaranth.

Seasonality: In Colorado this species is occasionally resident, although most leave the state from mid-October until early April. Wyoming records extend from March 16 to early October. In Montana the birds are usually present from mid-April to late September. Colorado egg records are from May 21 to June 27, while in Montana active nests have been observed from May 6 to the end of July. Alberta egg records are from May 27 to June 11.

Suggested Reading: Bent, 1968; Moriarty, 1965.

LATILONG STATUS

	M	s	S
			S
V			

s		s	S
		M	M
		M	M

		M	M
		M	s
		M	M

Snow Bunting (*Plectrophenax nivalis*)

Identification: Except during the breeding season, both sexes of this species are almost entirely white below and brownish white below, with extensive white patches on the wings and outer tail feathers. No other wintering sparrow of the region has so much white in the plumage as this one. Migrating birds utter a nearly constant twittering while in flight.

LATILONG STATUS

W	W	W	W
M	W	W	M
W	M	W	W

W	M	M	M
W	W	W	W
	W	W	W

W		W	W
	W	W	W
	W	W	W

WINTERING MIGRANT

Status: An overwintering migrant throughout the region, mainly at lower altitudes on grasslands. Generally uncommon to rare in the more southern montane parks, but fairly common in the Canadian parks. The nearest breeding areas are in northern Canada.

Habitats and Ecology: While in the Rocky Mountain region these birds are usually found along snow-free roads, in partially snow-free weedy fields, stubble fields, snow-free hilltops on cultivated lands, and similar areas where grain or weed seeds are likely to be found.

Seasonality: Colorado records are too few to estimate seasonality. In Wyoming the birds are usually present from early November until mid-March, and in Montana the range of dates is from October 26 to May 2, but the birds are mainly present from early November to mid-March. In Alberta they usually arrive in October and are mostly gone by the end of April, with a few stragglers remaining into May.

Comments: These are among the most northerly nesting of all land birds, and also among the most tolerant of winter cold of the small birds of the region.

Suggested Reading: Nethersole-Thompson, 1966; Tinbergen, 1939.

Bobolink *(Dolichonyx oryzivorus)*

Identification: Breeding males have an almost entirely black plumage except for white wing patches, a white rump, and a large patch of yellow on the nape and hindneck. Females closely resemble various female sparrows, but are somewhat larger and have buffy crown and "eyebrow" stripes, as well as slightly spotted flanks and a distinctly striped buffy and brown back. The male's song is a loud *bob-o-link.*

Status: A summer resident in grasslands mainly east of the mountains, but local in western Montana and northwestern Colorado. A rare migrant in most of the montane parks, but a reported breeder at Yellowstone N.P. and a possible rare breeder elsewhere.

LATILONG STATUS

s	S	M	?
	S	s	s
V	S	S	s

s	S	s	s
M	S		
S	s	M	s

		M	M
M	s	M	M
	S	s	S

Habitats and Ecology: Breeding occurs in tall-grass prairies, ungrazed or lightly grazed mid-grass prairies, wet meadows, hayfields, retired croplands, and similar habitats. Scattered bushes or other singing posts in the territory add to its attractiveness.

Seasonality: Colorado records are from May 4 to September 9. Wyoming records extend from May 13 to August 27, with migration peaks in May and September. In Montana the birds are usually present from mid-May to September, while in Alberta they often do not arrive on the breeding grounds until early June. Colorado nest records are from June 19 to July 1, while in Montana and Alberta active nest records extend from June 14 to the latter part of July.

Comments: Bobolinks have extremely long migratory routes, wintering in central South America south to northern Argentina.

Suggested Reading: Martin, 1974; Wittenberger, 1978.

Red-winged Blackbird (*Agelaius phoeniceus*)

Identification: Males of this abundant species are entirely black except for their yellow-lined red epaulets, which are evident on standing as well as flying birds. Immature males are less colorful, and are somewhat brown and streaked, but with some reddish color on the upper wing surface. Females are heavily streaked with brown and resemble large sparrows, but have sharply pointed tapering beaks. The male's song is a liquid *kong-ka-ree*, uttered with the epaulets raised and the wings partially spread.

LATILONG STATUS

S	S	S	s
s	R	S	S
S	R	R	S

R	S	S	R
S	S	S	S
R	R	S	S

S	S	S	S
s	S	S	S
R	R	R	R

Status: A seasonal or permanent resident in suitable habitats throughout the region, including both lowlands and montane areas, and a relatively common breeder in all the montane parks.

Habitats and Ecology: Typical breeding habitats are wetlands ranging from deep marshes or the emergent vegetation zones of lakes and reservoirs through variably drier habitats including wet meadows, ditches, brushy patches in prairies, hayfields, and weedy croplands or roadsides. Wetlands with bulrushes or cattails are especially favored for nesting, but sometimes shrubs or other woody vegetation are used for nest sites. Outside the breeding season the birds often stray far from water, and seek grainfields, city parks, pasturelands, and other habitats offering food sources.

Seasonality: Year-round residency is typical, at least locally, in Colorado and parts of Wyoming, although large seasonal changes in abundance are evident. In Montana the birds are usually present from February through November, and in southern Alberta the birds begin to arrive in late March or early April. They usually remain until late October or early November. Wyoming egg records are from April 20 to June 9, while in Montana nest records extend from mid-May to late June. In southern Alberta eggs have been reported from May 28 to July 18.

Suggested Reading: Peek, 1971; Orians, 1961; Holm, 1973; Payne, 1969.

Western Meadowlark (*Sturnella neglecta*)

Identification: Meadowlarks are grassland songbirds with bright yellow underparts except for a conspicuous black V on the breast region. The upperparts are spotted and striped with brown and buff, and the outer tail feathers are mostly white (apparent only during flight). Eastern and western meadowlarks are almost impossible to separate visually, but the male western meadowlark has a complex, fluty song of many syllables. Eastern meadowlarks are unreported for the region.

Status: A seasonal or permanent resident almost throughout the region, becoming rarer northwardly, and not known to nest in the higher mountain areas of Alberta, although common on the adjacent plains and foothills.

LATILONG STATUS

R	S	s	s
s	R	S	S
R	R	S	S

R	S	S	R
S	S	S	S
S	s	s	S

S	S	S	S
S	S	S	S
R	R	S	R

Habitats and Ecology: During the breeding season this species occupies mixed-grass to tall-grass prairies, wet meadows, hayfields, the weedy borders of croplands, retired croplands, and to some extent short-grass prairies and sage prairies as well as mountain meadows as high as about 7,000 feet in southern parts of the region.

Seasonality: In Colorado this species is a local year-round resident. In Wyoming there is also some year-round residency, but major migrations occur in March and October. In Montana the birds are usually present from February to November, with peaks in early April and September. In Alberta they usually arrive in March and are mostly gone by the end of October. Colorado nest records extend from late May to early July. In Wyoming there are egg records from May 10 to June 26, and in Montana and Alberta there are egg records from mid-May to June 20.

Comments: Western meadowlarks are birds of the Great Plains, overlapping only slightly with the eastern meadowlark, which is adapted to moister grasslands of eastern North America.

Suggested Reading: Lanyon, 1956, 1957; Falls & Krebs, 1975; Rowher, 1971.

Yellow-headed Blackbird (*Xanthocephalus xanthocephalus*)

Identification: Males of this large blackbird have bright yellow on the head, neck, and upper breast, and white wing-patches, but otherwise are entirely black. Females are less obviously striped than are female red-winged blackbirds, and have a yellowish throat and breast. The "song" of males is a grating croak resembling the sound made by a rusty hinge.

LATILONG STATUS

S	S	s	s
	S	s	S
S	S	S	s

S	S	S	S
S	S	S	S
S	S	s	S

S	s	S	S
M	S	S	S
S	S	S.	S

Status: A summer resident in wetland habitats throughout almost the entire region; rarer in montane areas and not known to breed in most of the Canadian montane parks, although common on the nearby prairie marshes. Breeds to about 8,000 feet in mountains at the southern end of the region.

Habitats and Ecology: Restricted during the breeding seasons to relatively permanent marshes, the marsh zones of lakes, and the shallows of river impoundments where there are good stands of cattails, bulrushes, or phragmites. Although sometimes breeding in the same areas as red-winged blackbirds, yellow-headed occupy the deeper areas adjacent to open water.

Seasonality: Colorado records extend largely from March 14 to October 3, with rare overwintering. Wyoming records are from April 14 to October, rarely to December. In Montana the birds are usually present from late April to late September or early October, and in Alberta they are typically present from early May to September. Colorado egg records are from May 18 to June 15, and in Wyoming there are egg records from May 20 to early July. Montana and Alberta egg records are from late May to June 19, with nestlings reported to late July.

Comments: This species is more dependent upon aquatic insects than is the red-winged blackbird, and thus is much more limited to relatively permanent water areas than is that species.

Suggested Reading: Willson, 1964; Orians & Christman, 1963, 1968.

Rusty Blackbird (*Euphagus carolinus*)

Identification: Male rusty blackbirds are very similar to Brewer's black-birds, but have less iridescent plumage that is somewhat greenish rather than purplish. Females are slightly more brownish than males, and have pale yellowish eyes. During fall the male has a rusty brown plumage with yellowish eyes, and females are a brighter rust. More often found near water than the Brewer's blackbird.

Status: A local summer resident in woodland wet-lands near the northern limits of the region, breeding south to Jasper N.P.; rare in summer and breeding unproven for Banff N.P. Otherwise an irregular to uncommon migrant over much of the region, mainly east of the mountains.

LATILONG STATUS

M	M	M
	M	?
V	M	M

M		M
	M	

	M	
	M	
		W

Habitats and Ecology: During the breeding season this species is largely limited to wooded wetlands including alder–willow bogs, the brushy borders of lakes and slow-moving streams, receding muskegs, forest edges, and the borders of beaver ponds. Nests are usually placed over water, either in bushes or low conifers. On migration and dur-ing winter they use a much bigger variety of habi-tats, but typically roost in marshy or swampy areas.

Seasonality: In Colorado these are overwintering migrants, with records from August 29 to May 19. In Wyoming and Montana they are spring and fall migrants, with most records for fall. Alberta records are from April 3 to December 3, but they typically arrive in mid-April and migrate south in October and November. Alberta egg records are from May 15 to June 30, with the majority between May 21 and June 6.

LOCAL MIGRANT

Comments: The close similarity of this species and the Brewer's blackbird no doubt helps to account for the relatively few available migration records, especially in spring, for the rusty blackbird in this region.

Suggested Reading: Kennard, 1920; Bent, 1958.

Brewer's Blackbird (*Euphagus cyanocephalus*)

Identification: Males of this blackbird are entirely a glossy black, with a purplish head sheen and more greenish on the body, and with a yellow eye. Females are a uniformly dark brown, with dark brown eyes. The male rusty blackbird is very similar in spring, but is less iridescent, with no greenish tinges. Females of the two species are also similar in plumage, but the difference in eye color helps to separate them.

LATILONG STATUS

S	S	s	s
s	S	s	S
S	S	S	S

S	S	S	S
s	S	S	S
S	S	S	S

S	s	S	S
s	S	S	S
S	S	S	R

Status: A summer resident virtually throughout the entire region, breeding in most habitats from plains to mountain meadows, and a fairly common breeder in nearly all the montane parks.

Habitats and Ecology: Low-stature grasslands are the primary breeding habitats of this species, including mowed or burned areas, farmsteads and residential areas, the edges of marshes, especially where scattered shrubs are present, aspen groves, the brushy edges of prairie creeks, and similar locations. Nesting occurs on the ground or in low shrubs, and shrubs or fenceposts also serve as singing posts where they are available. Outside the breeding season a wider array of open habitats are used, especially grainfields, orchards, berry farms, and similar agricultural lands.

Seasonality: Locally resident as far north as southern Colorado, and in Wyoming the records are mainly from early March to early December, with migration peaks in May and October. In Montana the birds are usually present from late April to October, and in Alberta they normally arrive in late April and leave by October, with a few laggards remaining until November. Colorado egg records are from May 16 to June 17, and in Montana records of active nests are from May 16 to June 6. Alberta egg records are from May 20 to July 2, with a peak in late May and early June.

Suggested Reading: Orians & Horn, 1969; Horn, 1968, 1970; Williams, 1952.

Common Grackle (*Quiscalus quiscula*)

Identification: This is a large, long-tailed blackbird that has pale yellow eyes in both sexes, a highly iridescent plumage with bronzy sheen on the back, a long, tapering beak, and a tail that is often bent upwards in a V (males) or U (females) while in flight. The male has a loud, wheezy call in spring. Juveniles are uniformly brownish, and have dark eyes.

Status: A summer resident in suitable plains or foothills habitats over most of the region east of Idaho, but generally rare in the mountains, and usually rare or absent from the montane parks. There is one Idaho breeding record.

Habitats and Ecology: Breeding habitats consist of woodland edges, areas partially planted to trees such as residential areas, farmsteads, shelterbelts, coniferous or deciduous woodlands of an open nature, woody shorelines around lakes, and riparian woodlands. Junipers, spruces, and other small and dense conifers are preferred for nesting, although hardwoods, shrubs, buildings, birdhouses, and even cattails are sometimes also used.

Seasonality: A local year-round resident as far north as Colorado, and in Wyoming the records extend from April 11 to December 18, with migration peaks in April and October. In Montana the birds are usually present from mid-April to late October, and in Alberta from the latter part of April until October. Active nest records from Colorado are from April to mid-June, and in Montana and Wyoming eggs have been found from May 15 to June 11. Alberta egg records are from May 12 to June 2.

Suggested Reading: Maxwell & Putnam, 1972; Maxwell, 1965, 1970; Ficken, 1963.

LATILONG STATUS

	M	s	s
	s	?	s
V	M		S

S	S	S	S
	V	S	S
	M	S	S

s	M	S	s
	s	S	M
S	s	S	S

Brown-headed Cowbird (*Molothrus ater*)

Identification: Unlike the other "blackbirds," males of this species have a short, strongly tapering beak, and a sharp contrast between the iridescent body and the non-glossy and brownish head and neck. Females are a very uniformly grayish brown without any distinctive field-marks, but they are stouter than the other blackbird females and have a much shorter and blunter beak. The male's song is squeaky and gurgling.

LATILONG STATUS

S	S	S	s
s	S	S	S
S	S	S	S

S	S	S	S
s	S	M	S
	S	S	S

s	s	S	s
s	S	S	S
S	S	S	S

Status: A summer resident throughout the region in most habitats; variably common in the montane parks and probably breeding in all.

Habitats and Ecology: Breeding occurs in a variety of woodland edge habitats, including brushy thickets, forest clearings, brushy creek bottoms in prairies, aspen groves, sagebrush, desert scrub, agricultural lands, and open coniferous forests at lower altitudes (up to about 7,000 feet in southern parts of the region).

Seasonality: Most Colorado records are from late March to December, with a few for January and February. Wyoming records are from February 6 to late fall, with a few winter records. In Montana and southern Alberta the birds are usually present from late April or early May to late August.

Comments: This species is a social parasite, laying its eggs in the nests of such common species as meadowlarks, red-winged blackbirds, American robins, and others.

Suggested Reading: Friedmann, 1963; Payne, 1965, 1973; Norman & Robertson, 1975; Mayfield, 1965.

Orchard Oriole (*Icterus spurius*)

Identification: Adult males have black heads and upperparts including the breast and upper back, chestnut-brown underparts and rump, and blackish wings and tail. They are smaller and much duller than northern orioles, and larger and less colorful than redstarts, which also have white underparts and orange patches on the tail. Females are mostly lime-green to yellowish, with whitish wing-bars, and first-year males are similar to females but have black chins and throats. The male's song is a long and complex melodious warbling, often ending in a descending slurred note.

Status: A local summer resident in eastern Montana and extreme eastern Wyoming, and a migrant farther south, with vagrants occasionally reaching the mountains.

Habitats and Ecology: Associated with lightly wooded river bottoms, scattered trees in open country, shelterbelts, farmsteads, residential areas, and orchards during the breeding season, and extending into sagebrush and juniper woodlands during the non-breeding season. Nests are built in small to moderately large trees, from 5 to 70 feet above ground.

Seasonality: Records in Colorado and Wyoming are from May 11 to late September. Too few records for Montana exist to estimate migration periods, but in North Dakota the birds are present from late May to early September. Colorado nest records are from June 5 to July 14, and North Dakota egg records are from May 21 to July 3.

Comments: Nests of this species are much smaller and rounder than those of the northern oriole, and are less easily found. However, kingbirds and orioles often share the same nesting trees, with the orioles presumably gaining protection from the highly territorial defensive behavior of the kingbirds.

Suggested Reading: Dennis, 1948; Bent, 1958.

LATILONG STATUS

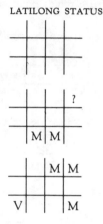

Northern Oriole (*Icterus galbula*)

Identification: Adult males are a brilliant golden to orange-red on the underparts, outer tail feathers, and white wing-patches, and with black upperpart coloration and a black throat. The race that breeds over most of the region (Bullock's oriole) has the orange-yellow color extending forward to the cheeks and has more white on the wings, while the more easterly race (Baltimore oriole) has a completely black head and less white but more yellow on the upper wing surface. Females and first-year males are yellowish to grayish on the underparts, with a brighter yellowish breast than occurs on the female orchard oriole, and immature males have varying amounts of black on the throat.

LATILONG STATUS

M	S	s	
s	S	s	s
S	S	S	S

S	S	S	S
M	S	M	S
s	S	s	S

S	M	S	S
M	s	s	S
S	S	s	S

Status: A local summer resident in wooded plains and foothills areas throughout most of the region, but becoming rare in the mountains, and absent from most of the northern montane parks. Breeds up to about 7,000 feet at the southern portions of the region.

Habitats and Ecology: During the breeding season males of the Bullock's race especially favor river bottom forests of willows and cottonwoods, but also occur in city parks, and plains or foothill slopes and valleys with aspen, poplars, birches, and similar vegetation. Later on in the summer the birds are attracted to trees and bushes that provide berries.

Seasonality: Colorado records extend from March 15 to November 15. Wyoming records are from April 30 to September 6, and those from Montana are from May 8 to September 8. Active nests in Colorado have been seen from June 2 to 27, and in Montana and Wyoming from May 29 to early July.

Comments: Originally considered two species, the Baltimore and Bullock's orioles are known to hybridize over a rather wide zone in the Great Plains.

Suggested Reading: Bent, 1958; Sibley & Short, 1964.

BALTIMORE

BULLOCK'S

Rosy Finch (*Leucosticte arctoa*)

Identification: These are small alpine finches with rose-tinted rumps, underparts, and wing coverts. In northern areas the birds ("gray-crowned rosy finches") have gray crowns and reddish brown back and breast colors. From central Idaho and west-central Montana south through central Wyoming the nape color is still gray, but the back and breast are dusky brown ("black rosy finch"). Finally, in Colorado and adjacent southeastern Wyoming the crown and nape are dark brownish, and the breast and back are grayish brown, becoming reddish on the belly ("brown-capped rosy finch"). Hybrids occur in some areas (Seven Devils area of western Idaho and Bitterroot Mountains of eastern Idaho), and during winter the northern forms occur south of their breeding ranges.

Status: All three forms are limited to alpine areas of high mountains in the region, probably occurring on all such montane areas, including all the montane parks.

Habitats and Ecology: During the breeding season these birds inhabit cirques, talus slopes, alpine meadows with nearby cliffs, and adjacent snow and glacial surfaces (where foraging for frozen insects occurs). Nesting is done in cliff crevices or among talus rocks. During fall and winter the birds move to lower elevations including mountain meadows, grasslands, sagebrush areas, and agricultural lands.

Seasonality: Collectively these birds are resident throughout the region, although on the foothills and plains areas they are present only as winter visitors. Colorado egg records are from June 28 to July 27, in Wyoming active nest records extend from July 1 to August 11, and in Montana and southern Alberta nestlings have been observed in July and early August.

Suggested Reading: Hendricks, 1978; Shreeve, 1980; Twining, 1940; Johnson, 1965; French, 1959.

LATILONG STATUS

S	R	R	–
s	M	S	M
R	S	S	M

r	S	S	M
r	R	R	–
M	R	R	M

W		W	
W	W	R	W
W	R	R	R

Pine Grosbeak (*Pinicola enucleator*)

Identification: This is the largest of the reddish finch group, or about the size of an American robin, and has the combination of a blunt, uncrossed beak and blackish wings with double white wing-bars. Males are reddish on the head, back, and breast, while females and young males are mostly grayish olive except for the white wing-bars and blackish wings. The male's song is a musical warbling.

LATILONG STATUS

r	R	r	M
M	r	r	M
R	r	r	M

r	R	r	M
s	R	R	
s	r	r	

r		M	
		R	R
W	W	S	R

Status: A permanent resident in coniferous forests throughout the region, including all of the montane parks, and probably breeding in all of them.

Habitats and Ecology: Breeding occurs in the upper levels of the coniferous forest, primarily the alpine fir, Engelmann spruce zone. Nesting usually occurs in such conifers, especially in open or scattered woods near meadows or streams. Outside the breeding season the birds descend to lower conifer zones, especially the pinyon–juniper zone, where the birds often feed on pinyon nuts. Some berries, grains, and other food sources are also used, but conifer seeds are primarily eaten.

Seasonality: A permanent resident throughout the region, although altitudinal movements cause the birds to be migrants or winter visitors at lower altitudes. Active nests in Colorado have been observed from June 23 to August 25. In Wyoming eggs have been reported in late June, and in Montana dependent young have been observed in early September.

Comments: The young of this species are fed regurgitated materials that are primarily of vegetable matter, including various berries and seeds, as well as some insect materials.

Suggested Reading: French, 1954; Bent, 1968.

Purple Finch (*Carpodacus purpureus*)

Identification: Males of this small wine-red colored finch are rather uniformly red over the entire head and upper breast, including the hindneck. The upper back is also distinctly tinted with reddish. Females resemble heavily streaked brown and white sparrows, but have notched tails and a conspicuous whitish eyebrow stripe. The male's song is a rich, bubbling warble that may last for several seconds.

Status: A local summer resident at the northern edge of the region, breeding south uncommonly to Jasper N.P. and occasionally to Banff N.P. Farther south it is a migrant and wintering visitor in Montana, an occasional winter visitor in Wyoming, and an accidental winter visitor in Colorado.

Habitats and Ecology: Breeding in this species occurs in natural conifer forests, mixed forests, and conifer plantings, especially where moist and shaded habitats occur. Preference seems to be for mixed forests, with the birds nesting in conifers but feeding in deciduous trees. Buds and blossoms of a variety of broad-leaved trees are favored in spring, while in summer they consume a variety of berries, fruit, and insects. During the winter period they eat a variety of weed and grass seeds, and thus have a broad winter habitat distribution.

Seasonality: In Alberta these birds are present from about the last week of April until early October. Alberta egg records are from June 2 to 12, and in British Columbia there are egg records from May 1 to July 25. Fledged young have been observed at Jasper N.P. by June 28.

Comments: In spite of their transcontinental Canadian breeding distribution, wintering in the Rocky Mountain area is strangely almost entirely lacking except apparently in Montana; even there it has been reported from only 11 latilongs.

Suggested Reading: Salt, 1952; Bent, 1968.

LATILONG STATUS

Cassin's Finch (*Carpodacus cassinii*)

Identification: This small finch resembles the purple finch and house finch, but males have a much more brownish and less reddish upper back and hindneck than the purple finch, and a whitish, unstreaked belly and flanks rather than the heavily streaked underparts of the Cassin's finch. Females very closely resemble females of the purple finch, but are more strongly streaked on the underparts and have a more streaked and obscured facial pattern, without nearly white striping present above the eye or in the mustache area. The male's call-note is a two-noted *kee-up*, unlike the purple finch's *tick* note.

LATILONG STATUS

S	S	s	
s	r	r	s
R	r	R	r

R	s	s	r
r	R	R	M
S	R	R	R

r	r	r	r
R	r	r	r
R	r	R	R

Status: A local summer or permanent resident in coniferous forests of the region north to extreme southern British Columbia and adjacent Alberta (breeding at Watertown Lakes N.P. and reported rarely north to Jasper, but not known to breed there).

Habitats and Ecology: Breeding typically occurs in open, rather dry coniferous forests, including ponderosa pine forests, with the nests placed at considerable heights in large conifers. Generally it occurs at rather higher altitudes than do either the house finch or the purple finch, sometimes almost to timberline. Throughout the year it is primarily vegetarian, feeding on buds, berries, and seeds, especially those of conifers.

Seasonality: In Colorado and Wyoming the birds are year-round residents, with substantial movements up and down the mountains with the seasons. In Montana the birds are more clearly migratory, with some seasonal movements out of northern regions, such as Glacier N.P. Seasonal movements probably also occur in Canada, but remain undocumented. Colorado egg records are from June 11 to July 30, and in Wyoming and Montana nest-building or incubation activities have been reported from May 20 to July 17.

Suggested Reading: Salt, 1952; Samson, 1976; Jones & Baylor, 1969.

House Finch (*Carpodacus mexicanus*)

Identification: These are small finches (size of house sparrows) in which the males are bright red on the rump, breast, throat, and crown, but with little or no red on the rear cheek area, and with distinctly streaked flanks and underparts. The females are streaked with brownish, lack pale markings on the head, and have only a very slightly notched tail. The male's song is a scrambled warbling, and the calls include a hoarse *weet* note and a sweeter, often repeated *cheep*.

Status: A local summer resident from northern Idaho to southeastern Wyoming, mainly at lower altitudes, including plains and foothills up to 9,000 feet at the southern end of this region. Generally rare or absent from the montane parks, but an uncommon nester in Rocky Mountain N.P.

Habitats and Ecology: Now generally associated with human habitations over most of its range, nesting on buildings in such areas. Otherwise it nests in open woods, river-bottom woodlands, scrubby desert or semi-desert vegetation such as sagebrush, and tree plantings. Deciduous underbrush, preferably close to water, is favored over dense coniferous woods, and sources of seeds, berries, or fruits are also needed throughout the year.

Seasonality: A year-round resident throughout the region, becoming more uncommon and local northwardly and at higher altitudes. Colorado egg records are from April 24 to July 22, with a peak during the second half of May. Egg records in Wyoming are from April 27 onward, with nestlings observed to August 16. In Idaho eggs have been seen as early as April 28.

Suggested Reading: Thompson, 1960; Van Riper, 1976; Evenden, 1957.

LATILONG STATUS

	M	?	
	M		
S	R	R	M

r	M	s	r
	V	R	M
	V	r	R

r	M	r	M
R	R	r	R
R	s	R	R

Red Crossbill (*Loxia curvirostra*)

Identification: Only this and the following species have beaks with crossed tips, and this species lacks the double white wing-bars typical of the white-winged crossbill. Males are variably reddish, depending on age, and females are mostly yellowish green, with a typical finch-like notched tail and a yellowish rump. The call is a repeated *kip* note, often uttered in flight.

LATILONG STATUS

r	R	r	?
r	r	r	s
R	r	R	r

R	s	s	R
s	R	r	
	r	r	r

r	M	r	r
M	r	r.	R
r	R	R	R

Status: A local resident in coniferous forest areas throughout the region, including all the montane parks, where it is a probable breeder in all.

Habitats and Ecology: Breeding is associated with coniferous forest habitats, especially those of pines, including ponderosa, lodgepole, and pinyon, but nesting in the region has also been observed in Engelmann spruces and subalpine firs, at elevations from 4,000 feet to as high as 10,000 feet or more. Breeding in the Rocky Mountain region seems associated with the higher levels of coniferous forests, but the pinyon zone is often used by wintering birds.

Seasonality: A year-round resident throughout the region, but altitudinal movements occur, with the birds moving to foothills and plains during winter. Nesting times are irregular and apparently dependent upon seed crop availability; in Colorado nesting has been observed from December to mid-September. In Wyoming nestlings have been noted from February 12 to June, and newly fledged young observed in August. In Montana active nests or newly fledged young have been seen from April 30 to July 27, and in Alberta there are egg records from March 3 to May 3. Probable nesting activities in Idaho have been noted from February 21 (carrying nesting materials) to July (singing males).

Suggested Reading: Griscom, 1937; Snyder, 1954; Nethersole-Thompson, 1953; Bailey et al., 1953.

White-winged Crossbill (*Loxia leucoptera*)

Identification: Like the previous species this one has uniquely crossed mandibles and black wing and tail coloration, but unlike the red crossbill this species has two broad white wing-bars on the upper wing surface. Otherwise, both species are nearly identical in plumage color. The songs and call-notes of the two species are also very similar, the song consisting of an extended series of trills, often given in flight, and the call a series of *wink* notes, often uttered in flight. The call of this species is somewhat less harsh than that of the red crossbill, and is inflected upwardly, rather than downwardly.

Status: A local resident in coniferous forests of the northernmost part of the region (Banff and Jasper N.P., Cypress Hills), has possibly bred in Glacier N.P., locally resident through the mountains of Montana south almost to Yellowstone, but no definite breeding records. There are summer observations for five Wyoming latilongs.

Habitats and Ecology: During the breeding season associated with coniferous forests and mixed forests containing spruces. Generally, spruces and tamaracks seem to be this species' prime food sources, as their beaks are too weak to handle the larger cones of pines. Although nesting occurs most commonly during spring and summer or early fall, like the red crossbill it can apparently occur almost any time a rich seed source becomes available. In spring, catkins of aspens and poplars are sometimes eaten, and large weed seeds may be eaten in fall and winter.

Seasonality: A permanent resident in the region. There are no specific nesting records for the area. Throughout Canada nesting has been reported for every month of the year.

Comments: Like the red crossbill this is a seemingly eruptive species that might occur in a particular area for several years, then disappear and not be seen again in the area for an indefinite period.

Suggested Reading: Bent, 1968; Newman, 1972.

LATILONG STATUS

r	r	s	?
	M	M	M
V	M	M	M

s	s	?	M
	r	W	
	W	W	

W		W	W
		W	W
		r	W

Common Redpoll (*Carduelis flammeus*)

Identification: This is a small and gregarious finch with a notched tail, a black chin, and a bright red cap. The only similar species of the region is the rare hoary redpoll, which is considerably paler throughout, including a whitish, unstreaked rump. The common redpoll's song is a rapid mixture of whistles, trills, and buzzy notes, and its call-notes are repeated *chit* notes as well as a goldfinch-like *sweeeet* note that rises in pitch.

LATILONG STATUS

W	W	W	
W	W	W	M
W	W	W	W

W	W	M	W
M	W	W	
W	W	W	

W		W	
			W
W		W	W

WINTERING MIGRANT

Status: A local wintering migrant almost throughout the region, both in montane areas and in plains or foothills, but probably commoner at lower altitudes in winter. The nearest breeding records are from central Alberta (Edmonton), but regular breeding occurs along the northern edges of the prairie provinces.

Habitats and Ecology: Breeding typically occurs in subarctic forests, typically nesting in dwarf spruces or in thickets of willows and alders. In the Rocky Mountain region the birds are associated with such open habitats as desert scrub, sagebrush, and grasslands. They also visit cities during winter to eat the seed cones of birches, visit bird feeders, and seek out weedy patches.

Seasonality: From central Alberta southward this species is a variably common overwintering migrant. In southern Alberta and Montana they usually arrive in mid-October, and remain until early April. In Wyoming the records extend from October 14 to April 14, but are primarily present from November to March. In Colorado the records extend from October 21 to April 25.

Comments: These are gregarious little finches that move south in winter only as far as their food supplies require, and are highly tolerant of snow, occasionally even spending the night in snow tunnels.

Suggested Reading: Bent, 1968.

Hoary Redpoll (*Carduelis hornemanni*)

Identification: This species closely resembles the common redpoll, but is noticeably paler, with only sparse streaking on the flanks and a pure white rump. Its song is apparently identical to that of the common redpoll, and its call-notes are extremely similar if not identical. Most likely to be seen as lone individuals within a flock of common redpolls.

Status: A rare wintering visitor in the region, with the nearest breeding areas in northeastern Manitoba, along the coast of Hudson Bay. In Alberta the species is generally but not invariably less common than the common redpoll in wintering flocks, but in Montana it has been reported in only about half as many latilongs as has the common redpoll. In Wyoming there are reports from four latilongs (vs. 16 for the common redpoll), and in Colorado it is as yet unreported. It is also unreported from Idaho, but is likely to occur in northern parts of the state.

LATILONG STATUS

M			
M		?	
V	M	M	M

M	?	M	M
	V		

		V	

Seasonality: A wintering visitor throughout the region, with a seasonality that probably coincides with that described for the common redpoll.

Comments: The taxonomic level of distinction of this form from the common redpoll is still in doubt, and some ornithologists consider them as no more than racially distinct.

Suggested Reading: Alsop, 1973; Bent, 1973.

WINTERING MIGRANT

Pine Siskin (*Spinus pinus*)

Identification: This small finch has a short, notched tail, a rather sharp but short beak, and a body that is mostly streaked with brownish and white, but with yellow markings at the base of the tail and the base of the inner flight feathers. Usually seen in groups, and almost always near conifers. The song is a goldfinch-like warbling, and the calls include a hoarse *teee* and a hoarse *jeeeah* note.

LATILONG STATUS

r	R	R	s
s	r	r	s
R	R	R	R

R	S	s	R
S	R	R	R
S	R	R	R

r	r	R	r
M	R	R	R
S	S	S	R

Status: A local resident in coniferous forests virtually throughout the region, including all the montane parks, where common to abundant, and probably breeding in all.

Habitats and Ecology: Breeding occurs in coniferous or mixed forests, and rarely in deciduous woodlands. Nesting preferentially occurs in conifers of almost any type, but has also been observed in cottonwoods, lilacs, and willows in the Rocky Mountain region. Foods are mainly conifer seeds, but also may include those of alders, birches, or various weeds, and seasonally feed on flower buds and insects.

Seasonality: A permanent resident almost throughout the region, although with considerable seasonal movements and with most birds moving out of Alberta from October to May. Colorado egg records are from April 21 to July 5, while in Wyoming and Montana there are egg records for late June and newly fledged young observed as late as the latter part of August.

Comments: Like many of the true finches, these birds often nest in small, loose colonies, and with the colonies often shifting in location irregularly.

Suggested Reading: Rodgers, 1937; Weaver & West, 1943.

Lesser Goldfinch (*Carduelis psaltria*)

Identification: Males in this part of the species' range have a plumage similar to that of an American goldfinch, but with the back, rump, and hindneck dark greenish rather than yellow. Females have a uniformly colored greenish back and rump color. The male's song is similar to that of an American goldfinch but more scratchy, while the calls are plaintive or questioning notes.

Status: A very local summer resident in the southeasternmost part of the region, breeding north to near Lyons, or 20 miles from Estes Park, and observed in Rocky Mountain N.P. as a rare visitor. Occasional at Dinosaur N.M. and a potential breeder there.

Habitats and Ecology: Breeding occurs in sagebrush and riparian thicket areas, as well as where scrub oaks merge with ponderosa pines. In Colorado it has been found breeding between 5,000 and 11,500 feet, but is certainly most common at the lower elevations in oak–pine woodlands. It also nests commonly in trees in cities and suburbs.

Seasonality: Migratory at the northern end of its breeding range, and usually present from about April to mid-September. Active nests in Colorado have been reported from May 10 to August 3. A few stragglers sometimes persist into December.

Comments: This species also occurs in a black-backed male plumage type that is variably common from Colorado southward into Mexico.

Suggested Reading: Linsdale, 1957; Coutlee, 1968a, 1968b.

LATILONG STATUS

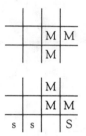

American Goldfinch (*Carduelis tristis*)

Identification: Breeding males have a bright lemon-yellow plumage except for a black forehead, a black notched tail, and mostly black wings except for white forewing patches. Winter males and females are much duller, but have white wing-bars, a short and notched tail, an unstreaked yellowish to brownish buff breast and underpart color, and a short, stubby beak. The usual call is a *per-chik-o-ree* or *ti-dee-di-di*, often given in flight while wing-beating between gliding phases of flight.

LATILONG STATUS

r	s	s	s
s	s	s	s
S	s	s	S

R	S	S	S
s	S	S	S
S	S	s	S

M	s	s	s
M	s	s	S
S	R	S	R

Status: A seasonal or permanent resident almost throughout the entire region, but absent from the northernmost montane areas, and apparently only a rare breeder in the montane parks.

Habitats and Ecology: Breeding occurs in open grazing country, especially where thistles are abundant, or where cattails are to be found. The seeds of thistles and other composites are used for feeding the young, and the "down" of thistles or cattails are used in nest construction. Riparian woodlands near weed-infested fields provide an ideal nesting situation. During winter the birds range widely over weedy fields and farmlands.

Seasonality: A permanent resident throughout much of the region, but migratory to the north, and usually absent from Alberta from November to late May. In Montana and Wyoming the birds are present mainly from May until September or October. Nesting records in Colorado are from July 31 to August 10, while in Wyoming and Montana active nests have been observed from mid-July to early September.

Comments: This is perhaps the most widespread and familiar of the true finches of the region, and is often called the "wild canary" by laymen.

Suggested Reading: Nickell, 1951; Coutlee, 1961; Stokes, 1950.

Evening Grosbeak (*Hesperiphona vespertina*)

Identification: This is a large, stocky finch with a short, massive beak, black wings and tail (the wings having white outer patches in females and white inner patches in males), and a bright yellow (males) to dingy gray (females) body color. In both sexes the beak is whitish to pale greenish. Like other finches it flies in a distinctive undulating flight, and a loud *cleeep* is the most common call. Males also utter a warbling song.

Status: A local resident in coniferous forests almost throughout the region. Present in all the montane parks and probably breeding in all of them.

Habitats and Ecology: During the breeding season this species is primarily associated with mature coniferous forests, although nesting has also been observed in riparian willow thickets and also in city parks and orchards. Nesting in elms, maples, and box elders has been reported, and seeds of the last-named tree appear to be a highly favored food. During fall and winter they often occur in flocks that feed on such large and nutritious seeds as maples, ashes, and sunflowers.

Seasonality: A permanent resident throughout the region, but with major seasonal wanderings or migrations. Nesting in Colorado has been reported from early June to late July. Dependent young have been seen in Montana in early July, and there are Manitoba egg records for mid-June.

Suggested Reading: Parks & Parks, 1963; Blair & Parks, 1964; Bent, 1958.

LATILONG STATUS

r	R	s	
s	r	r	s
R	R	R	r

R	R	s	s
s	R	R	M
	R	M	M

r		M	
M	r	M	
M	M	M	S

House Sparrow (*Passer domesticus*)

Identification: Males of this abundant sparrow have black beaks, throats, and breasts, while their cheeks are pale gray and their napes a chestnut brown. Females closely resemble several other sparrow species, but have an unstreaked breast, a buffy eyebrow line, and a back that is broadly streaked with bright buff. The most commonly uttered call is a loud, monotone chirp. Invariably found rather close to human habitations.

LATILONG STATUS

R	R	R	r
r	R	R	R
R	R	R	R

R	R	R	R
r	r	R	R
R	R	r	R

R	R	R	R
M	R	R	R
R	R	R	R

Status: A local permanent resident throughout the region in human-associated habitats. Generally locally common in the montane parks around developed areas, but rare or absent in more remote habitats.

Habitats and Ecology: Associated throughout the year with humans, and breeding occurs in cities, suburbs, farmsteads, ranches, developed campgrounds in parks, etc. Nesting is usually done on artificial structures such as buildings that offer cavities or crevices, such as vine-covered buildings, billboard braces, bird houses, or old nests of other species, but occasionally occurs in tree cavities.

Seasonality: A permanent resident throughout the region. Nesting occurs over an extended period, usually of at least four months, and at the latitudes concerned typically begins in early April and may last to mid-September. During this time single pairs may produce two or three clutches, rarely as many as four.

Comments: This introduced species, like the starling, is an aggressive nester that often excludes more desirable bird species from nesting boxes or nest cavities, and thus tends to reduce avian diversity in areas near humans.

Suggested Reading: Kendeigh, 1973; Sappington, 1977; Summers-Smith, 1963.

Regional and Local References

State and Multi-state References

Arvey, M. D., 1947. A check-list of the birds of Idaho. *University of Kansas, Museum of Natural History Publications* 1:193–216.

Burleigh, T. D., 1972. *Birds of Idaho.* Caldwell, Idaho: Caxton Printers.

Bailey, A. M., and Niedrach, R. J., 1965. *Birds of Colorado.* 2 vols. Denver: Denver Museum of Natural History.

Cary, M., 1917. Life zone investigations in Wyoming. U.S. Dept. of Agriculture, Bureau of Biological Survey, *North American Fauna* No. 40.

Chase, C. A., III, Bissell, S. J., Kingery, H. E., and Graul, W. D., 1982. Colorado Bird Latilong Study. Revised ed. Denver: Denver Museum of Natural History. (1st ed., 1978, edited by H. E. Kingery & W. D. Graul.)

Davis, C. V., 1961. A distributional study of the birds of Montana. Ph.D. dissertation, Oregon State University, Corvallis.

Dorn, J. L., 1978. Wyoming ornithology: A history and bibliography with species and Wyoming indexes. B.L.M. and Wyoming Game & Fish Dept.

Grave, B. H., and Walker, E. P., 1916. Wyoming birds. *University of Wyoming Bulletin* (Laramie) 12(6):1–137.

Johnsgard, P. A., 1979. *Birds of the Great Plains: breeding species and their distribution.* Lincoln: University of Nebraska Press.

Knight, W. C., 1902. The birds of Wyoming. *Wyoming Experiment Station Bulletin* (Laramie) 55:1–174.

Larrison, E. J., Tucker, J. L., and Jollie, M. T., 1967. *Guide to Idaho birds.* Rexburg, Idaho: Idaho Academy of Science, Ricks College.

Larrison, E. J., 1981. *Birds of the Pacific Northwest: Washington, Oregon, Idaho and British Columbia.* Moscow: University Press of Idaho.

McCafferty, C. E., 1930. An annotated and distributional list of the birds of Wyoming. M.S. thesis, University of Wyoming, Laramie.

418

McCreary, O., 1939. *Wyoming bird life.* Revised ed. Minneapolis: Burgess Publishing Co.

Matthews, W. H., III, 1968. *A guide to the national parks: their landscape and geology.* Vol. 1. The western parks. Garden City: Natural History Press.

Oakleaf, R., Downing, H., Raynes, B., Rayne, M., and Scott, O., 1982. Revised ed. Wyoming avian atlas. Wyoming Game & Fish Dept. and Bighorn Audubon Society. (Preliminary working draft published in 1978.)

Saunders, A. A., 1921. A distributional list of the birds of Montana. Cooper Ornithological Society, *Pacific Coast Avifauna* No. 14.

Skaar, P. D., 1980. *Montana bird distribution: mapping by latilong.* 2nd ed. Bozeman: P. D. Skaar.

Rocky Mountain National Park

Collister, A. E., 1965. A list of birds of Rocky Mountain National Park. 16 pp. Estes Park: Rocky Mountain Nature Association.

————, 1970. Annotated checklist of birds of Rocky Mountain National Park and Shadow Mountain Recreation Area in Colorado. 64 pp. *Denver Natural History Museum Pictorial* No. 18.

Gregg, H. R., 1938. Birds of Rocky Mountain National Park. 80 pp. Estes Park: Rocky Mountain Nature Association.

Kleinschnitz, F. C., 1937. Field manual of birds: Rocky Mountain National Park. 60 pp. National Park Service.

Packard, F. M., 1950. The birds of Rocky Mountain National Park. 81 pp. Estes Park: Rocky Mountain Nature Association. (See also *Auk* 62:371–94.)

Grand Teton National Park

Houston, D. B., 1969. The bird fauna of Grand Teton National Park. 14 pp. mimeo. Special study No. 1, Grand Teton National Park.

Johnsgard, P. A., 1982. *Teton wildlife: observations by a naturalist.* Boulder: Colorado Associated University Press.

Nye, D. L., Back, M., and Hinchman, H., undated. Birds of the Upper Wind River Valley. 34 pp. Shoshone National Forest Publication.

Raynes, B., 1984. *Birds of Grand Teton National Park and the surrounding area.* 90 pp. Moose: Grand Teton Natural History Association.

Yellowstone National Park

Brodrick, H. J., 1952. Birds of Yellowstone National Park. 58 pp. U.S. National Park Service, Yellowstone Interpretive Series No. 2.

Follet, D., undated. Birds of Yellowstone and Grand Teton National Parks. 71 pp. U.S. National Park Service and Yellowstone Library and Museum Association.

Komsies, E., 1930. Birds of Yellowstone National Park, with some recent additions. *Wilson Bulletin* 42:198–210.

———, 1935. Changes in the list of birds of Yellowstone National Park. *Wilson Bulletin* 47:68–70.

Meagher, M., 1963. Bird list of Yellowstone National Park. Unpublished manuscript (mimeo) in files of National Park Headquarters Library, Mammoth Hot Springs.

Skinner, M. P., 1925. The birds of Yellowstone National Park. *Roosevelt Wildlife Bulletin*, Syracuse University School of Forestry 3:11–189.

Glacier National Park

Anonymous, 1937. Check-list of birds of Glacier National Park. 29 pp. (mimeo). Wildlife Division, U.S. National Park Service.

Bailey, F. M., 1918. "The birds," pp. 103–199, in: *Wild animals of Glacier National Park*. U.S. National Park Service, Department of Interior.

Beaumont, G., 1978. *Many splendored mountains: the life of Glacier National Park*. 138 pp. U.S. National Park Service.

Parrat, L. P., 1970. Birds of Glacier National Park. 86 pp. U.S. National Park Service.

Alberta, General References

Anonymous, 1982. Checklist of Alberta birds. 9 pp. 4th ed. Edmonton: Provincial Museum of Alberta.

Hardy, W. G. (ed.), 1967. *Alberta, a natural history*. Edmonton: Hurtig.

Rand, A. L., 1948. The birds of southern Alberta. *National Museum of Canada Bulletin* 111:1–105.

Sadler, T. S., and Myres, M. T., 1976. Alberta birds, 1961–1970, with particular reference to migration. 314 pp. *Occasional Paper No. 1, Provincial Museum of Alberta*, Natural History Section, Edmonton.

Salt, W. R., and Salt, J. R., 1976. *The birds of Alberta*. Edmonton: Hurtig Publishers.

420

Salt, W. R., and Wilk, A. L., 1958. *The birds of Alberta*. 511 pp. Edmonton: Department of Economic Affairs.

Alberta Parks

Clarke, C. H. D., and Cowan, I. McT., 1945. Birds of Banff National Park, Alberta. *Canadian Field-Naturalist* 59:83–103.

Cowan, I. McT., 1955. Birds of Jasper National Park, Alberta, Canada. *Canadian Wildlife Service, Wildlife Management Bulletin*, Series 2, No. 8:1–67.

Kondla, N. G., 1978. The birds of Dinosaur Provincial Park, Alberta. *Blue Jay* 36:103–14.

Godfrey, W. E., 1950. Birds of the Cypress Hills and Flotten Lake region, Saskatchewan. *National Museums of Canada Bulletin* 120:1–96.

Soper, J. D., 1947. Observations on mammals and birds in the Rocky Mountains of Alberta. *Canadian Field-Naturalist* 61:143–73.

Van Tighem, K., and Holroyd, G., 1981. A birder's guide to Jasper National Park, Alberta. *Alberta Naturalist* 11:134–40.

References on Individual Species

Adkisson, C. S., 1966. The nesting and behavior of mockingbirds in northern Lower Michigan. *Jack-Pine Warbler* 44:102–16.

Aldrich, J. W., 1953. Habits and habitat differences in two races of Traill's flycatcher. *Wilson Bulletin* 65:8–11.

Alison, R. M., 1975. Breeding biology and behavior of the oldsquaw (*Clangula hyemalis* L.). American Ornithologists' Union, *Ornithological Monographs* No. 18.

Allen, A. A., 1924. A contribution to the life history and economic status of the screech owl (*Otus asio*). *Auk* 41:1–16.

Allen, D. L. (ed.), 1956. *Pheasants in North America*. Stackpole Co., Harrisburg, Pa., and Wildl. Manage. Inst., Washington, D.C.

Allen, R. P., 1952. *The whooping crane*. Research Report No. 2, New York: National Audubon Society.

Allen, R. P., and Mangels, F. P., 1940. Studies of the nesting behavior of the black-crowned night heron. *Proc. Linnean Soc. New York* 50–51:1–28.

Alsop, F. J., III, 1973. Notes on the hoary redpoll on its central arctic breeding grounds. *Wilson Bulletin* 85:484–85.

Alt, K. L., 1980. Ecology of the breeding bald eagle and osprey in the Grand Teton–Yellowstone Parks complex. M.S. thesis, Montana State Univ., Bozeman.

Ambrose, J. E., Jr., 1963. The breeding ecology of *Toxostoma curvirostre* and *T. bendirei* in the vicinity of Tucson, Arizona. M.S. thesis, University of Arizona, Tucson.

Anderson, D. R., Skaptason, P. A., Kahey, K. G., and Henny, C. J., 1974. Population ecology of the mallard. III. Bibliography of published research and management findings. Bur. Sport Fish. and Wildl., Resour. Publ. 119.

Anderson, S. H., 1976. Comparative food habits of Oregon nuthatches. *Northwest Sci.* 50(4):213–21.

Andersson, M., 1971. Breeding behaviour of the long-tailed skua. *Ornis Scandinavica* 2:35–54.

———, 1973. Behavior of the pomerine skua with comparative remarks on Stercorariinae. *Ornis Scandinavica* 4:1–6.

Angell, T., 1969. A study of the ferruginous hawk: adult and brood behavior. *Living Bird* 8:225–41.

Armitage, K. B., 1955. Territorial behavior in fall migrant rufous hummingbirds. *Condor* 57:239–40.

Armstrong, E. A., 1955. *The wren.* London: Collins.

———, 1956. Territory in the wren, *Troglodytes troglodytes. Ibis* 98(3):430–37.

Armstrong, J. T., 1965. Breeding home range in the nighthawk and other birds: its evolutionary and ecological significance. *Ecology* 46:619–29.

Armstrong, W. H., 1958. Nesting and food habits of the long-eared owl in Michigan. *Michigan State University Museum Publications, Biological Series* 1(2):63–96.

Ashmole, N. P., 1968. Competition and interspecific territoriality in *Empidonax* flycatchers. *Syst. Zool.* 17(2):210–12.

Austin, G. R., 1964. *The world of the red-tailed hawk.* Philadelphia and New York: Lippincott.

Bailey, A. M., Neidrach, R. J., and Bailey, A. L., 1953. The red crossbills of Colorado. *Denver Museum of Natural History, Museum Pictorial* 9:1–64.

Bailey, P. F., 1977. The breeding biology of the black tern (*Chlidonias niger surinamensis* Gmelin). M.S. thesis, State University of Wisconsin, Oshkosh.

Baird, P. A., 1976. Comparative ecology of California and ring-billed gulls (*Larus californicus* and *L. delawarensis*). Ph.D. diss., University of Montana, Missoula.

Bakus, G. J., 1959a. Observations on the life history of the dipper in Montana. *Auk* 76:190–207.

———, 1959b. Territoriality, movements, and population density of the dipper in Montana. *Condor* 61:410–25.

Balda, R. P., and Bateman, G. C., 1971. Flocking and annual cycle of the piñon jay, *Gymnorninus cyanocephalus. Condor* 73:287–302.

———, 1973. The breeding biology of the piñon jay. *Living Bird* 11:5–42.

Balda, R. P., Bateman, G. C., and Foster, G. F., 1972. Flocking associates of the piñon jay. *Wilson Bull.* 84:60–76.

Baldwin, P. H., and Hunter, W. F., 1963. Nesting and nest visitors of the Vaux's swift. *Auk* 80:81–85.

Baldwin, P. H., and Zaczkowski, N. K., 1963. Breeding biology of the Vaux's swift. *Condor* 65:400–6.

Balgooyen, T. G., 1976. Behavior and ecology of the American kestrel (*Falco sparverius* L.) in the Sierra Nevada of California. *Univ. Calif. Publ. Zool.* 103:1–83.

Banfield, A. W. F., 1947. A study of the winter feeding habits of the short-eared owl (*Asio flammeus*) in the Toronto region. *Can. J. Res. Sect. D.* 25(2):45–65.

Banko, W., 1960. The trumpeter swan. U.S. Fish and Wildlife Service, *North American Fauna* 63:1–214.

Barclay, R., 1977. Solitary vireo breeding behavior. *Blue Jay* 35:33–37.

Barlow, J. C., and Rice, J. C., 1977. Aspects of the comparative behavior of red-eyed and Philadelphia vireos. *Can. J. Zool.* 55:528–42.

Barlow, J. C., James, R. D., and Williams, N., 1970. Habitat co-occupancy among some vireos of the subgenus *Vireo* (Aves: Vireonidae). *Can. J. Zool.* 48:395–98.

Bartholomew, G. A., Howell, T. R., and Cade, T. J., 1957. Torpidity in the white-throated swift, Anna hummingbird and poor-will. *Condor* 59:145–155.

Baskett, T. S., 1947. Nesting and production of the ring-necked pheasant in north-central Iowa. *Ecological Monographs* 17:1–30.

Bateman, G. C., and Balda, R. P., 1973. Growth, development, and food habits of young piñon jays. *Auk* 90(1):36–61.

Baumann, S. A., 1959. The breeding cycle of the rufous-sided towhee *Pipilo erythropthalmus* (Linnaeus) in central California. *Wasmann Journal of Biology* 17:161–220.

Baxter, W. L., and Wolfe, C. W., 1973. *The ring-necked pheasant in Nebraska.* Lincoln: Nebraska Game and Parks Commission.

Beason, R. C., and Franks, E. C., 1974. Breeding behavior of the horned lark. *Auk* 91:65–74.

Beaver, D. L., and Baldwin, P. L., 1975. Ecological ovelap and the problem of competition and sympatry in the western Hammond's flycatchers. *Condor* 77:1–13.

Beecham, J. J., and Kichert, M. N., 1975. Breeding biology of the golden eagle in southwestern Idaho. *Wilson Bull.* 87:506–13.

424

Behle, W. H., 1942. Distribution and variations of the horned larks (*Otocoris alpestris*) of western North America. *Univ. Calif. Publ. Zool.* 46:205–316.

———, 1958. *The bird life of Great Salt Lake.* Salt Lake City: Univ. of Utah Press.

Bendell, J. F., and Elliott, P. W., 1967. *Behavior and the regulation of numbers in blue grouse.* Can. Wildl. Serv. Rep. Ser. 4.

Bengston, S. A., 1966a. Field studies on the harlequin duck in Iceland. Wildfowl Trust Ann. Rep. (1964–1965) 17:79–94.

———, 1966b. Breeding ecology of the harlequin (*Histrionicus histrionicus*) in Iceland. *Ornis Scandinavica* 3:25–43.

Bent, A. C., 1907. The marbled godwit on its breeding grounds. *Auk* 24: 160–67.

———, 1921. Life histories of North American gulls and terns. *United States National Museum Bulletin* 113:1–345.

———, 1926. Life histories of North American marsh birds. *U.S. Natl. Mus. Bull.* 135:1–490.

———, 1927. Life histories of North American shorebirds, I. *U.S. Natl. Mus. Bull.* 142:1–420.

———, 1929. Life histories of North American shorebirds, II. *U.S. Natl. Mus. Bull.* 146:1–412.

———, 1932. Life histories of North American gallinaceous birds. *U.S. Natl. Mus. Bull.* 162:1–490.

———, 1937. Life histories of North American birds of prey. Part 1. *U.S. Natl. Mus. Bull.* 167:1–409.

———, 1938. Life histories of North American birds of prey. Part 2. *U.S. Natl. Mus. Bull.* 170:1–428.

———, 1939. Life histories of North American woodpeckers. *U.S. Natl. Mus. Bull.* 174:1–322.

———, 1940. Life histories of North American cuckoos, goatsuckers, hummingbirds, and their allies. *U.S. Natl. Mus. Bull.* 176:1–506.

———, 1942. Life histories of North American flycatchers, larks, swallows, and their allies. *U.S. Natl. Mus. Bull.* 179:1–555.

———, 1946. Life histories of North American jays, crows, and titmice. *U.S. Natl. Mus. Bull.* 191:1–495.

———, 1948. Life histories of North American nuthatches, wrens, thrashers, and their allies. *U.S. Natl. Mus. Bull.* 195:1–475.

————, 1949. Life histories of North American thrushes, kinglets, and their allies. *U.S. Natl. Mus. Bull.* 196:1–454.

————, 1950. Life histories of North American wagtails, shrikes, vireos, and their allies. *U.S. Natl. Mus. Bull.* 197:1–411.

————, 1953. Life histories of North American wood warblers. *U.S. Natl. Mus. Bull.* 203:1–734.

————, 1958. Life histories of North American blackbirds, orioles, tanagers, and allies. *U.S. Natl. Mus. Bull.* 211:1–549.

————, 1968. Life histories of North American cardinals, grosbeaks, buntings, towhees, finches, sparrows, and allies. In three parts. *U.S. Natl. Mus. Bull.* 237:1–1889.

Bergman, C. A., 1983. Flaming owl of the ponderosa. *Aubudon* 85(6): 66–70.

Bergman, R. D., Swain, P., and Weller, M. W., 1970. A comparative study of nesting Forster's and black terns. *Wilson Bull.* 82:435–44.

Bertin, R. I., 1977. Breeding habitats of the wood thrush and veery. *Condor* 79(3):303–11.

Best, L. B., 1972. First-year effects of sagebrush control on two sparrows. *J. Wildl. Manag.* 36:534–44.

————, 1977. Territory quality and mating success in the field sparrow (*Spizella pusilla*). *Condor* 79(2):192–203.

————, 1978. Field sparrow reproductive success and nesting ecology. *Auk* 95(1):9–22.

Bibbee, P. C., 1947. The Bewick's wren, *Thryomanes bewickii* (Audubon). Ph.D. diss., Cornell University, Ithaca, N.Y.

Bicak, T. K., 1977. Some eco-ethological aspects of a breeding population of long-billed curlews (*Numenius americanus*) in Nebraska. M.A. thesis, University of Nebraska at Omaha.

Blackford, J. L., 1958. Territoriality and breeding behavior of a population of blue grouse in Montana. *Condor* 60(3):145–58.

Blais, J. R., and Parks, G. H., 1964. Interaction of evening grosbeak (*Hesperiphona vespertina*) and spruce budworm (*Choristoneura fumiferana* [Clem]) in a localized budworm outbreak treated with DDT in Quebec. *Can. J. Zool.* 42(6):1017–24.

Blanchard, B. D., 1941. The white-crowned sparrows (*Zonotrichia leucophrys*) of the Pacific seaboard: environment and annual cycle. *Univ. Calif. Publ. Zool.* 46:1–178.

Blanchard, B. D., and Erickson, M. M., 1949. The cycle in the Gambel's sparrow. *Univ. Calif. Publ. Zool.* 47:255–318.

Bock, C. E., 1969. Intra- vs. interspecific aggression in pygmy nuthatch flocks. *Ecology* 50:903–5.

————, 1970. The ecology and behavior of the Lewis woodpecker (*Asyndesmus lewis*). *Univ. Calif. Publ. Zool.* 92:1–100.

Bock, C. E., and Bock, J. H., 1974. On the geographical ecology of the three-toed woodpeckers, *Picoides tridactylus* and *P. arcticus*. *Amer. Midl. Natur.* 92:397–405.

Bock, C. E., Hadow, H. H., and Somers, P., 1971. Relations between Lewis' and red-headed woodpeckers in southeastern Colorado. *Wilson Bull.* 83(3):237–48.

Bowdish, B. S., and Philipp, P. B., 1916. The Tennessee warbler in New Brunswick. *Auk* 33:1–8.

Bowles, J. H., and Decker, F. R., 1934. Swainson's hawk in Washington state. *Auk* 51:446–50.

Braaten, D. J., 1975. Observations at three brown creeper nests in Itasca State Park. *Loon* 47:110–13.

Bradley, H. L., 1948. A lift history study of the indigo bunting. *Jack-Pine Warbler* 26:103–13.

Brakhage, G. K., 1965. Biology and behavior of tub-nesting Canada geese. *J. Wildl. Manag.* 29:751–71.

Braun, C. E., and Rogers, G. E., 1971. The white-tailed ptarmigan in Colorado. Colo. Div. Game, Fish, and Parks Tech. Publ. 27, Denver.

Breckenridge, W. J., 1956. Measurements of the habitat niche of the least flycatcher. *Wilson Bull.* 68:47–51.

Brenton, D. F., and Pittaway, R., Jr., 1971. Observations of the great gray owl on its winter range. *Can. Field Nat.* 85:315–22.

Brewer, R., 1963. Ecological and reproductive relationships of black-capped and Carolina chickadees. *Auk* 80:9–47.

Brown, J. L., 1964. The integration of agonistic behavior in the Steller's jay *Cyanocitta stelleri* (Gmelin). *Univ. Calif. Publ. Zool.* 60:223–328.

Brown, L., and Amadon, D., 1968. *Eagles, hawks and falcons of the world.* 2 vols. New York: McGraw-Hill.

Brown, P. W., 1981. Reproductive ecology and productivity of the white-winged scoter. Ph.D. diss., Univ. of Minnesota, Minneapolis.

Brown, P. W., and Brown, M. A., 1981. Nesting biology of the white-winged scoter. *J. Wildl. Manag.* 45:38–45.

427

Brown, R. G. B., Blurton Jones, N. G., and Hussell, D. J. T., 1967. The breeding behaviour of Sabine's gull. *Behaviour* 28:110–40.

Bull, E. L., 1975. Habitat utilization of the pileated woodpecker, Blue Mountains, Oregon. M.S. thesis, Oregon State Univ., Corvallis.

Bull, E. L., and Meslow, E. C., 1977. Habitat requirements of the pileated woodpecker in northeastern Oregon. *J. Forestry* 75:335–37.

Bump, G., Darrow, R., Edminster, F., and Crissey, W., 1947. *The ruffed grouse: life history, propagation, management.* Albany: New York State Conservation Department.

Bunni, M. K., 1959. The killdeer, *Charadrius v. vociferus* Linneaus, in the breeding season: ecology, behavior and the development of homiothermism. Ph.D. diss., University of Michigan, Ann Arbor.

Burger, J., 1974. Breeding adaptations of Franklin's gulls (*Larus pipixcan*) to a marsh habitat. *Animal Behaviour* 22:521–67.

———, 1977. Nesting behavior of herring gulls; invasion into *Spartina* salt marsh areas of New Jersey. *Condor* 79:162–69.

Burger, J., and Miller, L. M., 1977. Colony and nest site selection in white-faced and glossy ibises. *Auk* 94:664–75.

Butterfield, J. D., 1969. Nest site requirements of the lark bunting in Colorado. M.S. thesis, Colorado State University, Fort Collins.

Caccamise, D. F., 1974. Competitive relationships of the common and lesser nighthawks. *Condor* 76:1–20.

Cade, T. J., 1955. Experiments on winter territoriality of the American kestrel, *Falco sparvenius. Wilson Bull.* 67:5–17.

———, 1960. Ecology of the peregrine and gyrfalcons in Alaska. *Univ. Calif. Publ. Zool.* 63:151–267.

———, 1962. Wing movements, hunting, and displays of the northern shrike. *Wilson Bull.* 74(2):386–408.

———, 1967. Ecological and behavioral aspects of predation by the northern shrike. *Living Bird* 6:43–86.

———, 1982. *The falcons of the world.* Ithaca: Cornell University Press.

Calder, W. A., 1971. Temperature relationships and nesting of the calliope hummingbird. *Condor* 73:314–21.

———, 1973. Microhabitat selection during nesting of hummingbirds in the Rocky Mountains. *Ecology* 54:127–34.

Carter, B. C., 1958. The American goldeneye in central New Brunswick. Canadian Wildlife Service, *Wildlife Management Bulletin* 9(2):1–47.

Cartwright, B. W., Shortt, T. M., and Harris, R. D., 1937. Baird's sparrow. *Transactions of the Royal Canadian Institute* 21:153–97.

Catling, P. M., 1972. A study of the boreal owl in southern Ontario with particular reference to the irruption of 1968–1969. *Canadian Field-Naturalist* 86:223–32.

Chabreck, R. H., 1963. Breeding habits of the pied-billed grebe in an impounded coastal marsh in Louisiana. *Auk* 80:447–52.

Chamberlain, D. R., and Cornwell, G. W., 1971. Selected vocalizations of the common crow. *Auk* 88:613–34.

Chamberlain, M. L., 1977. Observations of the red-necked grebe nesting in Michigan.*Wilson Bull.* 89:33–46.

Chapman, L. B., 1935. Studies of a tree swallow colony. *Bird-Banding* 6(2):45–57.

Choate, T. S., 1963. Habitat and population dynamics of white-tailed ptarmigan in Montana. *J. Wildl. Manag.* 27:684–99.

Christensen, G. C., 1952. An ecological study of the chukar partridge in western Nevada. M.S. thesis, Univ. Nev., Reno.

———, 1970. *The chukar partridge: its introduction, life history, and management.* Nevada Dep. Fish and Game Biol. Bull. 4.

Clark, R. J., 1975. A field study of the short-eared owl *Otus flammeus* (Pontoppidan) in North America. *Wildlife Monographs*, vol. 47.

Cogswell, H. L., 1962. Territory size in three species of chaparral birds in relation to vegetation density and structure. Ph.D. diss., Univ. Calif. Library, Berkeley.

Coles, V., 1938. Studies in the life history of the turkey vulture (*Cathartes aura septentrionalis* Wied). Ph.D. diss., Cornell Univ., Ithaca.

———, 1944. Nesting of the turkey vulture in Ohio caves. *Auk* 61:219–28.

Combellack, C. R. B., 1954. A nesting of violet-green swallows. *Auk* 71: 435–42.

Cooch, F. G., 1958.The breeding biology and management of the blue goose (*Chen caerulescens*). Ph.D. diss., Cornell Univ., Ithaca.

Cornwell, G. W., 1963. Observations on the breeding biology and behavior of a nesting population of belted kingfishers. *Condor* 65: 426–31.

Coulombe, H. N., 1971. Behavior and population ecology of the burrowing owl, *Speotyto cunicularia*, in the Imperial Valley of California. *Condor* 73:162–76.

Coutlee, E. L., 1967. Agonistic behavior in the American goldfinch. *Wilson Bull.* 79:89–109.

———, 1968a. Comparative behavior of lesser and Lawrence's goldfinches. *Condor* 70:228–42.

———, 1968b. Maintenance behavior of lesser and Lawrence's goldfinches. *Condor* 70:378–84.

Coulter, M. W., and Miller, W. R., 1968. Nesting biology of black ducks and mallards in northern New England. *Vermont Fish and Game Department Bulletin* 68(2):1–74.

Cowan, J. B., 1952. Life history and productivity of a population of western mourning doves in California. *Calif. Fish and Game* 38:505–21.

Craighead, J., and Craighead, F., 1940. Nesting pigeon hawks. *Wilson Bull.* 52:241–48.

———, 1956. *Hawks, owls, and wildlife.* Harrisburg: Stackpole Publ.

Criddle, N., 1927. Habits of the mountain bluebird in Manitoba. *Can. Field-Nat.* 41(2):40–44.

Cripps, B. J., Jr., 1966. The nesting cycle of the chestnut-sided warbler. *Raven* 37:43–48.

Crockett, A. B., 1975. Ecology and behavior of the Williamson's sapsucker in Colorado. Ph.D. diss., Univ. Colo., Boulder.

Crockett, A. B., and Hadow, H. H., 1975. Nest site selection by Williamson and red-naped sapsuckers. *Condor* 77:365–68.

Crockett, A. B., and Hansley, P. L., 1978. Coition, nesting and postfledging behavior of Williamson's sapsucker in Colorado. *Living Bird* 16:7–20.

Crooks, M. P., 1948. Life history of the field sparrow (*Spizella pusilla pusilla*). M.S. thesis, Iowa State College, Ames.

Dane, C. W., 1966. Some aspects of breeding biology of the blue-winged teal. *Auk* 83:389–402.

Davis, C. M., 1979. A nesting study of the brown creeper. *Living Bird* 17: 237–63.

Davis, D. E., 1954. The breeding biology of Hammond's flycatchers. *Auk* 71:164–71.

———, 1959. Observations on territorial behavior of least flycatchers. *Wilson Bull.* 71:73–85.

Davis, J., 1951. Distribution and variation of the brown towhee. *Univ. Calif. Publ. Zool.* 52:1–120.

————, 1960. Nesting behavior of the rufous-sided towhee in coastal California. *Condor* 62:434–56.

————, 1973. Habitat preferences and competition of wintering juncos and golden-crowned sparrows. *Ecology* 54:174–80.

Davis, J., Fisher, G. F., and Davis, B. S., 1963. The breeding biology of the western flycatcher. *Condor* 65:337–82.

Dawson, W. L., 1919. The solitaires of Shasta. *Condor* 21:12–21.

Day, K. C., 1953. Home life of the veery. *Bird-Banding* 24:100–6.

Dennis, J. V., 1948. Observations on the orchard oriole in the lower Mississippi delta. *Bird-Banding* 19:12–20.

————, 1958. Some aspects of the breeding ecology of the yellow-breasted chat (*Icteria virens*). *Bird-Banding* 29:169–83.

Devillers, P., 1970. Identification and distribution in California of the *Sphyrapicus varius* group of sapsuckers. *Calif. Birds* 1:47–76.

Diem, K. L., 1979. White pelican reproductive failures in the Molly Islands breeding colony in Yellowstone National Park. *Proc. Res. in Natl. Parks Symp. U.S. Natl. Park Serv. Trans. and Proc.* Ser. No. 5. I:489–96.

Diem, K. L., and Condon, D. D., 1967. *Banding of water birds on the Molly Islands, Yellowstone Lake, Wyoming.* Yellowstone National Park: Yellowstone Library and Museum Association.

Dilger, W. C., 1956. Hostile behavior and reproductive isolating mechanisms in the avian genera *Catharus* and *Hylocichla*. *Auk* 73:313–53.

Dimmick, R. W., 1968. *Canada geese of Jackson Hole: their ecology and management.* Wyoming Game and Fish Comm. Bull. 11.

Dixon, C. L., 1978. Breeding biology of the savannah sparrow on Kent Island. *Auk* 95:235–46.

Dixon, J. B., 1934. Nesting of the Clark nutcracker in California. *Condor* 36:229–34.

Dixon, K. L., 1961. Habitat distribution and niche relationships in North American species of *Parus*. In W. F. Blair, ed., *Vertebrate speciation: a University of Texas symposium*, pp. 179–216. Austin: University of Texas Press.

Dixon, K. L., and Gilbert, J. D., 1964. Altitudinal migration in the mountain chickadee. *Condor* 66:61–64.

Dorn, J. L., 1972. The common raven in Jackson Hole, Wyoming. M.S. thesis, Univ. Wyoming, Laramie.

Dow, D. D., 1965. The role of saliva in food storage of the gray jay. *Auk* 82:139–54.

Drewien, R. C., 1973. Ecology of Rocky Mountain greater sandhill cranes. Ph.D. diss., Univ. of Idaho, Moscow.

Drewien, R. C., and Bizeau, E. G., 1974. Status and distribution of greater sandhill cranes in the Rocky Mountains. *J. Wildl. Manag.* 38(4):720–42.

Dunham, D. W., 1964. Reproductive displays of the warbling vireo. *Wilson Bull.* 76:170–73.

———, 1966. Territorial and sexual behavior in the rose-breasted grosbeak. *Zeitschrift für Tierpsychologie* 23:438–51.

Dunker, H., 1974. Habitat selection and territory size of the black-throated diver, *Gavia arctica* (L.) in south Norway. *Norw. J. Zool.* 22(1):15–29.

Dunkle, S. W., 1977. Swainson's hawks on the Laramie Plains, Wyoming. *Auk* 94:65–71.

Dunnett, G. M., 1955. The breeding of the starling (*Sturnus vulgaris*) in relation to its food supply. *Ibis* 97:619–62.

Dunstan, T. G., and Sample, S. D., 1972. Biology of barred owls in Minnesota. *Loon* 44(4):111–15.

Earhart, C. M., and Johnson, N. K., 1970. Size dimorphism and food habits of North American owls. *Condor* 72:251–64.

Eaton, S. W., 1957. A life history study of *Seiurus noveboracensis*. St. Bonaventure University, *Science Studies* 19:7–36.

Eckhardt, R. C., 1976. Polygyny in the western wood peewee. *Condor* 78:561–62.

Edson, J. M., 1943. A study of the violet-green swallow. *Auk* 60:396–403.

Emlen, J. T., Jr., 1942. Notes on a nesting colony of western crows. *Bird-Banding* 13:143–53.

———, 1952. Social behavior in nesting cliff swallows. *Condor* 54:177–99.

———, 1954. Territory, nest building and pair formation in the cliff swallow. *Auk* 71:16–35.

Emlen, S. T., Rising, J. D., and Thompson, W. L., 1975. A behavioral and morphological study of sympatry in the indigo and lazuli buntings of the Great Plains. *Wilson Bull.* 87:145–79.

Enderson, J. H., 1960. A population study of the sparrow hawk in east-central Illinois. *Wilson Bull.* 72:222–31.

———, 1964. A study of the prairie falcon in the central Rocky Mountain region. *Auk* 81:332–52.

432

Erickson, R. C., 1948. Life history and ecology of the canvasback, *Nyroca valisineria* (Wilson) in southeast Oregon. Ph.D. diss., Iowa State Univ., Ames.

Erpino, M. J., 1968. Nest related activities of the black-billed magpie. *Condor* 70:154-65.

Errington, P. L., Hamerstrom, F., and Hamerstrom, F. N., 1940. The great horned owl and its prey in the north-central United States. *Iowa State Agricultural Experiment Station, Research Bull.* 277:758-850.

Erskine, A. J., 1972. *Buffleheads.* Canadian Wildlife Service Monograph Series, No. 4. Ottawa: Information Canada.

Erwin, W. G., 1935. Some nesting habits of the brown thrasher. *Journal of the Tennessee Academy of Science* 10:179-204.

Evenden, F. G., 1957. Observations on nesting behavior of the house finch. *Condor* 59:112-17.

Eyer, L. E., 1963. Observations on golden-winged warblers at Itasca State Park, Minnesota. *Jack-Pine Warbler* 41:96-109.

Faaborg, J., 1976. Habitat selection and territorial behavior of the small grebes of North Dakota. *Wilson Bull.* 88:390-99.

Faanes, C. A., 1980. Breeding biology of the eastern phoebe in northern Wisconsin. *Wilson Bull.* 92:107-10.

Falls, J. B., and Krebs, J. R., 1975. Sequence of songs in repertoires of western meadowlarks (*Sturnella neglecta*). *Can. J. Zool.* 53(8):1165-78.

Fehon, J. H., 1955. Life-history of the blue-gray gnatcatcher (*Polioptila caerulea caerulea*). Ph.D. diss., Florida State University, Tallahassee.

Ficken, M. S., 1962. Agonistic behavior and territory in the American redstart. *Auk* 79:607-32.

―――, 1963. Courtship of the American redstart. *Auk* 80(3):307-17.

Ficken, M. S., and Ficken, R. W., 1962. The comparative ethology of the wood warblers: a review. *Living Bird* 1:103-22.

―――, 1966. Behavior of myrtle warblers in captivity. *Bird-Banding* 37: 273-79.

Ficken, R. W., 1963. Courtship and aggressive behavior of the common grackle (*Quiscalus quiscula*). *Auk* 80:52-72.

Ficken, R. W., Ficken, M. S., and Morse, D. H., 1968. Competition and character displacement in two sympatric pine-dwelling warblers (*Dendroica*, Parulidae). *Evolution* 22:307-14.

Finzel, J. E., 1964. Avian population of four herbaceous communities in southeastern Wyoming. *Condor* 66:496-510.

Fischer, R. B., and Gills, G., 1946. A cooperative study of the white-throated sparrow. *Auk* 63:402–18.

Fisher, R., 1958. The breeding biology of the chimney swift. *New York State Mus. Sci. Serv. Bull.* 368:1–41.

Fitch, F. W., Jr., 1950. Life history and ecology of the scissor-tailed fly-catcher. *Auk* 67:145–68.

Fitch, H. S., Swenson, F., and Tillotson, D. F., 1946. Behavior and food habits of the red-tailed hawk. *Condor* 48(5):205–37.

Fitzner, J. N., 1978. The ecology and behavior of the long-billed curlew (*Numenius americanus*) in southeastern Washington. Ph.D. diss., Washington State University, Pullman.

Fjeldså, J., 1973. Antagonistic and heterosexual behavior of the horned grebe, *Podiceps auritus*. *Sterna* 12:161–217.

Flack, J. A. D., 1976. Bird populations in aspen forests in western North America. *A.O.U. Ornith. Mono.* No. 19.

Forbes, J. A., and D. W. Warner, 1974. Behavior of a radio-tagged saw-whet owl. *Auk* 91:783–95.

Frederickson, L. H., 1970. Breeding biology of American coots in Iowa. *Wilson Bull.* 82:445–57.

———, 1971. Common gallinule breeding biology and development. *Auk* 88:914–19.

French, N. R., 1954. Notes on breeding activities and on gular sacs in the pine grosbeak. *Condor* 56:83–85.

———, 1959. Life history of the black rosy finch. *Auk* 76:159–80.

Friedmann, H., 1963. Host relations of the parasitic cowbirds. *U.S. Nat. Mus. Bull.* 233:1–276.

Frydendall, M. J., 1967. Feeding ecology and territorial behavior of the yellow warbler. Ph.D. diss., Utah State University, Logan.

Fuller, R. W., 1953. Studies in the life history and ecology of the American pintail, *Anas acuta tzitzihoa*, in Utah. M.S. thesis, Utah State Univ., Logan.

Galbreath, D. S., and Moreland, R., 1953. The chukar partridge in (*Aythya affinis* eyton) at West Medical Lake, Spokane County, Wash-

Gates, J. M., 1962. Breeding biology of the gadwall in northern Utah. *Wilson Bull.* 74(1):43–67.

Gaunt, A. S., 1965. Fossorial adaptations in the bank swallow, *Riparia riparia* (Linnaeus). *Univ. Kans. Sci. Bull.* 46(2):99–146.

Gehrman, K. H., 1951. An ecological study of the lesser scaup duck

(*Aythya affinis eyton*) at West Medical Lake, Spokane County, Washington. M.S. thesis, Wash. State Univ., Pullman.

Geis, M. P., 1956. Productivity of Canada geese in Flathead Valley, Montana. *J. Wildl. Manag.* 20:409–19.

Gibbon, R. S., 1966. Observations on the behavior of nesting three-toed woodpeckers, *Picoides tridactylus*, in central New Brunswick. *Canadian Field-Naturalist* 80:223–26.

Gibbs, R. M., 1961. Breeding ecology of the common goldeneye, *Bucephala clangula americana*, in Maine. M.S. thesis, Univ. Maine, Orono.

Gibson, F., 1971. The breeding biology of the American avocet (*Recurvirostra americana*) in central Oregon. *Condor* 73:444–54.

Girard, G. L., 1939. Notes on life history of the shoveller. *North American Wildlife Conference Transactions* 4:363–71.

Glover, F. A., 1953. Nesting ecology of the pied-billed grebe in northwestern Iowa. *Wilson Bull.* 65:32–39.

Glue, D. G., 1977. Breeding biology of long-eared owls. *British Birds* 70:318–31.

Goforth, W. R., and Baskett, T. S., 1971. Social organization of penned mourning doves. *Auk* 88:528–42.

Goodwin, D., 1967. *Pigeons and doves of the world*. London: British Museum (Natural History).

———, 1976. *Crows of the world*. Ithaca: Cornell University Press.

Goodwin, R. A., 1960. A study of the ethology of the black tern, *Chlidonias niger surinamensis*. Ph.D. diss., Cornell University, Ithaca.

Graber, J. W., Graber, R. R., and Kirk, E. L., 1983. Illinois birds: wood warblers. *Illinois Natural History Society Biological Notes* 118:1–144.

Grant, R. A., 1965. The burrowing owl in Minnesota. *Loon* 37:1–17.

Greenhalgh, C. M., 1952. Food habits of the California gull in Utah. *Condor* 54:302–8.

Grant, G. S., and Quay, T. L., 1977. Breeding biology of cliff swallows in Virginia. *Wilson Bull.* 89:286–90.

Graul, W. D., 1974. Adaptive aspects of the mountain plover social system. *Living Bird* 12:69–94.

———, 1975. Breeding biology of the mountain plover. *Wilson Bull.* 87(1):6–31.

Green, R., 1976. Breeding behaviour of ospreys *Pandion haliaetus* in Scotland. *Ibis* 118:475–90.

Grice, D., and Rogers, J. P., 1965. *The wood duck in Massachusetts*. Massachusetts Division of Fisheries and Game, Final Report, Project W-19-R.

Grinnell, J., and Storer, T. I., 1924. *Animal life in the Yosemite*. Berkeley: University of California Press.

Griscom, L., 1937. A monograph study of the red crossbill. *Boston Soc. Nat. Hist. Proc.* 41(5):77–210.

———— (ed.), 1957. *The warblers of North America*. New York: Devin-Adair.

Gullion, G. W., 1954. The reproductive cycle of American coots in California. *Condor* 71:366–412.

Gutierrez, R. J., and Koenig, W. D., 1978. Characteristics of storage trees used by acorn woodpeckers in two California woodlands. *J. Forestry* 86:162–64.

Gutierrez, R. J., Braun, C. E., and Zapatka, T. P., 1975. Reproductive biology of the band-tailed pigeon in Colorado and New Mexico. *Auk* 92:665–77.

Haecker, F. A., 1948. A nesting study of the mountain bluebird in Wyoming. *Condor* 50:216–19.

Hahn, H. W., 1937. Life history of the ovenbird in southern Michigan. *Wilson Bull.* 49:145–237.

————, 1950. Nesting behavior of the American dipper in Colorado. *Condor* 52:49–62.

Hamerstrom, F., 1969. A harrier population study. In J. J. Hickey, ed., *Peregrine falcon populations, their biology and decline*, pp. 367–85. Madison: University of Wisconsin Press.

Hamilton, R. C., 1975. Comparative behavior of the American avocet and the black-necked stilt (Recurvirostridae). *A.O.U. Monographs* 17:1–98.

Hammond, M. C., and Henry, C. J., 1949. Success of marsh hawk nests in North Dakota. *Auk* 66:271–74.

Hancock, J., and Elliott, H., 1978. *The herons of the world*. New York: Harper & Row.

Hann, H. W., 1950. Nesting behavior of the American dipper in Colorado. *Condor* 52:49–62.

Hanson, H. C., and Kossack, C. W., 1963. The mourning dove in Illinois. Illinois Department of Conservation Technical Bulletin No. 2, Urbana, Ill.

Harcus, J. L., 1973. Song studies in the breeding biology of the catbird, *Dumetella carolinensis* (Aves: Mimidae). Ph.D. diss., Univ. Toronto, Canada.

Hardy, J., 1961. Studies in behavior and phylogeny of certain New World jays (Garrulinae). *Univ. Kans. Sci. Bull.* 62:13–149.

Harjer, H. J., 1974. An analysis of some aspects of the ecology of dusky grouse. Ph.D. diss., Univ. of Wyoming, Laramie.

Harlow, R. C., 1922. The breeding habits of the northern raven in Pennsylvania. *Auk* 39:399–410.

Harrison, H. H., 1951. Notes and observations on the Wilson warbler. *Wilson Bull.* 63:143–48.

Hartshorne, J. M., 1962. Behavior of the eastern bluebird at the nest. *Living Bird* 1:131–49.

Hays, H., 1973. Polyandry in the spotted sandpiper. *Living Bird* 11:43–57.

Heckenlively, D. B., 1967. Role of song in territoriality of black-throated sparrows. *Condor* 69:429–30.

———, 1970. Song in a population of black-throated sparrows. *Condor* 72:24–36.

Henderson, A. D., 1926. Bonaparte's gull nesting in northern Alberta. *Auk* 48:288–94.

Hendricks, P., 1978. Notes on the courtship behavior of brown-capped rosy finches. *Wilson Bull.* 90:285–87.

Henny, J. J., and Wight, H. W., 1972. Population ecology and environmental population: red-tailed and Cooper's hawks. *In* Population ecology and migratory birds: a symposium. *Bur. Sport Fish. and Wildl., Wildl. Res. Rep.* 2:229–50.

Hebrand, J. J., 1974. Habitat partitioning in two species of *Spizella* (Aves: Emberizidae): a concurrent laboratory and field study. Ph.D. diss., Clemson Univ., South Carolina.

Hespenheide, H. A., 1964. Competition and the genus *Tyrannus*. *Wilson Bull.* 76:265–81.

Heydweiller, A. M., 1935. A comparison of winter and summer territories and seasonal variations of the tree sparrow (*Spizella arborea*). *Bird-Banding* 6(1):1–11.

Hickey, J. J. (ed.), 1969. *Peregrine falcon populations: their biology and decline.* Madison: University of Wisconsin Press.

Higgins, K. F., and Kirsch, L. M., 1975. Some aspects of the breeding biology of the upland sandpiper in North Dakota. *Wilson Bull.* 87:96–102.

Hildén, O., 1964. Ecology of duck populations in the island group of Valassaaret, Gulf of Bothnia. *Annales Zoologici Fennici* 1:1–279.

Hines, J. E., 1977. Nesting and brood ecology of lesser scaup at Waterhen Marsh, Saskatchewan. *Canadian Field-Naturalist* 91:248–55.

Hochbaum, Hans Albert, 1944. The canvasback on a prairie marsh. Am. Wildl. Inst., Washington, D.C.

Hofslund, P. B., 1959. A life history of the yellowthroat, *Geothlypis trichas*. *Proceedings of the Minnesota Academy of Science* 27:144–74.

Hoglund, N. H., and Lansgren, E., 1968. The great grey owl and its prey in Sweden. *Viltrevy* 5(7):364–421.

Höhn, E. O., 1967. Observations on the breeding biology of Wilson's phalarope (*Steganopus tricolor*) in central Alberta. *Auk* 84:220–44.

———, 1971. Observations on the breeding behavior of grey and red-necked phalaropes. *Ibis* 113(3):335–48.

Holcomb, L. C., 1972. Traill's flycatcher breeding biology. *Nebraska Bird Review* 40:50–67.

Holm, C. H., 1973. Breeding sex ratios, territoriality, and reproductive success in the red-winged blackbird (*Agelaius phoeniceus*). *Ecology* 54:356–65.

Horak, C. J., 1970. A comparative study of the foods of the sora and Virginia rail. *Wilson Bull.* 82:206–13.

Horn, H. S., 1968. The adaptive significance of colonial nesting in the Brewer's blackbird. *Ecology* 49:682–94.

———, 1970. Social behavior of nesting Brewer's blackbirds. *Condor* 72:15–23.

Hostetter, D. R., 1961. Life history of the Carolina junco, *Junco hyemalis* Brewster. *Raven* 32:97–170.

Houston, D. B., 1963. A contribution to the ecology of the band-tailed pigeon, *Columba fasciata*, Say. M.A. thesis, Univ. Wyo., Laramie.

Howell, J. C., 1942. Notes on the nesting habits of the American robin (*Turdus migratorius* L.). *Amer. Midl. Nat.* 28:529–603.

Howell, T. R., 1952. Natural history and differentiation in the yellow-bellied sapsucker. *Condor* 54:237–82.

Hoyt, S., 1957. The ecology of the pileated woodpecker. *Ecology* 38:246–56.

Hubbard, J. P., 1969. The relationships and evolution of the *Dendroica coronata* complex. *Auk* 86:393–432.

Huckabee, J. W., 1965. Population study of waterfowl in the Third Creek Area, Jackson Hole, Wyoming. M.S. thesis, Univ. of Wyo., Laramie.

Hunter, W. F., and Baldwin, P. H., 1962. Nesting of the black swift in Montana. *Wilson Bull.* 74:409–16.

Jackman, S. M., 1974. Woodpeckers of the Pacific Northwest; their characteristics and their role in the forests. M.S. thesis, Oreg. State Univ., Corvallis.

Jackson, J. A., 1970. A quantitative study of the foraging ecology of downy woodpeckers. *Ecology* 51(2):318–23.

James, R. D., 1973. Ethological and ecological relationships of the yellow-throated and solitary vireos (Aves: Vireonidae) in Ontario. Ph.D. diss., University of Toronto, Canada.

———, 1976. Foraging behavior and habitat selection of three species of vireos in southern Ontario. *Wilson Bull.* 88(1):62–75.

Jenni, D. A., 1969. A study of the ecology of four species of herons during the breeding season at Lake Alice, Alachua County, Florida. *Ecological Monographs* 39:245–70.

Johns, J. E., 1969. Field studies of Wilson's phalarope. *Auk* 86(4):660– 70.

Johnsgard, P. A., 1973. *Grouse and quails of North America*. Lincoln: University of Nebraska Press.

———, 1975a. *North American game birds of upland and shoreline*. Lincoln: University of Nebraska Press.

———, 1975b. *Waterfowl of North America*. Bloomington: Indiana University Press.

———, 1979. *Birds of the Great Plains: the breeding species and their distribution*. Lincoln: University of Nebraska Press.

———, 1981. *The plovers, sandpipers and snipes of the world*. Lincoln: University of Nebraska Press.

———, 1983a. *Hummingbirds of North America*. Washington, D.C.: Smithsonian Institution Press.

———, 1983b. *Grouse of the world*. Lincoln: University of Nebraska Press.

———, 1983c. *Cranes of the world*. Bloomington: Indiana University Press.

Johnson, N. K., 1963. Biosystematics of sibling species of flycatchers in the *Empidonax hammondii-oberholseri-wrightii* complex. *Univ. Calif. Publ. Zool.* 66:79–238.

———, 1966. Bill size, and the questions of competition in allopatric and sympatric populations of dusky and gray flycatchers. *Syst. Zool.* 15:70–87.

————, 1976. Breeding distribution of Nashville and Virginia's warblers. *Auk* 93:219–30.

Johnson, R. E., 1965. Reproductive activities of rosy finches, with special reference to Montana. *Auk* 82:190–205.

Johnston, D. W., and Forster, M. E., 1954. Interspecific relations of breeding gulls at Honey Lake, California. *Condor* 56:38–42.

Jones, R. E., 1960. Activities of the magpie during the breeding period in southern Idaho. *Northwest Sci.* 34:18–25.

Jones, V. E., and Baylor, L. M., 1969. Nesting of the Cassin's finch at Pocatello, Idaho. *Tebiwa* 12:64–68.

Joyner, D. W., 1975. Nest parasitism and brood-related behavior of the ruddy duck (*Oxyura jamaicensis rubida*). Ph.D. diss., University of Nebraska, Lincoln.

Kale, H. W., II, 1965. Ecology and bioenergetics of the long-billed marsh wren, *Telmatodytes palustris griseus* (Brewster), in Georgia salt marshes. Publ. Nuttall Ornithol. Club 5.

Kangarise, C. M., 1979. Breeding biology of Wilson's phalarope in North Dakota. *Bird-Banding* 50:12–22.

Karalus, K. E., and Eckert, A. W., 1974. *The owls of North America.* Garden City, New York: Doubleday.

Kaufmann, G. W., 1971. Behavior and ecology of the sora, *Porzana carolina*, and Virginia rail, *Rallus limicola*. Ph.D. diss., University of Minnesota, Minneapolis.

Keith, J. A., 1966. Reproduction in a population of herring gulls (*Larus argentatus*) contaminated by DDT. M.S. thesis, University of Wisconsin, Madison.

Kendeigh, S. C., 1941. Territorial and mating behavior of the house wren. *Illinois Biological Monographs* 18:1–120.

————, 1945. Nesting behavior of wood warblers. *Wilson Bull.* 57:145–64.

————, 1973. A symposium on the house sparrow (*Passer domesticus*) and European tree sparrow (*P. montanus*) in North America. A.O.U. Ornithol. Monogr. 14.

Kennard, F. H., 1920. Notes on the breeding habits of the rusty blackbird in northern New England. *Auk* 37:412–22.

Kessel, B., 1957. A study of the breeding biology of the European starling (*Sturnus vulgaris* L.) in North America. *Amer. Midl. Nat.* 58:257–331.

440

Kilham, L., 1959. Early reproductive behavior of flickers. *Wilson Bull.* 71:323–36.

———, 1962a. Reproductive behavior of downy woodpeckers. *Condor* 64:126–33.

———, 1962b. Breeding behavior of yellow-bellied sapsuckers. *Auk* 79:31–43.

———, 1966. Reproductive behavior of hairy woodpeckers. 1. Pair formation and courtship. *Wilson Bull.* 78:251–65.

———, 1968a, 1972. Reproductive behavior in white-breasted nuthatches. *Auk* 85:477–92, 89:115–29.

———, 1968b. Reproductive behavior of white-breasted nuthatches: I. Distraction display, bill-sweeping, and nest hole defense. *Auk* 85(3): 477–92.

———, 1971a. Roosting habits of white-breasted nuthatches. *Condor* 73(1):113–14.

———, 1971b. Reproductive behavior of yellow-bellied sapsuckers: I. Preference for nesting in *Fomes*-infected aspens and nest hole interactions with flying squirrels, raccoons, and other animals. *Wilson Bull.* 83(2):159–71.

———, 1972. Reproductive behavior of white-breasted nuthatches. II. Courtship. *Auk* 89:115–29.

———, 1973. Reproductive behavior in the red-breasted nuthatch. 1. Courtship. *Auk* 90:597–609.

———, 1974a. Early breeding season behavior of downy woodpeckers. *Wilson Bull.* 84:407–18.

———, 1974b. Biology of young belted kingfishers. *Amer. Midl. Nat.* 92(1):245–47.

———, 1977a. Early breeding season behavior of red-headed woodpeckers. *Auk* 94:231–39.

———, 1977b. Nesting behavior of yellow-bellied sapsuckers. *Wilson Bull.* 89:310–24.

Killpack, M. S., 1970. Notes on sage thrasher nestlings in Colorado. *Condor* 72:486–88.

King, J. R., 1955. Notes on the life history of Traill's flycatcher (*Empidonax traillii*) in southeastern Washington. *Auk* 72:148–73.

Kingsbury, E. W., 1933. The status and natural history of the bobolink *Dolichonyx oryzivorus*. Ph.D. diss., Cornell University, Ithaca, N.Y.

Kirby, R. E., 1976. Breeding chronology and interspecific relations of pied-billed grebes in northern Minnesota. *Wilson Bull.* 88(3):493–95.

Kitchen, D. W., 1968. Brood habitat selection of the hooded merganser (*Lophodytes cucullatus*) in northeastern Wisconsin. M.S. thesis, University of Michigan, Ann Arbor.

Knapton, R. W., 1979. Breeding ecology of the clay-colored sparrow. *Living Bird* 17:137–58.

Knapton, R. W., and Krebs, J. R., 1974. Settlement patterns, territory size, and breeding density in the song sparrow (*Melospiza melodia*). *Can. J. Zool.* 52(11):1413–20.

Knorr, O. A., 1961. The geographical and ecological distribution of the black swift in Colorado. *Wilson Bull.* 73:155–70.

Krause, H., 1965. Nesting of a pair of Canada warblers. *Living Bird* 4:5–11.

Krebs, J. R., 1974. Colonial nesting and social feeding as strategies for exploiting food resources in the great blue heron (*Ardea herodias*). *Behaviour* 51:99–131.

Kroodsma, D. E., 1973. Coexistence of Bewick's wrens and house wrens in Oregon. *Auk* 90:341–52.

———, 1975. Song patterning in the rock wren. *Condor* 77(3):294–303.

Kroodsma, R. L., 1970. North Dakota species pairs. I. Hybridization in buntings, grosbeaks and orioles, II. Species-recognition behavior of territorial male rose-breasted and black-headed grosbeaks (*Pheucticus*). Ph.D. diss., North Dakota State University, Fargo.

———, 1974. Species-recognition behavior of territorial male rose-breasted and black-headed grosbeaks (*Pheucticus*). *Auk* 91(1):54–64.

Kuchel, C. R., 1977. Some aspects of the behavior and ecology of harlequin ducks breeding in Glacier National Park, Montana. M.S. thesis, University of Montana, Missoula.

Kuerzi, R. G., 1941. Life history studies of the tree swallow. *Proc. Linnean Soc. New York* 52–53:1–52.

Lancaster, D. A., 1970. Breeding behavior of the cattle egret in Colombia. *Living Bird* 9:167–93.

Langvatn, R., and Moksnes, A., 1979. On the breeding ecology of the gyrfalcon *Falco rusticus* in central Norway 1968–74. *Cinclus* 2:27–39.

Lanyon, W. E., 1956. Territory of the meadowlarks, genus *Sturnella*. *Ibis* 98:485–89.

————, 1957. The comparative biology of the meadowlarks (*Sturnella*) in Wisconsin. *Publications of the Nuttall Ornithological Club*, no. 1, pp. 1–67.

————, 1961. Species limits and distribution of ashy-throated and Nutting flycatchers. *Condor* 63:421–29.

Laskey, A. R., 1962. Breeding biology of mockingbirds. *Auk* 79:596–606.

Laun, C. H., 1957. A life history study of the mountain plover, *Eupoda montana* Townsend, in the Laramie Plains, Albany County, Wyoming. M.S. thesis, University of Wyoming, Laramie.

Lawrence, G. E., 1950. The diving and feeding activity of the western grebe on the breeding grounds. *Condor* 52:3–16.

Lawrence, L. de K., 1948. Comparative study of the nesting behavior of chestnut-sided and Nashville warblers. *Auk* 65:204–19.

————, 1949. Notes on nesting pigeon hawks at Pimisi Bay, Ontario. *Wilson Bull.* 61:15–25.

————, 1953a. Nesting life and behavior of the red-eyed vireo. *Canadian Field-Naturalist* 67:47–87.

————, 1953b. Notes on the nesting behavior of the Blackburnian warbler. *Wilson Bull.* 65:135–44.

————, 1967. A comparative life-history study of four species of woodpeckers. *A.O.U. Ornithological Monographs* No. 5.

Lea, R. B., 1942. A study of the nesting habitats of the cedar waxwing. *Wilson Bull.* 54:225–37.

Lederer, J. R., 1977. Winter feeding territories in the Townsend's solitaire. *Bird-Banding* 48:11–18.

Lewis, J. C., 1973. *The world of the wild turkey*. Philadelphia and New York: Lippincott.

Ligon, J. D., 1973. Foraging behavior of the white-headed woodpecker in Idaho. *Auk* 90:862–69.

Linsdale, J. M., 1928. Variations in the fox sparrow with reference to natural history and osteology. *Univ. Calif. Publ. Zool.* 30:251–392.

————, 1937. *The natural history of magpies*. Pacific Coast Avifauna 25.

————, 1938. Environmental responses of vertebrates in the Great Basin. *Amer. Midl. Nat.* 19:1–206.

————, 1957. Goldfinches on the Hastings Natural History Reservation. *Amer. Mid. Nat.* 57:1–119.

Littlefield, C. D., and Ryder, R. A., 1968. Breeding biology of the greater sandhill crane on Malheur National Wildlife Refuge, Oregon. *Trans-*

443

actions of the North American Wildlife and Natural Resources Conference 33:444–54.

Low, J. B., 1941. Nesting of the ruddy duck in Iowa. Auk 58:506–17.

———, 1945. Ecology and management of the redhead, Nyroca americana, in Iowa. Ecol. Monogr. 15(1):35–69.

Ludwig, J. P., 1965. Biology and structure of the Caspian tern (Hydroprogne caspia) population of the Great Lakes from 1896–1964. Bird-Banding 36:217–33.

Lumsden, H. G., 1965. Displays of the sharptail grouse. Ontario Department of Lands and Forests Technical Series Research Report No. 66. Maple, Ontario.

Lundberg, A., 1979. Residency, migration and compromise: adaptations to nest-site scarcity and food specializations in three Fennoscandinavian owl species. Oecologia 41:273–81.

Lunk, W. A., 1962. The rough-winged swallow Stelgidopteryx ruficollis (Vieillot); a study on its breeding biology in Michigan. Publications of the Nuttall Ornithological Club, No. 4, pp. 1–155.

Luttich, S., Rusch, D. H., Meslow, E. C., and Keith, L. B., 1970. Ecology of red-tailed hawk predation in Alberta. Ecology 51(2):190–203.

Luttich, S. N., Keith, L. B., and Stephenson, J. D., 1971. Population dynamics of the red-tailed hawk (Buteo jamaicensis) at Rochester, Alberta. Auk 88:75–87.

MacQueen, P. M., 1950. Territory and song in the least flycatcher. Wilson Bull. 62:194–205.

MacRoberts, M. H., and MacRoberts, B. R., 1976. Social organization and behavior of the acorn woodpecker in central California. A.O.U. Ornithological Monographs No. 21.

Manuwal, D. A., 1970. Notes on the territoriality of Hammond's flycatcher in western Montana. Condor 72:364–65.

March, J. R., 1967. Dominance relations and territorial behavior of captive shovelers, Anas clypeata. M.S. thesis, University of Minnesota, Minneapolis.

Marshall, J. T., 1939. Territorial behavior of the flammulated screech owl. Condor 41:71–78.

———, 1960. Interrelations of Abert and brown towhees. Condor 62:49–64.

———, 1967. Parallel variation in North and Middle American screech owls. Monograph Western Foundation Vertebrate Zoology No. 1.

444

Marti, C. D., 1974. Feeding ecology of four sympatric owls. *Condor* 76:45–61.

Martin, D. J., 1973. Selective aspects of burrowing owl ecology and behavior. *Condor* 75:446–56.

Martin, S. G., 1970. The agonistic behavior of varied thrushes (*Ixoreus naevius*) in winter assemblages. *Condor* 72:452–59.

————, 1974. Adaptations for polygynous breeding in the bobolink, *Dolichonyx oryzivorus*. *Am. Zool.* 14(1):109–19.

Mason, C. F., and Macdonald, S. M., 1976. Aspects of the breeding biology of the snipe. *Bird Study* 23(1):33–38.

Matray, P. F., 1974. Broad-winged hawk nesting and ecology. *Auk* 91: 307–24.

Maxwell, G. R., II, 1965. Life history of the common grackle, *Quiscalus quiscula* (Linnaeus). Ph.D. diss., Ohio State University, Columbus.

————, 1970. Pair information, nest building and egg laying of the common grackle in northern Ohio. *Ohio Journal of Science* 70:284–91.

Maxwell, G. R., II, and Putnam, L. S., 1972. Incubation, care of young, and nest success of the common grackle (*Quiscalus quiscula*) in northern Ohio. *Auk* 89:349–59.

Mayfield, H., 1958. Nesting of the black-backed three-toed woodpecker in Michigan. *Wilson Bull.* 70:195–96.

————, 1965. The brown-headed cowbird, with old and new hosts. *Living Bird* 4:13–27.

Mayhew, W. W., 1958. The biology of the cliff swallow in California. *Condor* 60:7–37.

McAllister, N. M., 1955. Reproductive behavior of the eared grebe, *Podiceps caspicus nigricollis*. M.S. thesis, University of British Columbia, Vancouver.

————, 1958. Courtship, hostile behavior, nest establishment, and egg-laying in the eared grebe. *Auk* 75:290–311.

————, 1964. Ontogeny of behavior of five species of grebes. Ph.D. diss., University of British Columbia, Vancouver.

McCabe, R. A., and Hawkins, A. S., 1946. The Hungarian partridge in Wisconsin. *Amer. Midl. Nat.* 36:1–75.

McClelland, B. R., 1977. Relationships between hole-nesting birds, forest snags and decay in western larch–Douglas fir forests of the northern Rocky Mountains. Ph.D. diss., University of Montana, Missoula.

McClelland, B. R., Young, L. S., Shea, D. S., McClelland, P. T., Allen, H. L., and Spottigue, E. B., 1982. The bald eagle concentration in

Glacier National Park, Montana: origin, growth and variation in numbers. *Living Bird* 19:133-55.

McGahan, J., 1968. Ecology of the golden eagle. *Auk* 85(1):1-12.

McKinney, F., 1965. The displays of the American green-winged teal. *Wilson Bull.* 77:112-21.

McKinney, T. D., 1966. Survival, distribution and reproduction of an introduced gray partridge population. M.S. thesis, Colorado State University, Fort Collins.

McLaren, M. A., 1975. Breeding biology of the boreal chickadee. *Wilson Bull.* 87:344-54.

McNicholl, M. K., 1971. The breeding biology and ecology of Forster's tern (*Sterna forsteri*) at Delta, Manitoba. M.S. thesis, University of Manitoba, Winnipeg.

Mendall, H. L., 1937. Nesting of the bay-breasted warbler. *Auk* 54:429-39.

————, 1958. The ring-necked duck in the Northeast. *University of Maine Bulletin* 60.

Meng, H., 1952. The Cooper's hawk, *Accipiter cooperii* (Bonaparte). Ph.D. diss., Cornell University, Ithaca, New York.

Mewaldt, L. R., 1956. Nesting behavior of the Clark nutcracker. *Condor* 58:3-23.

Meyeriecks, A. J., 1960. Comparative behavior of four species of North American herons. *Publications of the Nuttall Ornithological Club* No. 2, pp. 1-158.

Michael, C. W., 1935. Nesting of the Williamson sapsucker. *Condor* 37(4):209-10.

Michener, H., and Michener, J. R., 1935. Mockingbirds, their territories and individualities. *Condor* 37:97-140.

Mickey, F. W., 1943. Breeding habits of McCown's longspur. *Auk* 60:181-209.

Mikkola, H., 1983. *Owls of Europe.* Berkhamsted, England: T. & A. D. Poyser.

Miller, A. H., 1931. Systematic revision and natural history of the American shrikes (*Lanius*). *Univ. Calif. Publ. Zool.* 38:11-242.

Miller, E. V., 1941. Behavior of the Bewick wren. *Condor* 43:81-99.

Miller, J. R., and Miller, J. T., 1948. Nesting of the spotted sandpiper at Detroit, Michigan. *Auk* 65:558-67.

Minock, M. E., 1971. Social relationships among mountain chickadees. *Condor* 73:118-20.

Mitchell, R. M., 1977. Breeding biology of the double-crested cormorant at Utah Lake. *Great Basin Naturalist* 37:1–23.

Mock, D. M., 1976. Pair-formation displays of the great blue heron. *Wilson Bull.* 88:185–230.

Moisan, G., Smith, R. I., and Martinson, R. K., 1967. The green-winged teal: its distribution, migration, and population dynamics. U.S. Dep. Interior, Fish and Wildl. Serv., Spec. Sci. Rep. 100.

Moldenhauer, R. R., and Wiens, J. A., 1970. The water economy of the sage sparrow, *Amphispiza belli nevadensis*. *Condor* 72(3):265–75.

Morehouse, E. L., and Brewer, R., 1968. Feeding of nestling and fledgling eastern kingbirds. *Auk* 85:44–54.

Moriarty, L. J., 1965. A study of the breeding biology of the chestnut-collared longspur (*Calcarius ornatus*) in northeastern South Dakota. *South Dakota Bird Notes* 17:76–79.

Morse, D. H., 1972. Habitat differences of Swainson's and hermit thrushes. *Wilson Bull.* 84:206–8.

———, 1979. Habitat use by the blackpoll warbler. *Wilson Bull.* 91:234–43.

———, 1980. Foraging and coexistence of spruce-woods warblers. *Living Bird* 18:7–26.

Morse, T. E., Jakabosky, J. L., and McCrow, V. P., 1969. Some aspects of the breeding biology of the hooded merganser. *J. Wildl. Manag.* 33:596–604.

Morton, M. L., Horstmann, J. L., and Osborn, J. M., 1972. Reproductive cycle and nesting success of the mountain white-crowned sparrow (*Zonotrichia leucophrys oriantha*) in the central Sierra Nevada. *Condor* 74:152–63.

Mumford, R. E., and Zusi, R. L., 1958. Notes on movements, territories and habitat of wintering saw-whet owls. *Wilson Bull.* 70:188–91.

Munro, J. A., 1939. Studies of waterfowl in British Columbia. Barrow's golden-eye, American golden-eye. *Trans. Royal Can. Inst.* 22:259–318.

Murray, B. G., Jr., 1969. A comparative study of the Le Conte's and sharp-tailed sparrows. *Auk* 86:199–231.

Murray, B. G., Jr., and Gill, F. B., 1976. Behavioral interactions between blue-winged and golden-winged warblers. *Wilson Bull.* 88:231–54.

Murray, G. A., 1976. Geographic variation in the clutch sizes of seven owl species. *Auk* 93:602–13.

Murton, R. K., and Carke, S. P., 1968. Breeding biology of rock doves. *British Birds* 61:429–48.

Murton, R. K., Coombs, C. F. B., and Therle, R. J. P., 1972. Ecological studies of the feral pigeon *Columba livia* var., II. Flock behaviour and social organization. *J. Appl. Ecol.* 9(3):875–89.

Mussehl, T. W., 1960. Blue grouse production, movements, and populations in the Bridger Mountains, Montana. *J. Wildl. Manag.* 24:60–68.

Myers, H. W., 1912. Nesting habits of the western bluebird. *Condor* 14(6):221–22.

Neff, J. A., 1947. Habits, food, and economic status of the band-tailed pigeon. *North Amer. Fauna* 58:1–76.

Nelson, T., 1939. The biology of the spotted sandpiper, *Actitis macularia* (Linn.). Ph.D. diss., University of Michigan, Ann Arbor.

Nero, R. E., 1956. A behavior study of the red-winged blackbird. *Wilson Bull.* 68:5–37, 129–50.

———, 1980. *The great gray owl: phantom of the northern forest.* Washington, D.C.: Smithsonian Institute Press.

Nethersole-Thompson, D., 1966. *The snow bunting.* London: Oliver and Boyd Ltd.

———, 1975. *Pine crossbills. A Scottish contribution.* Berkhamstead: T. & A. D. Poyser.

Newman, G. A., 1970. Cowbird parasitism and nesting success of lark sparrows in southern Oklahoma. *Wilson Bull.* 82:304–9.

Newman, I., 1972. *Finches.* London: Collins.

Nice, M. M., 1922. A study of the nesting of mourning doves (Part I). *Auk* 39:457–74.

———, 1937. Studies in the life history of the song sparrow. Part I. A population study of the song sparrow. *Trans. Linnean Soc. New York* 4:1–227.

———, 1939. Home life of the American bittern. *Wilson Bull.* 51:83–85.

———, 1943. Studies in the life history of the song sparrow. Part II. The behavior of the song sparrow and other passerines. *Trans. Linnean Soc. New York* 6:1–329.

Nice, M. M., and Collias, N. E., 1961. A nesting of the least flycatcher. *Auk* 78:145–49.

Nice, M. M., and Nice, L. B., 1932. A study of the two nests of the black-throated green warbler. *Bird-Banding* 3:95–105, 157–72.

Nickell, W. P., 1951. Studies of habitats, territory, and nests of the eastern goldfinch. *Auk* 68:447–70.

————, 1965. Habitats, territory and nesting of the catbird. *Amer. Midl. Nat.* 73:443–78.

————, 1966. The nesting of the black-crowned night heron and its associates. *Jack-Pine Warbler* 44(3):130–39.

Noble, G. K., Wurm, M., and Schmidt, M., 1938. Social behavior of the black-crowned night heron. *Auk* 55:7–40.

Norman, F. R., and Robertson, R. J., 1975. Nest-searching behavior in the brown-headed cowbird. *Auk* 92(3):610–11.

Norris, R. A., 1958. Comparative biosystematics and life history of the nuthatches *Sitta pygmaea* and *Sitta pusilla*. *Univ. Calif. Publ. Zool.* 56: 119–300.

Nowicki, T., 1973. A behavioral study of the marbled godwit in North Dakota. M.S. thesis, Central Michigan University, Mt. Pleasant.

Nuechterlein, G., 1975. Nesting ecology of western grebes on the Delta Marsh, Manitoba. M.S. thesis, Colorado State University, Fort Collins.

Odum, E. P., 1941–42. Annual cycle of the black-capped chickadee. *Auk* 58:314–33, 518–35, 59:499–531.

Oeming, A. F., 1955. A preliminary study of the great gray owl (*Scotiaptex nebulosa nebulosa* Forster) in Alberta, with observations on some other species of owls. M.S. thesis, University of Alberta, Edmonton.

Ohlendorf, H. M., 1974. Competitive relationships among kingbirds (*Tyrannus*) in Trans-Pecos Texas. *Wilson Bull.* 86:357–73.

————, 1976. Comparative breeding ecology of phoebes in Trans-Pecos Texas. *Wilson Bull.* 88:255–71.

Ohlendorf, R. R., 1975. *Golden eagle country*. New York: Alfred A. Knopf.

Olson, D. P., 1964. A study of canvasback and redhead breeding populations: nesting habitats and productivity. Ph.D. diss., University of Minnesota, Minneapolis.

Olson, S. T., and Marshall, W. H., 1952. The common loon in Minnesota. *Occasional Papers of the Minnesota Museum of Natural History* No. 5, pp. 1–77.

Orians, G. H., 1961. The ecology of blackbird (*Agelaius*) social systems. *Ecol. Monogr.* 31:285–312.

Orians, G. H., and Christman, G. M., 1968. A comparative study of the behavior of red-winged, tricolored, and yellow-headed blackbirds. *Univ. Calif. Publ. Zool.* 84:1–85.

Orians, G. H., and Horn, H. S., 1969. Overlap in foods and foraging of four species of blackbirds in the potholes of central Washington. *Ecology* 50:930–38.

Oring, L. W., 1969. Summer biology of the gadwall at Delta, Manitoba. *Wilson Bull.* 81:44–54.

Oring, L. W., and Knudson, M. L., 1973. Monogamy and polyandry in the spotted sandpiper. *Living Bird* 11:59–73.

Palmer, R. S., 1941. A behavior study of the common tern. *Proceedings of the Boston Society of Natural History* 42:1–119.

Palmer, R. S., ed., 1962. *Handbook of North American birds*, Vol. I. New Haven and London: Yale University Press.

———, 1976. *Handbook of North American birds*, Vols. 2 and 3. Waterfowl. New Haven and London: Yale University Press.

Parks, G. H., and Parks, H. C., 1963. Some notes on a trip to an evening grosbeak nesting area. *Bird-Banding* 34:22–30.

Patterson, R. L., 1952. *The sage grouse in Wyoming.* Denver: Sage Books.

Payne, R. B., 1965. Clutch size and numbers of eggs laid by brown-headed cowbirds. *Condor* 67:44–60.

———, 1969. Breeding season and reproductive physiology of tricolored and redwinged blackbirds. *Univ. Calif. Publ. Zool.* 90:1–114.

———, 1973. The breeding season of a parasitic bird, the brown-headed cowbird, in central California. *Condor* 75:80–99.

Paynter, R. A., Jr., 1954. Interrelations between clutch-size, brood-size, prefledging survival, and weight in Kent Island tree swallows. *Bird-Banding* 25:35–58, 101–10, 136–48.

Peakall, David B., 1970. The eastern bluebird: its breeding season, clutch size and nesting success. *Living Bird* 9:239–55.

Peek, F. W., 1971. Seasonal change in the breeding behavior of the male red-winged blackbird. *Wilson Bull.* 83:383–95.

Perdeck, A. C., 1953. The early reproductive behavior of the arctic skua. *Ardea* 51:1–15.

Peterson, A. J., 1955. The breeding cycle in the bank swallow. *Wilson Bull.* 67:235–86.

Peterson, M. R., 1976. Breeding biology of arctic and red-throated loons. M.S. thesis, University of California, Davis.

———, 1979. Nesting ecology of arctic loons. *Wilson Bull.* 91:608–15.

Pettingill, O. S., Jr., 1930. Observations on the nesting activities of the hermit thrush. *Bird-Banding* 1:72–77.

450

Phillips, R. S., 1972. Sexual and agonistic behavior in the killdeer (*Charadrius vociferus*). *Animal Behavior* 20:1–9.

Pickens, A. L., 1936. Notes on nesting ruby-throated hummingbirds. *Wilson Bull.* 48:80–85.

Pickwell, G. B., 1931. The prairie horned lark. *St. Louis Acad. Sci. Trans.* 27:1–153.

———, 1947. The American pipit in its arctic–alpine home. *Auk* 64:1–14.

Pitelka, F. A., 1940. Breeding behavior of the black-throated green warbler. *Wilson Bull.* 52:2–18.

———, 1951. Speciation and ecological distribution in American jays of the genus *Aphelocoma*. *Univ. Calif. Publ. Zool.* 50:195–464.

Pitelka, F. A., Tomich, P. Q., and Treichel, G. W., 1955. Ecological relations of jaegers and owls as lemming predators near Barrow, Alaska. *Ecol. Monogr.* 25:85–117.

Planck, R. J., 1967. Nest site selection and nesting of the European starling in California. Ph.D. diss., University of California, Davis.

Platt, J. B., 1976. Sharp-shinned hawk nesting and nest site selection in Utah. *Condor* 78:102–3.

Porter, D. K., Strong, M. S., Giezentanner, J. B., and Ryder, R. A., 1975. Nest ecology, productivity and growth of the loggerhead shrike on the short-grass prairie. *Southwestern Naturalist* 19:429–36.

Porter, R. D., and White, C. M., in collaboration with Erwin, R. J., 1973. The peregrine falcon in Utah, emphasizing ecology and competition with the prairie falcon. *Brigham Young Univ. Sci. Bull. Biol. Ser.* 18:1–74.

Portnoy, J. W., and Dodge, W. E., 1979. Red-shouldered hawk nesting ecology and behavior. *Wilson Bull.* 91:104–17.

Pospahala, R. S., Anderson, D. R., and Henny, C. J., 1974. Population ecology of the mallard II: breeding habitat conditions, size of breeding populations and production indices. *Bur. Sport Fish. and Wildl. Res. Publ.* 115.

Pospichal, L. B., and Marshall, W. H., 1954. A field study of the sora rail and Virginia rail in central Minnesota. *Flicker* 26:2–32.

Poston, H. J., 1974. Home range and breeding biology of the shoveler. *Can. Wildl. Serv. Rep. Ser.* 25.

Potter, E., 1980. Notes on nesting yellow-billed cuckoos. *J. Field Ornithol.* 51:17–29.

Potter, P. E., 1972. Territorial behavior in savannah sparrows in southeastern Michigan. *Wilson Bull.* 72:48–59.

Power, H. W., III, 1966. Biology of the mountain bluebird in Montana. *Condor* 68:351–71.

Pratt, H. M., 1970. Breeding biology of great blue herons and common egrets in central California. *Condor* 72:407–16.

Preble, N. A., 1957. The nesting habits of the yellow-billed cuckoo. *Amer. Midl. Nat.* 57:474–82.

Prescott, K. W., 1965. The scarlet tanager. New Jersey State Museum, Investigations, No. 2, Trenton, N.J.

Pulliainen, E., 1978. Nesting of the hawk owl, *Surnia ulula*, and short-eared owl, *Asio flammeus*, and the food consumed by owls on the island of Uklokrunni in the Bothnian Bay in 1977. *Aquilo* (Zool.) 18:17–22.

Putnam, L. S., 1949. The life history of the cedar waxwing. *Wilson Bull.* 61:141–82.

Randle, W., and Austing, R., 1952. Ecological notes on the long-eared and saw-whet owls in southwestern Ohio. *Ecology* 33:422–26.

Raper, E. L., 1976. Influence of the nesting habitat on the breeding success of California gulls (*Lanus californicus*), Banforth Lake, Albany County, Wyoming. M.S. thesis, University of Wyoming, Laramie.

Rawls, C. K., Jr., 1949. An investigation of life history of the white-winged scoter (*Melanitta fusca deglandi*). M.S. thesis, University of Minnesota, Minneapolis.

Reller, A. W., 1972. Aspects of behavioral ecology of red-headed and red-bellied woodpeckers. *Amer. Midl. Nat.* 88:207–90.

Rea, A. M., 1970. Winter territoriality in a ruby-crowned kinglet. *West. Bird-Bander* 45:4–7.

Rees, W. E., 1973. Comparative ecology of three sympatric sparrows of the genus *Zonotrichia*. Ph.D. diss., University of Toronto, Ontario.

Reynolds, R. T., 1978. Food and habitat partitioning in two groups of coexisting Accipiters. Ph.D. diss., Oregon State University, Corvallis.

Reynolds, T. D., and Rich, T. D., 1978. Reproductive ecology of the sage thrasher (*Oreoscoptes montanus*) on the Snake River Plain in south-central Idaho. *Auk* 95:580–82.

Richardson, L., 1980. Ospreys in the Bridger-Teton National Forest, Wyoming. M.S. thesis, Yale School of Forestry & Environmental Studies, New Haven.

Richmore, M. L., DeWeese, L. R., and Pillmore, R. E., 1980. Brief observations on the breeding biology of the flammulated owl in Colorado. *Western Birds* 11:35–46.

Rice, J., 1978. Ecological relationships of two interspecifically territorial vireos. *Ecology* 59:526–38.

Ritter, L. V., 1972. The breeding biology of scrub jays. M.A. thesis, California State University, Chico.

Robbins, S., 1974. The willow and alder flycatcher in Wisconsin: a preliminary description of summer range. *Passenger Pigeon* 36(4):147–52.

Robertson, F. D., 1957. The flocking habits of the golden-crowned sparrow in a winter society. *West. Bird-Bander* 32:29–31.

Robinson, T. S., 1957. *The ecology of bobwhites in south-central Kansas.* University of Kansas Museum of Natural History and State Biological Survey Miscellaneous Publication No. 15.

Robinson, W. L., 1980. *Fool hen: the spruce grouse on the Yellow Dog Plains.* Madison: University of Wisconsin Press.

Rogers, T. L., 1937. Behavior of the pine siskin. *Condor* 39:143–49.

Rogers, J. P., 1962. The ecological effect of drought on reproduction of the lesser scaup, *Aythya affinis* (Eyton). Ph.D. diss., University of Missouri, Columbia.

Root, R. B., 1967. The niche exploitation pattern of the blue-gray gnatcatcher. *Ecol. Monogr.* 37:317–50.

———, 1969. The behavior and reproductive success of the blue-gray gnatcatcher. *Condor* 71:16–31.

Rosene, W., 1969. *The bobwhite quail: its life and management.* New Brunswick, N.J.: Rutgers University Press.

Rohwer, S. A., 1971. Systematics and evolution of Great Plains meadowlarks, genus *Sturnella*. Ph.D. diss., University of Kansas, Lawrence.

Rohwer, S., Ewald, P. W., and Rohwer, F. C., 1981. Variation in size, appearance and dominance within and among the age and sex classes of Harris sparrow. *J. Field Ornithol.* 52:291–303.

Rutter, R. J., 1969. A contribution to the biology of the gray jay (*Perisoreus canadensis*). *Can. Field Nat.* 83(4):300–16.

Ryder, J. P., 1972. Biology of nesting Ross's geese. *Ardea* 60(3–4):185–215.

Ryder, R. A., 1967. Distribution, migration and mortality of the white-faced ibis (*Plegadis chihi*) in North America. *Bird-Banding* 38(4):257–77.

Sabine, W. S., 1959. The winter society of the Oregon junco: intolerance, dominance, and the pecking order. *Condor* 61(2):110–35.

Salomonson, M. G., and Balda, R. P., 1977. Winter territoriality of Townsend's solitaires (*Myadestes townsendi*) in a piñon–juniper–ponderosa pine ecotone. *Condor* 79:148–61.

Salt, G. W., 1952. The relation of metabolism to climate and distribution in three finches of the genus *Carpodacus*. *Ecol. Monogr.* 22:121–52.

———, 1957a. Observation on fox, Lincoln and song sparrows at Jackson Hole, Wyoming. *Auk* 74:258–59.

Salt, W. R., 1966. A nesting study of *Spizella pallida*. *Auk* 83:274–81.

Salyer, J. C., and Lagler, K. F., 1946. The eastern belted kingfisher, *Megaceryle alcyon alcyon* (Linnaeus), in relation to fish management. *Trans. Amer. Fish. Soc.* 76:97–117.

Samuel, D. E., 1971. The breeding biology of barn and cliff swallows in West Virginia. *Wilson Bull.* 83:284–301.

Samson, F. B., 1976. Territory, breeding density, and fall departure in Cassin's finch. *Auk* 93:477–97.

Sanderson, G. C., and Shultz, H. C. (eds.), 1973. *Wild turkey management.* Proc. Second Natl. Wild Turkey Symp., Columbia, Missouri, 1970. Missouri Chapter of the Wildl. Soc. and University of Missouri Press, Columbia.

Sanderson, G. C. (ed.), 1977. *Management of migratory shore and upland game birds in North America.* Washington, D.C.: U.S. Dept. of Interior and International Association of Fish and Wildlife Agencies. (Reprinted in 1980 by the University of Nebraska Press, Lincoln.)

Santee, R., and Granfield, W., 1939. Behavior of the saw-whet owl on its nesting grounds. *Condor* 41:3–9.

Sappington, J. N., 1977. Breeding biology of house sparrows in north Mississippi. *Wilson Bull.* 89:300–9.

Schaller, G. B., 1964. Breeding behavior of the white pelican at Yellowstone Lake, Wyoming. *Condor* 66:3–23.

Schnell, G. D., 1968. Differential habitat utilization by wintering rough-legged and red-tailed hawks. *Condor* 70:373–77.

Schnell, J. H., 1958. Nesting behavior and food habits of goshawks in the Sierra Nevada of California. *Condor* 60:377–403.

Schrantz, F. G., 1943. Nest life of the yellow warbler. *Auk* 60:367–87.

Schukman, J. N., 1974. Comparative nesting ecology of the eastern phoebe (*Sayornis phoebe*) and Say's phoebe (*Sayornis saya*) in west-central Kansas. M.S. thesis, Fort Hays State College, Fort Hays, Kansas.

454

Schwartz, C. W., 1945. The ecology of the prairie chicken in Missouri. *University of Missouri Studies* 20:1–99.

Scott, M. D., 1982. Distributions and habitat use of white-tailed ptarmigan in Montana. *Proc. Montana Academy of Science* 41:57–66.

Sealy, S. G., 1974. Ecological segregation of Swainson's and hermit thrushes on Langara Island, British Columbia. *Condor* 76:350–51.

Seanstedt, T. R., and MacLean, S. F., 1979. Territory size and composition in relation to seasonal abundance in Lapland longspurs breeding in arctic Alaska. *Auk* 96:131–42.

Sedgwick, J. A., 1975. A comparative study of the breeding biology of Hammond's and dusky flycatchers. M.S. thesis, University of Montana, Missoula.

Selander, R. K., 1954. A systematic review of the booming nighthawks of western North America. *Condor* 56:57–82.

Semple, J. B., and Sutton, G. M., 1932. Nesting of Harris' sparrow at Churchill, Manitoba. *Auk* 49:166–83.

Shea, R. E., 1979. The ecology of the trumpeter swan in Yellowstone National Park and vicinity. M.S. thesis, University of Montana, Missoula.

Sherrod, S. K., White, C. M., and Williamson, F. S. L., 1976. Biology of the bald eagle on Amchitka Island, Alaska. *Living Bird* 15:143–82.

Short, L. L., Jr., 1965. Hybridization in the flickers (*Colaptes*) of North America. *Bulletin of the American Museum of Natural History* 129:309–428.

———, 1971. Systematics and behavior of some North American woodpeckers, genus *Picoides* (Aves). *Bulletin of the American Museum of Natural History* 145:1–118.

———, 1974. Habits and interactions of North American three-toed woodpeckers (*Picoides arcticus* and *P. tridactylus*). *Amer. Mus. Novitates* 2547:1–42.

———, 1983. *Woodpeckers of the world.* Delaware Museum of Natural History, Monograph Series No. 4.

Shreeve, D. F., 1977. Comparative behavior of Aleutian gray-crowned and brown-capped rosy finch. Ph.D. diss., Cornell University, Ithaca, N.Y.

———, 1980. Behaviour of the Aleutian grey-crowned and brown-capped rosy finches *Leucosticte tephrocotis*. *Ibis* 122:146–65.

Sibley, C. G., Short, L. L., Jr., 1959. Hybridization in the buntings (*Passerina*) of the Great Plains. *Auk* 76:443–63.

————, 1964. Hybridization in orioles of the Great Plains. *Condor* 66: 130–50.

Sibley, C. G., and West, D. A., 1959. Hybridization in the rufous-sided towhees of the Great Plains. *Auk* 76:326–38.

Smith, D. G., and Murphy, J. T., 1973. Breeding ecology of raptors in the eastern Great Basin of Utah. *Brigham Young Univ. Sci. Bull.*, Biol. Ser. 19.

————, 1978. Biology of the ferruginous hawk in central Utah. *Sociobiology* 3:79–95.

Smith, R. I., 1968. The social aspects of reproductive behavior in the pintail. *Auk* 83(3):381–96.

Smith, R. L., 1963. Some ecological notes on the grasshopper sparrow. *Wilson Bull.* 75:159–65.

Smith, S. M., 1973. A study of prey-attack behaviour in young loggerhead shrikes, *Lanius ludovicianus* L. *Behaviour* 44(1–2):113–41.

————, 1974. Nest site selection in black-capped chickadees. *Condor* 76(4):478–79.

Smith, W. J., 1966. Communication and relationships in the genus *Tyrannus*. *Publ. Nuttall Ornithol. Club* 6:1–250.

————, 1969. Displays of *Sayornis phoebe* (Aves, Tyrannidae). *Behaviour* 33:283–322.

————, 1970. Courtship and territorial displaying in the vermilion flycatcher, *Pyrocephala rubinus*. *Condor* 72:488–91.

Smith, W. P., 1934. Observations on the nesting habits of the black-and-white warbler. *Bird-Banding* 5:31–36.

Snapp, B. D., 1976. Colonial breeding in the barn swallow (*Hirundo rustica*) and its adaptive significance. *Condor* 78:471–80.

Snow, C., 1972. American peregrine falcon (*Falco peregrinus anatum*) and arctic peregrine falcon (*Falco peregrinus tundrius*). Habitat Management Series for Endangered Species, U.S. Bur. Land Manage., Tech Note 167. Denver, Colorado.

————, 1973a. Habitat management series for unique or endangered species: golden eagle *Aquila chrysaetos*. U.S. Dept. Inter., Bur. Land Manage. Tech. Note 239, Rep. 7. Denver, Colo.

————, 1973b. Southern bald eagle (*Haliaeetus leucocephalus leucocephalus*) and northern bald eagle (*Haliaeetus leucocephalus alascanus*). Habitat Management Series for Endangered Species, U.S. Dept. Inter., Bur. Land Manage. Tech. Note 171. Denver, Colo.

————, 1974a. Habitat management series for unique or endangered species: ferruginous hawk *Buteo regalis*. U.S. Dept. Inter., Bur. Land Manage. Tech. Note 255, Rep. 13. Denver, Colo.

———, 1974b. Prairie falcon (*Falco mexicanus*). Habitat Management Series for Endangered Species, U.S. Dept. Inter., Bur. Land Manage. Tech. Note 240.

Snyder, D. E., 1954. Nesting study of red crossbills. *Wilson Bull.* 66:32–37.

Southern, W. E., 1958. Nesting of the red-eyed vireo in the Douglas Lake region, Michigan. *Jack-Pine Warbler* 36:105–30, 185–207.

Sowls, L. K., 1955. *Prairie ducks: a study of the behavior, ecology and management.* Washington, D.C.: Wildlife Management Institute; Harrisburg, Pa.: Stackpole Company (reprinted 1978, University of Nebraska Press).

Spencer, H., Jr., 1953. The cinnamon teal, *Anas cyanoptera* (Vieillot): its life history, ecology and management. M.S. thesis, Utah State University, Logan.

Spencer, O. R., 1943. Nesting habits of the black-billed cuckoo. *Wilson Bull.* 55:11–22.

Springer, A. M., 1975. Observations on the summer diet of rough-legged hawks in Alaska. *Condor* 77:338–39.

Staebler, A. E., 1949. A comparative life history study of the downy and hairy woodpeckers (*Dendrocopos pubescens* and *Dendrocopos villosus*). Ph.D. diss., University of Michigan, Ann Arbor.

Stallcup, P. L., 1968. Spatio-temporal relationships of nuthatches and woodpeckers in ponderosa pine forests of Colorado. *Ecology* 49(5):831–43.

Stein, R. C., 1962. A comparative study of songs recorded from five closely related warblers. *Living Bird* 1:61–71.

Stiehl, R. B., 1978. Aspects of the biology of the common raven in Harney Basin, Oregon. Ph.D. diss., Portland State University, Portland.

Stewart, R. E., 1949. Ecology of a nesting red-shouldered hawk population. *Wilson Bull.* 61:26–35.

———, 1953. A life history study of the yellow-throat. *Wilson Bull.* 65:99–115.

Stewart, R. M., 1973. Breeding behavior and life history of the Wilson warbler. *Wilson Bull.* 85:21–30.

Stewart, R. M., Henderson, R. P., and Darling, K., 1977. Breeding ecology of the Wilson's warbler in the high Sierra Nevada, California. *Living Bird* 17:83–102.

Stocek, R. F., 1970. Observations on the breeding biology of the tree swallow. *Cassinia* 52:3–20.

Stokes, A. W., 1950. Breeding behavior of the goldfinch. *Wilson Bull.* 62:107–27.

Stout, G. D. (ed.), 1967. *The shorebirds of North America*. New York: Viking Press.

Sturman, W. A., 1968a. The foraging ecology of *Parus atricapillus* and *P. rufescens* in the breeding season, with comparisons with other species of *Parus*. *Condor* 70:309–22.

———, 1968b. Description and analysis of breeding habitats of the chickadees, *Parus atricapillus* and *P. rufescens*. *Ecology* 49:418–31.

Summers-Smith, D., 1963. *The house sparrow*. London: Collins.

Sumner, L., and Dixon, J. S., 1953. *Birds and mammals of the Sierra Nevada, with records from Sequoia and Kings Canyon National Parks*. Berkeley: University of California Press.

Sutherland, C. A., 1963. Notes on the behavior of common nighthawks in Florida. *Living Bird* 2:31–39.

Sutton, G. M., 1949. *Studies of the nesting birds of the Edwin S. George Reserve. Part I. The vireos.* University of Michigan Museum of Zoology Miscellaneous Publication No. 74.

Sutton, G. M., and Parmelee, D. F., 1955. Summer activity of the Lapland longspur on Baffin Island. *Wilson Bull.* 67:110–27.

Swearingen, E. M., 1977. Group size, sex ratio, reproductive success and territory size in acorn woodpeckers. *Western Birds* 8:21–24.

Swenson, J. E., 1975. Ecology of the bald eagle and osprey in Yellowstone National Park. M.S. thesis, Montana State University, Bozeman.

Tanner, W. D., Jr., and Hendrickson, G. O., 1954. Ecology of the Virginia rail in Clay County, Iowa. *Iowa Bird Life* 24:65–70.

Tate, D. G., 1973. Habitat usage by the chipping sparrow (*Spizella passerina*) in northern Lower Michigan. Ph.D. diss., University of Nebraska, Lincoln.

Tate, J., Jr., 1970. Nesting and development of the chestnut-sided warbler. *Jack-Pine Warbler* 48:57–65.

Taylor, D. L., and W. L. Barmore, Jr., 1980. Post-fire succession of avivauna of coniferous forests of Yellowstone and Grand Teton National

Parks, Wyoming. In *Management of Western Forests and Grasslands for Nongame Birds*, USDA Forest Service General Technical Report INT-86.

Taylor, P. S., 1974. Breeding behavior of the snowy owl. *Living Bird* 12:137–54.

Taylor, W. K., and Hanson, H., 1970. Observations on the breeding biology of the vermilion flycatcher in Arizona. *Wilson Bull.* 82:315–19.

Thomas, R. H., 1946. A study of eastern bluebirds in Arkansas. *Wilson Bull.* 58:143–83.

Thompson, C. F., and V. Nolan, Jr., 1973. Population biology of the yellow-breasted chat (*Icteria virens* L.) in southern Indiana. *Ecol. Monogr.* 43:145–71.

Thomsen, L., 1971. Behavior and ecology of burrowing owls at the Oakland Municipal Airport. *Condor* 73:177–92.

Thompson, W. L., 1960. Agonistic behavior in the house finch, Part 1: annual cycle and display patterns. *Condor* 62:245–71.

Threlfall, W., and Blacquiere, J. R., 1982. Breeding biology of the fox sparrow in Newfoundland. *J. Field Ornithol.* 53:235–39.

Tinbergen, N., 1939. The behavior of the snow bunting in spring. *Trans. Linn. Soc. N.Y.* 5:1–94.

————, 1959. Comparative studies of the behaviour of gulls (Laridae): A progress report. *Behaviour* 15:1–70.

Tomback, D. F., 1977. The behavioral ecology of the Clark's nutcracker (*Nucifraga columbiana*) in the eastern Sierra Nevada. Ph.D. diss., University of California, Santa Barbara.

Tomlinson, I. R., 1965. The willets of Georgia and South Carolina. *Wilson Bull.* 77:151–67.

Tomlinson, D. N. S., 1976. Breeding behaviour of the great white egret. *Ostrich* 47:161–78.

Tramontano, J. P., 1964. Comparative studies of the rock wren and the canyon wren. M.S. thesis, University of Arizona, Tucson.

————, 1971. Summer foraging behavior of sympatric Arizona grassland sparrows. Ph.D. diss., University of Arizona, Tucson.

Trauger, D. L., 1971. Population ecology of lesser scaup (*Aythya affinis*) in subarctic taiga. Ph.D. diss., Iowa State University, Ames.

Trefethen, J. B. (ed.), 1966. *Wood duck management and research: a symposium. Theme: emphasizing management of forests for wood ducks.* Wildl. Manage. Inst.

Tuck, L. M., 1972. *The snipes: a study of the genus Capella.* Canadian Wildlife Service, Monograph Series No. 5. Ottawa.

Tvrdik, G. M., 1971. Pendulum display by the olive-sided flycatcher. *Auk* 88:174.

———, 1977. Aggressive and courtship behavior in two races of the brown towhee (*Pipilo fuscus*). Ph.D. diss., University of California, Berkeley.

Twining, H., 1940. Foraging behavior and survival in the Sierra Nevada rosy finch. *Condor* 42:64–72.

Twomey, A. C., 1934. Breeding habits of Bonaparte's gull. *Auk* 51:291–96.

Tyrell, W. B., 1945. A study of the northern raven. *Auk* 62:1–7.

van Riper, C., III, 1976. Aspects of house finch breeding biology in Hawaii. *Condor* 78:224–29.

Verbeck, N. A. M., 1967. Breeding biology and ecology of the horned lark in alpine tundra. *Wilson Bull.* 79:208–18.

———, 1970. Breeding ecology of the water pipit. *Auk* 87(3):425–51.

———, 1975. Comparative feeding behavior of three coexisting tyrannid flycatchers. *Wilson Bull.* 87:231–40.

Vermeer, K., 1970. *Breeding biology of California and ring-billed gulls: a study of ecological adaptations to the inland habitat.* Canadian Wildlife Service, Report Series No. 12. Ottawa.

Verner, J., 1965. Breeding biology of the long-billed marsh wren. *Condor* 67:6–30.

———, 1975a. Complex song repertoire of male long-billed marsh wrens in eastern Washington. *Living Bird* 14:300–463.

———, 1975b. Interspecific aggression between yellow-headed blackbirds and long-billed marsh wrens. *Condor* 77(3):328–31.

Walker, L. W., 1974. *The book of owls.* New York: Alfred A. Knopf.

Walkinshaw, L. H., 1944. The eastern chipping sparrow in Michigan. *Wilson Bull.* 56:193–205.

———, 1953. Life history of the prothonotary warbler. *Wilson Bull.* 65:152–68.

———, 1966. Summer biology of Traill's flycatcher. *Wilson Bull.* 78:31–46.

———, 1975. Savannah sparrow breeding and territoriality on a Nova Scotia beach. *Auk* 92:235–51.

Walkinshaw, L. H., and Henry, C. J., 1957. Yellow-bellied flycatcher nesting in Michigan. *Auk* 74:293–304.

Wallace, G. J., 1939. Bicknell's thrush: its taxonomy, distribution, and life history. *Proc. Boston Soc. Nat. Hist.* 41:211–402.

Watson, A. T., 1977. *The hen harrier.* Berkhamstead, England: A. & A. D. Poyser.

Wattel, J., 1973. Geographic differentiation in the genus *Accipiter. Publ. Nuttal Ornithol. Club* 13:1–231.

Watts, C. R., and Stokes, A. W., 1971. The social order of turkeys. *Scientific American* 224(6):112–18.

Weaver, R. L., and West, F. H., 1943. Notes on the breeding of the pine siskin. *Auk* 60:492–504.

Webster, J. D., 1961. Revision of Grace's warbler. *Auk* 78:554–66.

Weeden, J. S., 1965. Territorial behavior of the tree sparrow. *Condor* 67(3):193–209.

Weeden, R. B., 1960. The ecology and distribution of ptarmigan in western North America. Ph.D. diss., University of British Columbia, Vancouver.

Weidmann, U., 1955. Some reproductive activities of the common gull. *Ardea* 43:85–132.

Weller, M. W., 1958. Observations on the incubation behavior of a common nighthawk. *Auk* 75:48–59.

Welsh, D. A., 1971. Breeding and territoriality of the palm warbler in a Nova Scotia bog. *Canadian Field-Naturalist* 65:99–115.

———, 1975. Savannah sparrow breeding and territoriality on a Nova Scotia dune beach. *Auk* 92:235–51.

Welter, W. A., 1935. The natural history of the long-billed marsh wren. *Wilson Bull.* 47:3–34.

Wemmer, C., 1969. Impaling behavior of the loggerhead shrike. *Zeit. für Tierpsychologie* 26:208–24.

West, D. A., 1962. Hybridization in grosbeaks (*Pheucticus*) of the Great Plains. *Auk* 79:399–424.

Weston, H. G., Jr., 1947. Breeding behavior of the black-headed grosbeak. *Condor* 49:54–73.

Weston, J. B., 1969. Nesting ecology of the ferruginous hawk *Buteo regalis. Brigham Young Univ. Sci. Bull. Biol. Ser.* 10(4):25–36.

Whedon, A. D., 1938. Nesting behavior of kingbirds. *Wilson Bull.* 50:288–89.

White, C. M., and W. H. Behle, 1960. Birds of Flaming Gorge Reservoir basin, pp. 185–208. In *Ecological studies of the flora and fauna of Flaming Gorge Reservoir basin, Utah and Wyoming.* University of Utah, Dept. of Anthropology, Anthropology Paper No. 48.

White, H. C., 1953. The eastern belted kingfisher in the Maritime Provinces. *Fish. Res. Bd. Can. Bull.* 97:1–44.

———, 1957. Food and natural history of mergansers on salmon waters in the Maritime Provinces of Canada. *Fish. Res. Bd. Can. Bull.* 116:1–63.

White, J. M., 1973. Breeding biology and feeding patterns of the Oregon junco in two Sierra Nevada habitats. M.S. thesis, University of California, Berkeley.

Wiens, J. A., 1973. Interterritorial behavior variation in grasshopper and savannah sparrows. *Ecology* 54:877–84.

Wiese, J. H., 1975. Courtship and pair formation in the great egret. *Auk* 93:709–24.

Wiley, J. W., 1975. The nesting and reproductive success of red-tailed hawks and red-shouldered hawks in Orange County, California, 1973. *Condor* 77:133–39.

Williams, L., 1942. Interrelations in a nesting group of four species of birds. *Wilson Bull.* 54:238–49.

———, 1952. Breeding behavior of the Brewer blackbird. *Condor* 54:3–47.

Williamson, P., 1971. Feeding ecology of the red-eyed vireo (*Vireo olivaceus*) and associated foliage-gleaning birds. *Ecol. Monogr.* 41(2):129–52.

Willoughby, E. J., and Cade, T. J., 1964. Breeding behavior of the American kestrel (sparrow hawk). *Living Bird* 3:75–96.

Willson, M. F., 1964. Breeding ecology of the yellow-headed blackbird. *Ecological Monogr.* 36:51–77.

Willson, M. F., and Orians, G. H., 1963. Comparative ecology of red-winged and yellow-headed blackbirds during the breeding season. *Proc. 16th Int. Congr. Zool.* 3:342–46.

Winter, J., 1974. The distribution of the flammulated owl in California. *Western Birds* 5:25–44.

Wittenberger, J. F., 1978. The breeding biology of an isolated bobolink population in Oregon. *Condor* 80:355–71.

Wood, N. A., 1974. The breeding behaviour and biology of the moorhen. *British Birds* 67:104–15, 137–58.

Woolfenden, G., 1975. Florida scrub jay helpers at the nest. *Auk* 92: 1–15.

Wythe, M. W., 1938. The white-throated sparrow in western North America. *Condor* 40(3):110–17.

Appendix

Abundance and Breeding Information for U.S. and Canadian National Parks

		Red-throated Loon	Arctic Loon	Common Loon	Pied-billed Grebe	Horned Grebe	Red-necked Grebe	Eared Grebe	Western Grebe	American White Pelican	Double-crested Cormorant
Page Number		55	56	57	58	60	61	62	63	64	65
Banff NP	SP			O	O	O	U	O	U	V	
	SU			U*	R*	O	O*				
	F	V		O	U	O	O	O	U	V	
	W										
	G										
Yoho NP	G		V	U			C	U	R	U	
Kootenay NP	G			R			R	R	R		
Waterton Lakes NP	G	V		U	U*	U	U	C	C		V
Glacier NP	SP			C	R	C	U	C	C	R	
	SU			C*	R	C*	U	C	C*		R
	F	V	V	C	R	U	U	U	C		R
	W				?	?		?			
	G										
Yellowstone NP	SP			R	O	C		R	R	C	O
	SU	V		R*	O*	O*		R*	R*	C*	U*
	F			R	O	O		O	R	O	
	W										
	G	*?					V				
Grand Teton NP	SP			O	O	R	V	C	R	O	O
	SU			R*	U*	R		O	R*	O	U
	F		V	O	O	O	V	O	O	O	O
	W			V							
	G										
Rocky Mountain NP	SP			U	U			U	U		
	SU			O	R				R	R	
	F			O	U			U	U		
	W										
	G										
Dinosaur NM	G			R	R*	V		R*	O*	R	

Legend: **A**=abundant; **C**=common; **U**=uncommon; **O**=occasional; **R**=rare; **V**=vagrant (accidental); **?**=inadequate information; *****=breeds (or has bred) in area.

Abundance and Breeding Information for U.S. and Canadian National Parks

		American Bittern	Least Bittern	Great Blue Heron	Great (Common) Egret	Snowy Egret	Cattle Egret	Green-backed Heron	Black-crowned Night Heron	White-faced Ibis	Tundra Swan
Page Number		66	67	68	70	71	72	73	74	75	76
Banff NP	SP	R		R							U
	SU	R*		V					V		V
	F			R							U
	W										
	G										
Yoho NP	G	R		R							R
Kootenay NP	G			R							V
Waterton Lakes NP	G	U		U							C
Glacier NP	SP	R		C							C
	SU	R		C*					V		
	F			C							C
	W										
	G										
Yellowstone NP	SP	R*		C					R		O
	SU	R		C*	V	R			R*		
	F	R		C					R		O
	W			R							
	G									V	
Grand Teton NP	SP	O		C	V	R	V	V	V	O	V
	SU	O*		C*		R					
	F	O		C		R			V		O
	W			O							V
	G										
Rocky Mountain NP	SP	V	V	R		V	V		V	R	R
	SU	V							V		
	F			R						V	R
	W										
	G										
Dinosaur NM	G	R*		C*		O			R*	R*	O

Abundance and Breeding Information for U.S. and Canadian National Parks

		Trumpeter Swan	Greater White-fronted Goose	Snow Goose	Ross's Goose	Canada Goose	Wood Duck	Green-winged Teal	American Black Duck	Mallard	Northern Pintail
Page Number		77	78	79	80	81	82	83	84	85	86
Banff NP	SP	V		R		C	U	U		C	U
	SU					C*	R*	O*		C*	O*
	F					C	V	U		C	U
	W							V			V
	G										
Yoho NP	G			R		U*	R	U*		C*	U
Kootenay NP	G			V		R		U*		U*	R
Waterton Lakes NP	G	U		U		C*	R	U		C*	C
Glacier NP	SP	R		U	R	C	U	U		A	U
	SU			V		C*	U*	U		A*	U*
	F			U		C	U	U		A	U
	W					U				C	R
	G										
Yellowstone NP	SP	C		R		C		C		A	C
	SU	C*				C*	R*	O*		C*	O*
	F	C		R		C	R	C		A	C
	W	C				C		O		C	R
	G								V		
Grand Teton NP	SP	C		O		C	R	C		A	O
	SU	C*				C*	R	O*		C*	O*
	F	C	V	O		C	R	C		A	C
	W	C		R		C	R	O		C	C
	G										
Rocky Mountain NP	SP					U		U		U	R
	SU					R		R		C*	
	F					U		U		U	R
	W		V	V		R	V	V		O	V
	G										
Dinosaur NM	G			R		C*		O*		C*	O*

Abundance and Breeding Information for U.S. and Canadian National Parks

		Blue-winged Teal	Cinnamon Teal	Northern Shoveler	Gadwall	Eurasian Wigeon	American Wigeon	Canvasback	Redhead	Ring-necked Duck	Greater Scaup
Page Number		87	88	89	90	91	92	93	94	95	96
Banff NP	SP	O	O	R	O		U	R	R	U	
	SU	O*	O				V*	V	V	O*	
	F	O	V	R	R		U	R	V	U	
	W										
	G										
Yoho NP	G	U	R	U	R		U	V	R	U*	
Kootenay NP	G	U		R			R			U*	R
Waterton Lakes NP	G	U*	R*	U	R*		U	U	U	U	
Glacier NP	SP	U	U	U	U	V	C	U	U	C	V
	SU	U	U	U	U*	V	C		R	U*	
	F	U	U	U	U		C	U	U	U	
	W			R	R		R				
	G										
Yellowstone NP	SP	O	O	O	C		C	R	R	O	O
	SU	O*	O*	O*	O*		O*		R*	O*	
	F	O	O	O	C		C	R	O	O	O
	W				R		O			R	V
	G										
Grand Teton NP	SP	C	O	O	C	V	C	O	R	O	
	SU	O*	O*	R*	O*	V	R*	R*	R*	C*	V
	F	C	R	O	C		C	O	C	C	
	W	V	V	V	O		R			R	
	G										
Rocky Mountain NP	SP	C	R	U	U		R	R	U	O	
	SU	O*		R*	O*						
	F	C		U	O		R	R	O		
	W									R	
	G										
Dinosaur NM	G	O*	O*	C*	O*		O*	O*	O*	R*	

Abundance and Breeding Information for U.S. and Canadian National Parks

		Lesser Scaup	Harlequin Duck	Oldsquaw	Black Scoter	Surf Scoter	White-winged Scoter	Common Goldeneye	Barrow's Goldeneye	Bufflehead	Hooded Merganser
Page Number		97	98	99	100	101	102	103	104	105	106
Banff NP	SP	U	U	R		U	U	U	C	U	R
	SU	O*	O*						U*	O	
	F	U				R	R	U	U	O	U
	W							C	R	V	
	G										
Yoho NP	G	U	U*				U	U	U	U*	U	U
Kootenay NP	G	R	U*					R	R	U	U*	
Waterton Lakes NP	G	U	U*	R				R	U	U*	U*	U
Glacier NP	SP	U	C					U	C	C	C	U
	SU	U	C*						C*	C*	C*	U*
	F	U	R					U	C	C	C	U
	W								C	U	U	R
	G											
Yellowstone NP	SP	C	O						C	C	O	R
	SU	C*	O*						C*	C*	O*	R
	F	O	O						C	C	O	R
	W	R		R					C	C	C	R
	G				V		V				
Grand Teton NP	SP	O	O						O	C	C	R
	SU	O	O*						O*	C*	O*	
	F	O	O				V		O	C	C	R
	W					V			O	O	O	O
	G										
Rocky Mountain NP	SP	U							C	C	R	R
	SU											
	F	U		V			V		C	C	R	
	W	U		V					C	C	R	
	G										
Dinosaur NM	G	R*		V						V	O	

Abundance and Breeding Information for U.S. and Canadian National Parks

		Common Merganser	Red-breasted Merganser	Ruddy Duck	Turkey Vulture	Osprey	Bald Eagle	Northern Harrier	Sharp-shinned Hawk	Cooper's Hawk	Northern Goshawk
Page Number		107	108	109	110	111	112	113	114	115	116
Banff NP	SP	U	R	R	V	O	R	U	R	R	R
	SU	O*		V		U*	O*	O	R*	R*	R*
	F	U	V	R		U	O	U	R	R	R
	W	V					R		R	V	R
	G										
Yoho NP	G	U*	R	R		C*	U	U	U	U*	R
Kootenay NP	G	R*	R	R		R	R	R	U	R	R
Waterton Lakes NP	G	C*	R	U		R*	U*	U*	R*	R*	R*
Glacier NP	SP	A	U	U	R	C	U	C	C	U	U
	SU	A*	U*	U		C*	U*	C	C*	U*	U*
	F	A	U	U		C	A	C	U	U	U
	W	C					U		U	U	U
	G										
Yellowstone NP	SP	C	R	O	R	C	O	O	O	O	O
	SU	C*	?	O*	R	C*	O*	O*	O*	O*	O*
	F	C	R	O		C	O	O	O	O	O
	W	C	O				C	R	R	?	O
	G										
Grand Teton NP	SP	C	O	O	R	C	C	O	O	O	U
	SU	C*		O*	R	C*	C*	O*	O*	O*	U*
	F	C	O	O	R	C	C	O	O	O	U
	W	C		V		O	C	R			U
	G										
Rocky Mountain NP	SP	C	R	O	U	U	R	U	U	U	U
	SU	C*			U	U*	R	C	U	U	U*
	F	C		O		U	R	U	R	U	U
	W	C					R	U	R	R	O
	G										
Dinosaur NM	G	O		O*	A*	R	C	C*	O	O*	R

Abundance and Breeding Information for U.S. and Canadian National Parks

		Red-shouldered Hawk	Broad-winged Hawk	Swainson's Hawk	Red-tailed Hawk	Ferruginous Hawk	Rough-legged Hawk	Golden Eagle	American Kestrel	Merlin	Peregrine Falcon
Page Number		117	118	119	120	121	122	123	124	125	126
Banff NP	SP		V	V	.U		V	R	U	R	V
	SU		V	V	U*			O*	U*	R	V
	F				U		V	O	U	R	
	W						R	R			
	G										
Yoho NP	G			V	C*	V	R	U*	U*	U	
Kootenay NP	G			V	C*			U	U*	R	
Waterton Lakes NP	G		R	U	U	R	U	U*	C*	U	V
Glacier NP	SP			U	C		R	C	C	R	R
	SU			U	C*	R	R	C*	C*	R	
	F	V		U	C		R	C	C	R	
	W						R	C			
	G										
Yellowstone NP	SP			C	C	R	R	O	C	R	R
	SU			C*	C*	R*	R	O*	C*	R*	R*
	F			C	C	R	O	O	C	R	R
	W				R		C	O			
	G										
Grand Teton NP	SP		V	C	C	R	R	O	C	O	R
	SU			C*	C*	R*		O*	C*	R	R*
	F			C	C	R	C	O	C	O	R
	W				R		C	O	R	V	V
	G										
Rocky Mountain NP	SP				C		O	U	C	U	R
	SU			R	C		R	U*	C*	R*	R*
	F			O	U	U	O	U	C	U	
	W				O		U	U	R		
	G										
Dinosaur NM	G		V	O	C*	R	C	C*	C*	R*	R

Abundance and Breeding Information for U.S. and Canadian National Parks

		Gyrfalcon	Prairie Falcon	Gray Partridge	Chukar	Ring-necked Pheasant	Spruce Grouse	Blue Grouse	Willow Ptarmigan	White-tailed Ptarmigan	Ruffed Grouse
Page Number		127	128	130	131	132	133	134	135	136	137
Banff NP	SP		V			V	O	O		U	O
	SU		V				O*	O*		U	U*
	F		R			V	O	O		U	U
	W						O	O		U	O
	G										
Yoho NP	G						U*	U*		U*	U*
Kootenay NP	G		R				U*	U*		U*	C*
Waterton Lakes NP	G		R	V		V	U*	U*		U*	C*
Glacier NP	SP		R	R		R	A	C		C	A
	SU		R*				A*	C*		C*	A*
	F		R				A	C		C	A
	W		R				A	C	V	C	A
	G	V									
Yellowstone NP	SP		R					C			O
	SU		R*					C*			O*
	F		R					C			O
	W							C			O
	G			X			V?			V	
Grand Teton NP	SP	V	O	R	R			C			C
	SU		O*	O*	R			C*			C*
	F	V	O	O	R			C			C
	W	V	V	O	R			C			C
	G										
Rocky Mountain NP	SP		U					C		C	
	SU		U					C*		C*	
	F		U					C		C	
	W							C		C	
	G				X						
Dinosaur NM	G		O		O	O		C			

Abundance and Breeding Information for U.S. and Canadian National Parks

		Sage Grouse	Sharp-tailed Grouse	Wild Turkey	Bobwhite	Virginia Rail	Sora	Common Moorhen	American Coot	Sandhill Crane	Whooping Crane
Page Number		138	139	140	141	142	143	144	145	146	148
Banff NP	SP					V	O		U		
	SU						O*		O		
	F						O		U	V	
	W										
	G										
Yoho NP	G						U		U		
Kootenay NP	G						U		U*		
Waterton Lakes NP	G		U*				R		C		V
Glacier NP	SP						U		C	V	
	SU		R				U*		C		
	F								C	V	
	W										
	G				X						
Yellowstone NP	SP	R	R				O		C	O	
	SU	R*	R				O*		C*	O*	V
	F	R	R	V			O		C	O	
	W	R	R								
	G										
Grand Teton NP	SP	C					O		O	C	V
	SU	C*	V				O*		O*	U*	V
	F	C	V			V	O		C	C	V
	W	C	V						R		
	G										
Rocky Mountain NP	SP	R					U		U		
	SU	R*				R*	U*	V	U*		
	F	R					U		U	R	
	W	R							R		
	G										
Dinosaur NM	G	C*				R*	R*		O*	O	

Abundance and Breeding Information for U.S. and Canadian National Parks

		Black-bellied Plover	Lesser Golden Plover	Semi-palmated Plover	Killdeer	Mountain Plover	Black-necked Stilt	American Avocet	Greater Yellowlegs	Lesser Yellowlegs	Solitary Sandpiper
Page Number		149	150	151	152	153	154	155	156	157	158
Banff NP	SP				O			V	O		U
	SU				O*				O*		U*
	F	V		R	O			V	O	V	U
	W				R						
	G										
Yoho NP	G	V	V	V	C*			V		U	U
Kootenay NP	G				U				R	U	R*
Waterton Lakes NP	G	R			C*			R	U	R	R
Glacier NP	SP	V			C			U	R	R	U
	SU				C*					V	V
	F				C			U			U
	W				R						
	G										
Yellowstone NP	SP				C	R		O	O	R	R
	SU				C*			R*	V	V	
	F	O			C	R		O	R	R	U
	W				U						
	G										
Grand Teton NP	SP	R	V	R	C			O	O	O	R
	SU				C*	V	V	O	O	V	V
	F	R		R	C		V	O	O	O	R
	W				R						
	G										
Rocky Mountain NP	SP			V	C			R	R	R	R
	SU				C*						R
	F				C				R	R	R
	W										
	G										
Dinosaur NM	G				C*	V	V	O		O	V

Abundance and Breeding Information for U.S. and Canadian National Parks

		Willett	Wandering Tattler	Spotted Sandpiper	Upland Sandpiper	Long-billed Curlew	Hudsonian Godwit	Marbled Godwit	Ruddy Turnstone	Black Turnstone	Sanderling
Page Number		159	160	161	162	163	164	165	166	167	168
Banff NP	SP			C	V						
	SU			C*							
	F			C		R			V		V
	W										
	G										
Yoho NP	G			C*	V		V				
Kootenay NP	G			C*							
Waterton Lakes NP	G		V	C*	R						
Glacier NP	SP	R		C	R	U		R			
	SU	R		C*						V	
	F			C							
	W										
	G										
Yellowstone NP	SP	R		C		R		R			R
	SU	R		C*		R*		V			
	F	R		C				R			R
	W										
	G								V		
Grand Teton NP	SP	O		C		O		O			V
	SU	R		C*	V	O*		R			V
	F	O		C		O		R			V
	W										
	G										
Rocky Mountain NP	SP	R		C		R		R			
	SU	R		C*							
	F			C		R					
	W										
	G										
Dinosaur NM	G	R		C*							

Abundance and Breeding Information for U.S. and Canadian National Parks

		Semi-palmated Sandpiper	Western Sandpiper	Least Sandpiper	White-rumped Sandpiper	Baird's Sandpiper	Pectoral Sandpiper	Dunlin	Stilt Sandpiper	Short-billed Dowitcher	Long-billed Dowitcher
Page Number		169	170	171	172	173	174	175	176	177	178
Banff NP	SP	V				V	V			R	V
	SU										
	F	U		V		U	R			U	O
	W										
	G										
Yoho NP	G	V	R	U		R	R		R	R	
Kootenay NP	G		R	R		R				R	
Waterton Lakes NP	G			R		R	R				R
Glacier NP	SP						R				R
	SU										
	F					R					
	W										
	G										
Yellowstone NP	SP										
	SU										
	F			U		R	R				
	W										
	G								V		
Grand Teton NP	SP	V	V	O		R			R		O
	SU	R	R	R		O		V			O
	F	O	O	O		O	R				O
	W										
	G										
Rocky Mountain NP	SP	V	V								V
	SU					V					
	F		V								
	W										
	G										
Dinosaur NM	G										

Abundance and Breeding Information for U.S. and Canadian National Parks

		Common Snipe	Wilson's Phalarope	Red-necked Phalarope	Red Phalarope	Pomarine Jaeger	Parasitic Jaeger	Long-tailed Jaeger	Franklin's Gull	Bonaparte's Gull	Mew Gull
Page Number		179	180	181	182	183	184	185	186	187	188
Banff NP	SP	U	R	R	V				V	R	V
	SU	O*	V	R					V	V	V
	F	U					V	V	V	R	V
	W	O									
	G										
Yoho NP	G	C*	R	U		/		V		R	R
Kootenay NP	G	U	R	U				V		V	V
Waterton Lakes NP	G	U	U*	R			R		R	R	
Glacier NP	SP	C	U	R					U		
	SU	C*		V					U		
	F	U	U	R						R	
	W										
	G										
Yellowstone NP	SP	O	O	R					O	R	
	SU	O*	O*						R*		
	F	O	O	R						R	
	W	R									
	G										
Grand Teton NP	SP	C	C	O					O	O	
	SU	C*	O*						O		
	F	C	R	R			V		O	R	
	W	O									
	G										
Rocky Mountain NP	SP	C	R						U	R	
	SU	U*	R						R		
	F	C		R		V			C		
	W										
	G										
Dinosaur NM	G	O*		O					R		

Abundance and Breeding Information for U.S. and Canadian National Parks

		Ring-billed Gull	California Gull	Herring Gull	Sabine's Gull	Caspian Tern	Common Tern	Forster's Tern	Black Tern	Rock Dove	Band-tailed Pigeon
Page Number		189	190	191	192	193	194	195	196	197	198
Banff NP	SP	O	O	O			O	V	R	O	
	SU	U	U		V			V		O*	
	F	C	C	O	V		O			O	
	W	O	O							O	
	G										
Yoho NP	G	U	U	U			V		V	U	
Kootenay NP	G	R	R							V	
Waterton Lakes NP	G	C	C	U				R	C*	R	V
Glacier NP	SP	C	C			V	R		U		
	SU	C	C						U*	R	
	F	C	C	R			R	R			
	W									R	
	G										
Yellowstone NP	SP	O	C			R	R	O	R	R	
	SU	O*	C*	R		R*	?		R*	R	
	F	O	C			R	R	O	R	R	
	W									R	
	G				V						
Grand Teton NP	SP	R	C				R	V	R	R	V
	SU		C				R	R	R	R	V
	F	R	C		V	O	R	R	R	R	
	W									R	
	G										
Rocky Mountain NP	SP	R	C						R		U
	SU	R	C					R	R		U*
	F	R	C	V							U
	W	R		V							
	G										
Dinosaur NM	G	V					O			C	

Abundance and Breeding Information for U.S. and Canadian National Parks

		Mourning Dove	Black-billed Cuckoo	Yellow-billed Cuckoo	Flammulated Owl	Western Screech Owl	Great Horned Owl	Snowy Owl	Northern Hawk-Owl	Northern Pygmy Owl	Burrowing Owl
Page Number		199	200	201	202	203	204	205	206	207	208
Banff NP	SP	O				V	O	V	R	R	U
	SU	O*				V*	O*		R*	R*	R*
	F	V					O	V	R	R	R
	W						O	V	R	R	R
	G										
Yoho NP	G	U					R		R	U	
Kootenay NP	G	R					U*		R*	R	
Waterton Lakes NP	G	U	R			R	U*			R	
Glacier NP	SP	C				R	C			C	
	SU	C				R*	C*			C*	
	F	C				R	C			C	
	W					R	C	R	V	C	
	G										
Yellowstone NP	SP	O				R	C			R	R
	SU	O*	V			R*	C*			R*	
	F	O				R	C		V	R	R
	W					R	C	R	V	R	
	G										
Grand Teton NP	SP	O	O			R	O			R	R
	SU	O*	O	V		R	O*			R	R
	F	O	O	V	V	R	O	V		R	R
	W					R	O	V		R	
	G										
Rocky Mountain NP	SP	U				U	U			R	
	SU	U*		V		U	U*			R*	
	F	U				U	U			R	
	W					U	U			R	
	G				X*						
Dinosaur NM	G	A*					C*			R	R*

Abundance and Breeding Information for U.S. and Canadian National Parks

		Barred Owl	Great Gray Owl	Long-eared Owl	Short-eared Owl	Boreal Owl	Northern Saw-whet Owl	Common Nighthawk	Common Poor-will	Black Swift	Chimney Swift
Page Number		209	210	211	212	213	214	216	217	218	219
Banff NP	SP	R	V	V		R	R				
	SU	R*	V	V*	V	O*	O*	O*		U*	
	F	R	V		V	R	R	O		U	
	W	R	V		V	R	R				
	G										
Yoho NP	G		U*	V	V	R	U	R		U	
Kootenay NP	G		R*	R*		V	R	C*		U	
Waterton Lakes NP	G		R			R	R	C*			
Glacier NP	SP	U	U			R	R	C		R	
	SU	U*	U*	R	R	R*	R*	C*		R*	
	F	U	U	R		R	R	O			
	W	U	U			R	R				
	G										
Yellowstone NP	SP		O	R	O		R	C			
	SU		O*	R*	O*		R*	C*			
	F		O	R	O		R	O			
	W		R	R	R						
	G					V				V	
Grand Teton NP	SP		O		R	R	R	C	V		
	SU		O*	O*	O*	R	R	C*			
	F	V	O	O	O	R	R	C			
	W		O	V	R	R	R				
	G	V									
Rocky Mountain NP	SP					V	U				
	SU			R*			U*	C	U*	U*	V
	F			R			U	C	R	U	
	W						U				
	G										
Dinosaur NM	G			R	R*			A*	O*		

Abundance and Breeding Information for U.S. and Canadian National Parks

		Vaux's Swift	White-throated Swift	Magnificent Hummingbird	Ruby-throated Hummingbird	Black-chinned Hummingbird	Calliope Hummingbird	Broad-tailed Hummingbird	Rufous Hummingbird	Belted Kingfisher	Lewis's Woodpecker
Page Number		220	221	222	223	224	225	226	228	229	230
Banff NP	SP									R	R
	SU						O		U*	O*	R*
	F						O		U	O	
	W									R	
	G										
Yoho NP	G	R					R		C*	U*	
Kootenay NP	G		V				R*		U	U*	
Waterton Lakes NP	G					R	U*		C	C*	R
Glacier NP	SP	C	R				C	R	C	C	U
	SU	C*				V	C*	R	C*	C*	U
	F									C	
	W									U	
	G										
Yellowstone NP	SP		O				R	R	R	O	R
	SU		O*				O*	R*	R*	C*	R*
	F		O							C	R
	W									R	
	G										
Grand Teton NP	SP						C	O	O	O	O
	SU		R			R	C*	O*	U*	C*	O*
	F					R	C	O	O	C	R
	W									O	
	G										
Rocky Mountain NP	SP		U						C	R	U	R
	SU		U*	V			R		C*	O	U?	U
	F		U						C	C	U	U
	W									R	
	G										
Dinosaur NM	G		C*			C*		C	O	O	

Abundance and Breeding Information for U.S. and Canadian National Parks

		Red-headed Woodpecker	Acorn Woodpecker	Yellow-bellied Sapsucker	Williamson's Sapsucker	Downy Woodpecker	Hairy Woodpecker	White-headed Woodpecker	Three-toed Woodpecker	Black-backed Woodpecker	Northern Flicker
Page Number		231	232	233	234	235	236	237	238	239	240
Banff NP	SP			O		R	O		O	R	U
	SU			O*		O*	O*		U*	R	U*
	F			O		O	O		O	R	O
	W					R	O		O	R	
	G										
Yoho NP	G			U*		U	U		U*	R	C*
Kootenay NP	G			U*		R	U*		C*	R*	C*
Waterton Lakes NP	G	R		U*	V	C*	C*		U*	R*	C*
Glacier NP	SP			C	U	C	C		C	U	C
	SU	V		C*	U*	C*	C*		C*	U*	C*
	F			U		C	C		C	U	C
	W					C	C		C	U	R
	G										
Yellowstone NP	SP			U	R	C	C		O	R	C
	SU	R		U*	R*	C*	C*		O*	R*	C*
	F			U	R	C	C		O	R	C
	W					C	C		R	R	R
	G										
Grand Teton NP	SP	V		C	O	C	C	V	O	O	C
	SU	V	V	C*	O*	C*	C*	V	O*	O*	C*
	F			C	R	C	C		R	O	C
	W			V		C	C			R	O
	G										
Rocky Mountain NP	SP	R		C	U	C	C		U		C
	SU	R		C*	C*	C*	C*		U*		C*
	F	R		C	U	C	C		U		C
	W					U	U		U		
	G										
Dinosaur NM	G	R		O		O					R*

Abundance and Breeding Information for U.S. and Canadian National Parks

		Pileated Woodpecker	Olive-sided Flycatcher	Western Wood Pewee	Yellow-bellied Flycatcher	Alder Flycatcher	Willow Flycatcher	Least Flycatcher	Hammond's Flycatcher	Dusky Flycatcher	Gray Flycatcher
Page Number		241	242	243	244	245	246	247	248	249	250
Banff NP	SP	R	U	U	V	U	U	U	U	O	
	SU	R*	U*	O*		U*	U*	U*	U*	O*	
	F	R									
	W	R									
	G										
Yoho NP	G		U*	C*	U*	U	C*	U	C	U	
Kootenay NP	G		U*	C	U	U	U	U	C	R	
Waterton Lakes NP	G		U	U	U	U	U	U	U		
Glacier NP	SP	C	C	U			C		C	U	
	SU	C*	C*	U*			C*	R	C*	U*	
	F	C									
	W	C									
	G										
Yellowstone NP	SP		O	C			C		O	O	
	SU		O*	C*			C*		O*	O*	
	F		R	R			C		O	O	
	W	V									
	G							?			
Grand Teton NP	SP		C	C			O		O	C	
	SU	V	C*	C*			C*	V	O	C*	
	F		C	C			O		O	C	
	W										
	G										
Rocky Mountain NP	SP		C	C			U	R	U	C	
	SU		C	C*			C*	R	U*	C*	
	F		C	C			U		U	C	
	W										
	G										
Dinosaur NM	G		R*	O*					V		O

Abundance and Breeding Information for U.S. and Canadian National Parks

		Western Flycatcher	Eastern Phoebe	Say's Phoebe	Vermilion Flycatcher	Ash-throated Flycatcher	Cassin's Kingbird	Western Kingbird	Eastern Kingbird	Scissor-tailed Flycatcher	Horned Lark
Page Number		251	252	253	254	255	256	257	258	259	260
Banff NP	SP	R	V	V				V	U		C
	SU	R	R	R				V	U*		C*
	F		V	R					O		C
	W										V
	G										
Yoho NP	G	U	V	V					U		R
Kootenay NP	G							V	U*		
Waterton Lakes NP	G	U		R	V			R	C*		U
Glacier NP	SP	R						U	C	V	C
	SU			R				U*	C*		C*
	F			R		R					C
	W										C
	G										
Yellowstone NP	SP	R		R				O	O		C
	SU	R*		R*	V			O*	O*		O*
	F								O		O
	W										O
	G						V				
Grand Teton NP	SP	R		R				R	O		O
	SU	R		R				R	O		O
	F	R		R				R	O		O
	W										O
	G										
Rocky Mountain NP	SP	C		U				U	U		C
	SU	C*		R		V			R*		C
	F	C		U				U	U		O
	W										R
	G						X				
Dinosaur NM	G	R		C*		C*		C*	R		O*

Abundance and Breeding Information for U.S. and Canadian National Parks

		Purple Martin	Tree Swallow	Violet-green Swallow	Rough-winged Swallow	Bank Swallow	Cliff Swallow	Barn Swallow	Gray Jay	Steller's Jay	Blue Jay
Page Number		262	263	264	265	266	267	268	269	270	271
Banff NP	SP		C	U	U	O	C	U	U	O	V
	SU		C*	U*	U*	O*	C*	U*	U*	R	V
	F		U					R	U	R	V
	W								U	O	V
	G										
Yoho NP	G		U*	C*	U*	R	C*	C*	C*	U*	
Kootenay NP	G		U*	U*	C*	U*	U*	C*	C*	U	
Waterton Lakes NP	G		C*	C	C	U	C*	C*	C*	C*	R
Glacier NP	SP		A	C	U	C	C	C	C	C	R
	SU		A*	C*	U*	C*	C*	C*	C*	C*	
	F		U	U	U	U	U	U	C	C	R
	W								C	C	
	G										
Yellowstone NP	SP		C	C	O	O	A	O	C	O	
	SU		C*	C*	O*	O*	A*	O*	C*	O*	
	F		C	R			O	O	C	O	
	W								C	O	
	G										
Grand Teton NP	SP		A	U	R	C	C	C	C	C	
	SU		A*	U*	R	C*	C*	C*	C*	C*	
	F		A				C	C	C	C	
	W								C	C	
	G										
Rocky Mountain NP	SP		C	U	R		C	C	C	C	
	SU		C*	A*	R		C*	C*	C*	C*	
	F		C	C			C	C	C	C	V
	W								C	C	
	G	V									
Dinosaur NM	G	R	O*	C*	R*	O*	A*	O*		O	

Abundance and Breeding Information for U.S. and Canadian National Parks

		Scrub Jay	Piñon Jay	Clark's Nutcracker	Black-billed Magpie	Common Crow	Common Raven	Black-capped Chickadee	Mountain Chickadee	Boreal Chickadee	Chestnut-backed Chickadee
Page Number		272	273	274	275	276	277	278	279	280	281
Banff NP	SP			U	U	R	C	C	C	C	
	SU			C*	U*	C*	C*	C*	C*	C*	
	F			U	U	C	C	C	C	C	
	W			U	U	R	C	C	C	C	
	G										
Yoho NP	G			C*	U	C*	C*	U*	U*	C*	
Kootenay NP	G			C*	U	R	C*	C*	C	U	
Waterton Lakes NP	G			U*	C*	C*	C*	C	C	U	V
Glacier NP	SP			C	C	C	C	A	A	U	R
	SU			C*	C*	C*	C*	A*	A*	U*	R*
	F			C	C	C	C	A	A	U	R
	W			C	C	C	C	A	A	U	R
	G										
Yellowstone NP	SP		R	C	U	O	C	U	C		
	SU		R	C*	U*	O*	C*	U*	C*		
	F		R	C	U	O	C	U	C		
	W			C	U	R	C	U	C		
	G										
Grand Teton NP	SP			C	C	O	C	C	C		
	SU		R	C*	C*	O*	C*	C*	C*		
	F		R	C	C	O	C	C	C		
	W			C	C	O	C	C	C		
	G										
Rocky Mountain NP	SP		R	C	C	U	C	C	C		
	SU			C*	C*	U	C*	C*	C*		
	F	R	U	C	C	U	C	C	C		
	W		U	C	C	U	C	C	C		
	G										
Dinosaur NM	G	O	C	C	A*	R	O	O	O		

Abundance and Breeding Information for U.S. and Canadian National Parks

		Plain Titmouse	Common Bushtit	Red-breasted Nuthatch	White-breasted Nuthatch	Pygmy Nuthatch	Brown Creeper	Rock Wren	Canyon Wren	Bewick's Wren	House Wren
Page Number		282	283	284	285	286	287	288	289	290	291
Banff NP	SP			R	R		O				
	SU			U*	V		O*	R			V
	F			U	V		O	R			
	W			R	R	V	O				
	G										
Yoho NP	G			C*	V		U*				
Kootenay NP	G			C*	R		R				
Waterton Lakes NP	G			C*	U		U	R			U*
Glacier NP	SP			C	R		C	U			U
	SU			C*	R*		C*	U*			U*
	F			C	R		C	U			
	W			C	R		C				
	G										
Yellowstone NP	SP			C	O		O	U			O
	SU			C*	O*	R	O*	U*			O*
	F			C	O	R	O	U			O
	W			C	O		R				
	G								?		
Grand Teton NP	SP			O	C		O	O			C.
	SU			O*	C*		O*	O*			C*
	F			O	C	V	O	O			C
	W			O	C		O				
	G										
Rocky Mountain NP	SP			O	C	C	C	U	U	V	C
	SU			R*	C*	C*	C*	C*	C*		C
	F			O	C	C	C	U	U		C
	W			U	C	C	C		R		
	G										
Dinosaur NM	G	C	O	O	V	V		A	C	O	C*

Abundance and Breeding Information for U.S. and Canadian National Parks

		Winter Wren	Marsh Wren	American Dipper	Golden-crowned Kinglet	Ruby-crowned Kinglet	Blue-gray Gnatcatcher	Eastern Bluebird	Western Bluebird	Mountain Bluebird	Townsend's Solitaire
Page Number		292	293	294	295	296	297	298	299	300	301
Banff NP	SP	U	V	U	U	C			V	V	U	O
	SU	U*	V	U*	C*	C*			V		U*	U*
	F	V	V	U	C	C					O	U
	W			U	U							R
	G											
Yoho NP	G	U*		U*	C*	C					U*	U*
Kootenay NP	G	C*		C*	C	C					U*	C*
Waterton Lakes NP	G	U*	R	C*	U	C					U*	U*
Glacier NP	SP	C	V	C	A	C					C	C
	SU	C*		C*	A*	C*			R		C*	C*
	F	U		C	C	C					C	C
	W	U		C	U							U
	G											
Yellowstone NP	SP		R	C	R	C					C	O
	SU		R*	C*	R*	C*					C*	O*
	F			C	R	C					O	O
	W			C	R							R
	G								V			
Grand Teton NP	SP	V	O	C	O	C				R	C	C
	SU	R	O*	C*	O	C*					C*	C*
	F		O	C	O	O	V			R	C	O
	W			C	R							O
	G											
Rocky Mountain NP	SP	V		C	R	U	V			U	C	C
	SU			C*	U*	C*				U*	C*	C*
	F	V		C	U	U			R	U	C	C
	W			C	C							U
	G											
Dinosaur NM	G		R*	C	V	O	C			O	C	O

Abundance and Breeding Information for U.S. and Canadian National Parks

		Veery	Gray-cheeked Thrush	Swainson's Thrush	Hermit Thrush	American Robin	Varied Thrush	Gray Catbird	Northern Mockingbird	Sage Thrasher	Brown Thrasher
Page Number		302	303	304	305	306	307	308	309	310	311
Banff NP	SP			U	O	U	U		V		
	SU	V	V	C*	U*	U*	C*	V			
	F		V	U	U	U	C				
	W					R					
	G										
Yoho NP	G	V		U*	U*	C*	C*	V			
Kootenay NP	G			C*	C*	C*	C*				
Waterton Lakes NP	G	U		U	U	C*	U*	U*			R
Glacier NP	SP	U		C	C	A	C	R			
	SU	U*		C*	C*	A*	C*	R*			
	F	U		C	C	C	C	R			
	W						U				
	G										
Yellowstone NP	SP	O		O	C	A		R		R	
	SU	O*		O*	C*	A*		R*		R*	
	F	R		R	O	A		R		R	
	W					R					
	G						V				
Grand Teton NP	SP	O		C	C	A		O		O	
	SU	O		C*	C*	A*	V	R	V	O*	
	F	O		O	O	A	V	R	V	O	
	W					R					
	G										
Rocky Mountain NP	SP	R		C	C	A		U		U	
	SU	R		C*	C*	A*		U	V	U	R
	F			U	C	A		U	V	U	R
	W					U			V		R
	G						V				
Dinosaur NM	G			R	R	C*	V			O*	V

Abundance and Breeding Information for U.S. and Canadian National Parks

		Water Pipit	Sprague's Pipit	Bohemian Waxwing	Cedar Waxwing	Northern Shrike	Loggerhead Shrike	European Starling	Solitary Vireo	Warbling Vireo	Philadelphia Vireo
Page Number		312	313	314	315	316	317	318	319	320	321
Banff NP	SP	C		C	V	O		U	U	U	V
	SU	C*		U	U			U*	U*	C*	
	F	C		C	U	O		U	O	U	V
	W			C		V		U			
	G										
Yoho NP	G	C*		U	U*	U		C*	U	C	V
Kootenay NP	G	C*		U*	R	R		R*	C	C	
Waterton Lakes NP	G	U	R	U	C*	R	R	C*	R	R	
Glacier NP	SP	C		C	C	R	U	U	C	C	
	SU	C*		C*	C*		U	U*	C*	C*	
	F	C		C	C	R	U	U			
	W			C		R		U			
	G										
Yellowstone NP	SP	O		O	R	O		U	O	C	
	SU	O*			U		O	U*	R	C*	
	F	C		O	R	C		U	O	O	
	W			O		O		O			
	G										
Grand Teton NP	SP	C	V	O	O	O	O	C	O	A	
	SU	C*			O		R	C*	V	A*	
	F	C	V	O	O	O		C	O	O	
	W	R		O	O	O	O	O			
	G										
Rocky Mountain NP	SP	C	V	U	V		U	C	U	C	
	SU	C*		O	V		R	C*	U*	C*	
	F	C		O	R	R	U	C	U	C	
	W			U	V	U		U			
	G										
Dinosaur NM	G	R			O	R	O	O*	O	O	

Abundance and Breeding Information for U.S. and Canadian National Parks

		Red-eyed Vireo	*Golden-winged Warbler*	*Tennessee Warbler*	*Orange-crowned Warbler*	*Nashville Warbler*	*Virginia's Warbler*	*Northern Parula*	*Yellow Warbler*	*Chestnut-sided Warbler*	*Magnolia Warbler*
Page Number		322	323	324	325	326	327	328	329	330	331
Banff NP	SP	O		O	U	V			O		V
	SU	O*		O*	C*				U*		R
	F	R		O	U	V			O		V
	W										
	G										
Yoho NP	G	R		R	C*				U		V
Kootenay NP	G	R		C	U				U		R
Waterton Lakes NP	G	R		R	R				C		
Glacier NP	SP	C		R	U	V					
	SU	C*		R	U*				C*		
	F								C		
	W										
	G										
Yellowstone NP	SP	O		R	R				C	V	
	SU	R			R*				C*		
	F	O			R				O		
	W										
	G					V					
Grand Teton NP	SP	O			O	V			A		
	SU	O		R	O				A*	V	
	F	O		V	O	V			C		
	W										
	G										
Rocky Mountain NP	SP	R		V	U		C		C		
	SU	R	V		U*		C*	V	C*	V	
	F	R		V	C	R	C		C		V
	W										
	G										
Dinosaur NM	G						O		C*		

Abundance and Breeding Information for U.S. and Canadian National Parks

		Cape May Warbler	Yellow-rumped Warbler	Black-throated Gray Warbler	Townsend's Warbler	Black-throated Green Warbler	Blackburnian Warbler	Grace's Warbler	Palm Warbler	Bay-breasted Warbler	Blackpoll Warbler
Page Number		332	333	334	335	336	337	338	339	340	341
Banff NP	SP		C		R	V					U
	SU		C*		C*						C*
	F		C	V	U						U
	W										
	G										
Yoho NP	G		A*		A*				V	V	U
Kootenay NP	G		C		C						R
Waterton Lakes NP	G		C*		U	R			R		R
Glacier NP	SP		C		C						
	SU		C*		C*						
	F		U								
	W										
	G										
Yellowstone NP	SP		C								
	SU		C*		R						
	F		C		R						
	W										
	G	V									V
Grand Teton NP	SP		A								
	SU		A*		R		V		V	V	
	F		C		R						
	W										
	G								V		
Rocky Mountain NP	SP		C	V							
	SU		C*				V	V		V	
	F		C	V	U	V		V	V		
	W										
	G										
Dinosaur NM	G	V	C	C	R						

Abundance and Breeding Information for U.S. and Canadian National Parks

		Black-and-white Warbler	American Redstart	Prothonotary Warbler	Worm-eating Warbler	Ovenbird	Northern Waterthrush	MacGillivray's Warbler	Common Yellowthroat	Hooded Warbler	Wilson's Warbler
Page Number		342	343	344	345	346	347	348	349	350	351
Banff NP	SP	O	U				O	U	C		U
	SU	O	C*				O*	U*	C*		C*
	F	O	U					O	U		U
	W										
	G										
Yoho NP	G		U				U	U*	C*		C*
Kootenay NP	G		U				U	C	C*		C
Waterton Lakes NP	G		U			R	U	U*	U*		U
Glacier NP	SP	V	C				C	C	C		C
	SU		C*				C*	C*	C*		C*
	F										
	W										
	G										
Yellowstone NP	SP		R				R	O	C		O
	SU		R*				R	O*	C*		O*
	F						R	O	C		O
	W										
	G				V						
Grand Teton NP	SP		O				R	C	C		C
	SU		O	V			R	C*	C*		C*
	F						V	O	C		C
	W										
	G										
Rocky Mountain NP	SP	V	R				V	C	R	V	C
	SU	V			V	V		C*	R		C*
	F		V				V	O	R		C
	W										
	G										
Dinosaur NM	G	R	R		V			R	R*		O

Abundance and Breeding Information for U.S. and Canadian National Parks

		Canada Warbler	Yellow-breasted Chat	Hepatic Tanager	Scarlet Tanager	Western Tanager	Rose-breasted Grosbeak	Black-headed Grosbeak	Blue Grosbeak	Lazuli Bunting	Indigo Bunting
Page Number		352	353	354	355	356	357	358	359	360	361
Banff NP	SP					O	V	V			
	SU	V				O*					
	F					O	V	V			
	W										
	G										
Yoho NP	G					U	R			R	
Kootenay NP	G					U*		V		R	
Waterton Lakes NP	G					C		U		U	R
Glacier NP	SP		V			C		R		C	
	SU					C*		R*		C*	
	F										
	W										
	G										
Yellowstone NP	SP					C		R		O	
	SU					C*		R		O*	
	F					O		R			
	W										
	G				V						
Grand Teton NP	SP		V			C	O	O		O	V
	SU		V			C*	V	C*		O*	V
	F					O		R		R	V
	W										
	G										
Rocky Mountain NP	SP		V			C	V	C		V	V
	SU			V	V	C*	R	C*		R	
	F					C		C		V	
	W										
	G										
Dinosaur NM	G		O*			O*	R	O	R	C	R

Abundance and Breeding Information for U.S. and Canadian National Parks

		Dickcissel	Green-tailed Towhee	Rufous-sided Towhee	Brown Towhee	American Tree Sparrow	Chipping Sparrow	Clay-colored Sparrow	Brewer's Sparrow	Field Sparrow	Vesper Sparrow
Page Number		362	363	364	365	366	367	368	369	370	371
Banff NP	SP					R	U	O	O		O
	SU			V			C*	U*	U*		O*
	F					R	U	O	O		O
	W										
	G										
Yoho NP	G					U	A*	R	R		R
Kootenay NP	G						C*	R			U
Waterton Lakes NP	G			R		R	C		R		C*
Glacier NP	SP			U		R	C		R		C
	SU			U		V	C*		R		C*
	F		V			R					U
	W					R					
	G									V	
Yellowstone NP	SP		U	R		R	C		U		C
	SU		U*	R*			C*		U*		C*
	F		U			R	C		C		C
	W					C					
	G										
Grand Teton NP	SP		O	R		O	C		C		C
	SU		C*	R			C*	R*	C*		C*
	F		C	R		O	C		C		C
	W					O					
	G										
Rocky Mountain NP	SP		C	R		U	C				C
	SU		C*	R			C*	V	R*		C*
	F		C	R		U	C	V	U		C
	W					U					
	G										
Dinosaur NM	G		O	C	R	O	C		V		O*

Abundance and Breeding Information for U.S. and Canadian National Parks

		Lark Sparrow	Black-throated Sparrow	Sage Sparrow	Lark Bunting	Savannah Sparrow	Baird's Sparrow	Grasshopper Sparrow	LeConte's Sparrow	Fox Sparrow	Song Sparrow
Page Number		372	373	374	375	376	377	378	379	380	381
Banff NP	SP				V	U	V	V		O	U
	SU					C*				U*	U*
	F					U				O	O
	W										V
	G										
Yoho NP	G	V				C*				U	U
Kootenay NP	G				V	U			V	C	R
Waterton Lakes NP	G				R	C			U	U*	U*
Glacier NP	SP	R				C			R	C	C
	SU				R	C*			R	C*	C*
	F									U	C
	W										R
	G										
Yellowstone NP	SP	O			O	C		V		R	C
	SU	O*		R		C*		V*		R*	C*
	F				O	C		V		R	C
	W										
	G										
Grand Teton NP	SP		V	V	R	C				O	C
	SU			V	R	C*		V		O*	C*
	F	O				C				O	C
	W	O									O
	G										
Rocky Mountain NP	SP	R		V	R	C				R	C
	SU	R*			R	C*				R*	C*
	F	R			R	C				R	C
	W										
	G										
Dinosaur NM	G	A*	O	O*		C*		V			O*

Abundance and Breeding Information for U.S. and Canadian National Parks

		Lincoln's Sparrow	Swamp Sparrow	White-throated Sparrow	Gold-crowned Sparrow	White-crowned Sparrow	Harris's Sparrow	Dark-eyed Junco	McCown's Longspur	Lapland Longspur	Chestnut-collared Longspur
Page Number		382	383	384	385	386	387	388	389	390	391
Banff NP	SP	U	V	O	O	C		C		V	V
	SU	C*	V	O*	U*	C*		C*			
	F	U	V	V	O	U	V	O		V	
	W					V		V		V	
	G										
Yoho NP	G	U				C*	R	C*		R	
Kootenay NP	G	U*				U*		C*			
Waterton Lakes NP	G	U		R	R	C		C*	V		R
Glacier NP	SP	U				C	R	A	R	U	U
	SU	U*				C*		A*			
	F	U				C		A		U	
	W			R				C		U	
	G										
Yellowstone NP	SP	O				C	R	A			
	SU	O*				C*		A*			
	F	O		R		C		C		O	
	W						R	O		O	
	G								V		
Grand Teton NP	SP	O		R		A	R	A		V	
	SU	C*	V			A*		A*	V		
	F	C	V	R		A	R	C			
	W					R	R	O		V	
	G										
Rocky Mountain NP	SP	C		V		C		C			
	SU	C*				C*		C*			
	F	C				C		C			
	W			V		R	R	C			
	G										
Dinosaur NM	G					C*	R	C	V		

Abundance and Breeding Information for U.S. and Canadian National Parks

		Snow Bunting	Bobolink	Red-winged Blackbird	Western Meadowlark	Yellow-headed Blackbird	Rusty Blackbird	Brewer's Blackbird	Common Grackle	Brown-headed Cowbird	Orchard Oriole
Page Number		392	393	394	395	396	397	398	399	400	401
Banff NP	SP	C	V	C	O	U	O	U	V	C	
	SU			C*		V	R	U*	V	C*	
	F	U		V	O	V	O	U	V	C	
	W	C						V			
	G										
Yoho NP	G	U	R	C*	U	U	U	U*	V	C*	
Kootenay NP	G	R		U*	R	R		U*		R	
Waterton Lakes NP	G	U	R	C*	C*	U*		U*	R	U	
Glacier NP	SP	U		C	U	R		U		U	
	SU		R	C*	U*		R	U	R	U*	
	F	U			U			U		U	
	W	U						R			
	G										
Yellowstone NP	SP		R	C	C	U		C		O	
	SU		R*	C*	C*	U*		C*		O*	
	F	O	R	C	C	U		C		O	
	W	O		R	R			R		R	
	G								V		
Grand Teton NP	SP	V	R	C	O	C		C	O	C	V
	SU		R	C*	O	C*		C*	O	C*	
	F	R		C	O	C	V	A	O	C	
	W	O		R	V	V		R			
	G										
Rocky Mountain NP	SP		R	C	U	R		C	R	R	
	SU			C*	U*	R		C*	R*	R*	
	F			C	U	R	V	C	R		
	W			R							
	G										
Dinosaur NM	G			O*	C*	O*		C*	R*	O*	

Abundance and Breeding Information for U.S. and Canadian National Parks

		Northern (Bullock's) Oriole	Rosy Finch	Pine Grosbeak	Purple Finch	Cassin's Finch	House Finch	Red Crossbill	White-winged Crossbill	Common Redpoll	Hoary Redpoll
Page Number		402	403	404	405	406	407	408	409	410	411
Banff NP	SP	V	C	U	O	V		U	U	U	V
	SU		C*	U*	O*	V		U*	U*	V	
	F		C	U		V		U	U	V	
	W		V	U				U	U	U	R
	G										
Yoho NP	G		U*	U*	R	R		U	U	U	
Kootenay NP	G		U	U*	U			U	R	R	
Waterton Lakes NP	G	R	U	U	U	U*	R	R	U	U	R
Glacier NP	SP	R	C	C		C		C	U	C	
	SU	R*	C*	C*		C*		C*	U		
	F		C	C		C		C	U	C	
	W		C	C				C	U	C	
	G										
Yellowstone NP	SP	R	O	C		C		O	R		
	SU	R*	O*	C*		C*		O*	R		
	F	V	O	C		C		O	R	R	
	W		O	O		O		R	R	R	
	G						V				
Grand Teton NP	SP	O	C	O		C	V	O	V	C	V
	SU	O*	C*	O		C*	V	O	V		
	F	O	O	O		C	V	O		C	
	W		O	O		R	V	O	V	O	
	G										V
Rocky Mountain NP	SP	R	U	U		C		U		U	
	SU	R	C*	U*		C*	U*	U*	R		
	F	R	U	U		C	V	U			
	W		U	U		C		U	R	U	
	G										
Dinosaur NM	G	C*	R	R		R	C	V		R	

Abundance and Breeding Information for U.S. and Canadian National Parks

Page Number		Pine Siskin	Lesser Goldfinch	American Goldfinch	Evening Grosbeak	House Sparrow
		412	413	414	415	416
Banff NP	SP	U		V	U	C
	SU	C*		V	O*	C*
	F	C			O	C
	W				U	C
	G					
Yoho NP	G	C*		R	U	C*
Kootenay NP	G	C			U	V
Waterton Lakes NP	G	C		C	U	C
Glacier NP	SP	C		U	C	
	SU	C*		U	C*	
	F	C		U	C	R
	W	U			C	
	G					
Yellowstone NP	SP	A		R	O	R
	SU	A*		R*	O*	R
	F	A		R	O	R
	W	O			O	R
	G					
Grand Teton NP	SP	C		O	C	C
	SU	C*		O*	O	C*
	F	C		O	C	C
	W	O			C	C
	G					
Rocky Mountain NP	SP	C	R	O	R	
	SU	C*	R	U	R*	
	F	C		R	R	
	W	U			U	
	G					
Dinosaur NM	G	O*	O	O*	O	O

Index

This index includes the common names of bird species mentioned in this book, with the principal species account indicated by italics. Some names that previously have been used as common names but which no longer are used, or apply only to forms now considered subspecies, have also been indexed.